# Choosing THE RIGHT STUFF

**Recent Titles in**
**Human Evolution, Behavior, and Intelligence**

Staying Human in the Organization: Our Biological Heritage and the Workplace
*J. Gary Bernhard and Kalman Glantz*

The End of the Empty Organism: Neurobiology and the Sciences of Human Action
*Elliott White*

Genes, Brains, and Politics: Self-Selection and Social Life
*Elliott White*

How Humans Relate: A New Interpersonal Theory
*John Birtchnell*

# Choosing THE RIGHT STUFF

## The Psychological Selection of Astronauts and Cosmonauts

PATRICIA A. SANTY

Human Evolution, Behavior, and Intelligence
SEYMOUR W. ITZKOFF, Series Editor

Westport, Connecticut
London

**Library of Congress Cataloging-in-Publication Data**

Santy, Patricia A.
Choosing the right stuff : the psychological selection of astronauts and cosmonauts / Patricia A. Santy.
p. cm. — (Human evolution, behavior, and intelligence, ISSN 1063-2158)
Includes bibliographical references and index.
ISBN 0-275-94236-8 (alk. paper)
1. Space flight—Psychological aspects. 2. Astronauts—Psychology. I. Title. II. Series.
TL856.S26 1994
155.9'66—dc20 93-22834

British Library Cataloguing in Publication Data is available.



Library of Congress Catalog Card Number: 93-22834
ISBN: 0-275-94236-8
ISSN: 1063-2158

First published in 1994

Praeger Publishers, 88 Post Road West, Westport, CT 06881
An imprint of Greenwood Publishing Group, Inc.

Printed in the United States of America

The paper used in this book complies with the Permanent Paper Standard issued by the National Information Standards Organization (Z39.48-1984).

10 9 8 7 6 5 4 3 2 1

**Copyright Acknowledgments**

The author and publisher are grateful for permission to use excerpts from the following:

Paul E. Meehl, *Clinical vs. Statistical Prediction: A Theoretical Analysis and a Review of the Evidence.* © 1954 by the University of Minnesota Press. Reprinted by permission of the University of Minnesota Press.

William H. Sledge and James A. Boydstun, "The Psychiatrist's Role in Aerospace Operations," *American Journal of Psychiatry* 137 (1980): 956–959.

This book is dedicated to two individuals without whom the work reported on would have been impossible:

To Carlos J. G. Perry, M.D. (1932–1989)

Thank you for your inspiration, your unselfish sharing of knowledge, and your friendship with a newcomer.

To Manley "Sonny" Carter, M.D. (1947–1991)

An extraordinary person who gave all of us the benefits of his many talents. Thank you for your support, your encouragement, and your efforts on behalf of our work.

*Little remains: but every hour is saved*
*From that eternal silence, something more,*
*A bringer of new things; and vile it were*
*For some three suns to store and hoard myself,*
*and this gray spirit yearning in desire*
*To follow knowledge like a sinking star,*
*Beyond the utmost bound of human thought.*

—Alfred, Lord Tennyson
*Ulysses*

# Contents

# Tables and Figures

## TABLES

## FIGURES

# Acknowledgments

Many people helped and supported me during the writing of this book, and I would like to give them my sincere thanks. They include J. J. Schulte, Stan Mohler, and Molly Hall at Wright State University and Wright Patterson Air Force Base, who searched out and obtained for me much of the archival material on the original Project *Mercury* selection; Dave Jones, John Patterson, and Herminio Cuervo, who assisted me in obtaining information from the Brooks Air Force Base archives and supported me through many trials and tribulations; George Ruff, Bryce Hartman, and Don Flinn, all of whom graciously allowed me to interview them about their roles in astronaut selection; Bob Rose and Bob Helmreich, who believed in the work I was doing when I was at NASA and who have supported me and continued the work after I left; Ernie Barratt at the University of Texas Medical Branch, whose extensive archives on psychological material and tests were invaluable in my research and who also donated his time, advice, and support; and Cooky and Jim Oberg, Natalie Karakulko, Mike Bungo, Dean Faulk, and Regina North, my close friends and colleagues without whom I would have never been able to finish; Nita Brannon and Judy Vanover, my very, very patient secretaries; my loving husband, Norman Richert; and, last, but not least, my daughter, Alexandra Ann, who was born at the same time I completed this manuscript and who—for me—already has the right stuff.

# Introduction

> Or a man could go for a routine physical one fine day, feeling like a million dollars, and be grounded for fallen arches. It happened!—Just like that! (And try raising them.) Or for breaking his wrist and losing only part of its mobility. Or for a minor deterioration of eyesight, or for any of hundreds of reasons that would make no difference to a man in an ordinary occupation. As a result all fighter jocks began looking upon doctors as their natural enemies. Going to see a flight surgeon was a no-gain proposition; a pilot could only hold his own or lose in the doctor's office. To be grounded for a medical reason was no humiliation, looked at objectively. But it was a humiliation, nonetheless!—for it meant you no longer had that indefinable, unutterable, integral stuff.
>
> —Tom Wolfe, *The Right Stuff*[1]

Flight surgeons are frequently viewed with a combination of suspicion and scorn by most pilots because of the enormous power they have to interrupt and even terminate a pilot's career. If being grounded for fallen arches is humiliating, imagine the humiliation the aviator experienced if he was grounded for *psychological* reasons. Unlike the case of fallen arches, a psychiatric disqualification carries an additional cultural burden of shame.

There is a tendency in our culture to be somewhat "anti-psychological." The American experience and heritage have always been oriented more toward seeking control over the outer world than toward understanding the inner one. On the whole, we eschew introspection and prefer action. Thus, the pilot—and by extension, the astronaut—is culturally revered and is symbolic of the American preference for exertion over emotion.

One effective defense mechanism aviators utilize to deal with emotional problems is to trivialize them. Thus, psychology and psychiatry become "pseudoscientific"[2] (Who understands that mumbo-jumbo, anyway?); mental health practitioners are considered useless, so there is no purpose in consulting one (What does a shrink know, anyway?); and even one's own emotions are downplayed as immaterial and beside the point, lest they threaten the carefully cultivated veneer of invulnerability (So what if my wife left me? Who cares? I can still fly that jet.). The capacity to rationalize and intellectualize can provide support to the battered self-esteem of an aviator who fears that he is "only human," after all.

In the person of the flight surgeon—or any physician—the pilot finds a concrete external enemy to focus on so he can avoid the internal enemy—himself and his own limitations. While the attitude of hostility toward flight surgeons in general and psychological issues in particular is not unanimously exhibited by all aviators, the typical pilot is expected by that "indefinable, unutterable, integral" mythology to sneer at the doctor's grim prognosis, knowing that someone who really has the "right stuff" would not be impeded by anything as mundane as medical or psychological problems.

In the U.S. military services, this attitude on the part of pilots is effectively compensated for by the administrative support given to flight surgeons (including psychiatrists) by the command hierarchy. Lacking an administrative command hierarchy to provide the same kind of checks and balances, the National Aeronautics and Space Administration (NASA) is an institution where the pilot has been elevated to the rank of "astronaut," and pilots' attitudes and beliefs have been elevated to agency policy. While pilots (and astronauts) have many fine qualities, their particular biases about physicians—psychiatrists in particular—and life scientists in general have had significant repercussions.

Contrary to the general public's perception, NASA is not an organization of scientists, but rather an organization of pilots and engineers. That is a significant difference. Members of either occupation, as a general rule, have always had difficulty in dealing with concepts from the psychological universe. Their professions require that their superior intellectual and technical skills be directed externally, to the outside world. It has taken many years for engineers to come to appreciate the valuable information that a knowledge of human factors contributes to their undertakings. Few pilots ever learn to appreciate the human side of a flying career. Why is this so? The minimization of personal emotions may actually be necessary to achieve peak performance. How a pilot minimizes emotion, however, is critical to his physical and psychological health. There is an important difference between putting aside emotion in order to accomplish a particular task (suppression) and denying emotion exists (repression). The first psychological technique is healthy and necessary for anyone to be able to function in the world; employing the second inevitably results in a buildup of psychological tension that ultimately threatens health and safe performance. The denial of emotion in an organization can be similarly destructive.

The repression of psychological factors by NASA explains why there is such a tremendous gap between our understanding of the human mind in space and the knowledge learned in every other medical discipline. NASA managers have maintained that they have not needed psychological input until recently, when the agency's focus shifted from short to long space missions. But the fact is that there have been interpersonal and behavioral problems from the beginning of the U.S. Space Program, and NASA managers know it—and have known it all along. There is a strong motivation on the part of these managers to deny such problems and protect the agency from negative public opinion. Hence, there is an institutional emphasis on how special astronauts are, both physically and mentally, and almost a universal denial of the very human problems they have. The high divorce rate and the family problems expressed by spouses and children attest to the astronauts' inability to escape from the real-life difficulties all of us have to confront. If one needs evidence to support this contention, I would suggest reading a few of the many biographies written in recent years by former astronauts, who detail some of the emotional problems and conflicts they and their families actually experienced during their NASA careers.

But NASA believes that the organization—and especially astronauts—must always appear in a positive light to the American public and to Congress. The agency's experience with the media, combined with waxing and waning support from Congress, has led its managers to expect that openness about any problems—particularly emotional or behavioral ones—will erode their public support. Anything considered to be potentially "damaging" to the agency must be carefully controlled or minimized to circumvent the possibility that funding might be cut. Since this fear has some justification, it has led to an unhappy situation where NASA actually believes in its own invulnerability. The *Challenger* disaster only momentarily pierced the defense.

## AN ANTI-PSYCHOLOGICAL BIAS

I began my career at NASA at the Johnson Space Center in 1984 in the Shuttle program. Even then, some of the original glitz had worn off the Space Program in the eyes of the public, but as the first full-time civil service psychiatrist (others had been paid as consultants) hired to be a flight surgeon to astronauts and their families, I had rather grandiose fantasies of following in the footsteps of the behavioral scientists who had come before me. Ruff, Levy, Flinn, Hartman, and Perry—these were the names of the psychiatrists and psychologists whose work on astronaut selection had always fascinated me. Their few papers on this subject influenced my choice of psychiatry as a medical specialty. What interested me most were the personalities of those extraordinary people who wanted to live and work in the dangerous environment of space or on distant planets.

During my years at NASA, I frequently encountered the anti-psychological bias that was rampant in the agency. The most blatant example of this bias occurred in the fall of 1988. I had conducted a rather fruitless search for the original *Mercury, Gemini,* and *Apollo* psychological data and had also discovered that no documentation of psychiatric procedures existed for the entire Shuttle period. This caused me to suggest to management that NASA might want to review its current psychological selection processes, since they seemed to be inadequate. To that end, I made a presentation to an outside review committee formed by NASA and chaired by Chris Kraft, former director of the Manned Spacecraft Center—as the Johnson Space Center used to be called. The committee was charged with the task of evaluating NASA's selection methods, and as part of that, it requested that I present a history of the psychological aspects of selection. During my presentation, I stated that no performance data were available to validate the psychological selection criteria used in the *Mercury* program and thereafter and that the documentation of the psychological procedures currently used was nonexistent. Kraft became agitated and interrupted me after only a few slides had been shown.

"Young lady, you are a dangerous person and are out to destroy NASA! I will not permit that to happen."

I responded that if the data or the documentation existed, I had been unable to find it.

He angrily replied, "You never will because it's [the data] in here [pointing to his head], and it's going to stay there so that people like you can't use it against NASA."

Years later, I was finally able to obtain a copy of the original *Mercury, Gemini,* and *Apollo* psychiatric evaluations and psychometric test results from a source outside of NASA. I cannot help but speculate that Kraft's—and NASA's—behavior suggested that there was something he believes needed to be hidden. I wondered what it was as I studied the original reports.

Many forces—nations, institutions, and individuals—in the world today work to limit the dissemination of information. Sometimes the purpose of the suppression is genuinely malicious: to *suppress* knowledge entirely; sometimes it is merely a practical matter: a desire to *control* knowledge and information. Lewis Thomas wrote in his book *The Fragile Species:*

> The scientific community has emerged in this century as the only genuine world community that I can think of. It has had nothing to connect it to the special interests of nation-states, it carries out its work of inquiry without respect for national borders, it passes information around as though at a great party; and because of these habits, science itself has grown and prospered. Every researcher, in whatever laboratory, depends for the work's progress on the cascades of information coming in from other laboratories everywhere, and sends out his laboratory's latest findings on whatever wind is, at the moment, blowing.
>
> I do not believe that science can be done in any other way. . . .[3]

## THE EXCLUSION OF BEHAVIORAL SCIENCE

This analysis is written, in part, to remedy a historical omission in the history of space medical research. Due to a fear of psychological issues, not free scientific enquiry, the behavioral sciences have, until very recently, been largely ignored by the U.S. Space Program. Thirty years of space flight experience in this country have yielded a gold mine of data and knowledge about the human body and its response to the space environment, but no objective data on the human psyche in that same environment has been produced—and many scientists consider psychological issues to be a limiting factor in the human exploration of the universe.

I came to realize slowly and gradually that I was the "figurehead" for all psychological issues at NASA. When accused of not taking psychological issues seriously, NASA officials could point to me and maintain that I was clear proof of their commitment to such issues. But they never provided much support (either financial or moral) in my endeavors (to be fair, many other medical specialties were also seriously underfunded), and any psychological input that I might have had on any specific issue was generally ignored. It was clear that while it was convenient for them to be able to point to me as representing their commitment to understanding the psychological factors of space flight, I was never really expected to *act* like a psychiatrist.

I discovered that all the work of my predecessors had disappeared into a black hole. No records existed at NASA about their work. There was no record of psychological selection on the medical charts of astronauts. There were no archives that housed all the data collected; there was, in truth, a complete absence of behavioral sciences in *any* recognizable form in every part of the Space Program from selection and training to flight. When I questioned why this was so, I was accused of wanting to "destroy" the agency and described as "dangerous."

If the paranoia was due to fears that medical confidentiality would be violated, NASA had no real justification for its stance. All ethical psychiatrists and psychologists, including those doing research, have the strongest respect for individuals who participate in research. Researchers generally deal with this issue by one of two methods: Either they publish only group data, or they disguise the individual data in such a way as to ensure that it could never be connected to any particular individual.

To this day, for example, George Ruff will not discuss any individual astronaut data. The possibility of revealing anything to the press about the psychological state of an astronaut or of revealing anything that might be detrimental to a mission was, as far as Ruff was concerned, not a possibility.

The tragedy of it was, that all we ever saw in our flights was very positive. This recent talk about Carpenter using a lot of fuel for his flight and not having the "right stuff" was not done 25 years ago. At the time, that was within the requirements of the mission. We psychiatrists sat in on all the debriefings, long and short. Nobody then felt that such a thing was outside the range of what was allowed. What we had to say was very positive.[4]

This lack of openness on the part of NASA management has plagued the agency throughout its life span, even if it only became obvious after the *Challenger* disaster. It has caused and causes rumors, innuendo, and fantasies about what is really going on. Particularly in the case of the psychological status of astronauts, there is a considerable amount of anecdotal—and often completely false—information that has been spread around for decades. "I remember reading once about the well-known case of the 'chicken' astronaut who went psychotic in space. Well, unless somebody later on knows something I don't know, that certainly didn't happen back in our day. I think it would have been very positive if more of it had been released."[5]

The exclusion of behavioral research began shortly after the *Mercury* program was under way. While psychiatrists and psychologists were still involved in astronaut selection, their role was ambiguous and became even more so as the years progressed. It is hardly surprising that psychological research in the U.S. Space Program is somewhere between twenty and thirty years behind all other medical disciplines. And yet the agency has repeatedly stated that psychological adaptation will be (along with cardiovascular and musculoskeletal adaptation) one of the most significant problems of long-duration space flight. Intuitively, this seems clear to most of the general public, who often ask me about psychological factors when I give talks and who are surprised and dismayed to learn that NASA has never done any research on the subject.

This book represents the accumulated work of many people over a period of thirty years of space history. It covers material that is not a well-known part of that history—a history that has celebrated a few courageous male, and even fewer female, astronauts who have had the privilege of flying in space. The record has tended to ignore the thousands of dedicated men and women who made the dream come true for those fortunate few astronauts. This book recognizes some of those lesser known but nonetheless dedicated individuals.

I believe that the destiny of the human species lies in the exploration and colonization of the universe. I felt admiration and awe as a teenager when *Apollo 11* landed on the Moon—without question one of the highlights of my young life. But it was a magical TV series called "Star Trek" that made me understand the role I wanted to play. Many of the show's episodes dealt with the psychological and philosophical foundations of mankind's need to explore space. What motivates the Captain Kirks? The Ensign Uhurus? The Doctor McCoys? I wanted to understand what it was that made those characters want to explore strange, new worlds. It was not until 1979 that Tom Wolfe coined the term the "right stuff" to describe the quality that so intrigued me. Later, as a psychiatrist, I was always bothered by the pilots' and the astronauts' concept of the right stuff. Was it really necessary to be so uninsightful, insensitive, and unaware of one's emotions to be able to "boldly go where no one had gone before"?

The concept of the right stuff, as it is perceived by pilots and astronauts, lies at the heart of understanding the denial of psychological issues in the U.S. Space Program. The most important contribution of the behavioral sciences to space exploration will be to provide a healthier and more useful perspective on that elusive and illusionary ideal.

My purpose in writing this book was simple. I felt that the story of psychiatry in the Space Program should be told and that the early work of Ruff and others should be resurrected and analyzed in light of plans for longer missions of exploration. Their early work eventually became the basis for the development of operational and research procedures in the behavioral sciences at the Johnson Space Center. I wanted to make sure that the efforts of the psychiatrists and the psychologists who assisted me in making these changes at NASA would not get "lost" again and that a behavioral scientist in the future would not have to go to the extremes I did in order to reconstruct the historical data.

## NOTES

1. Tom Wolfe, *The Right Stuff* (New York: Farrar, Straus & Giroux, 1979), 30.
2. C. Cordes, "Mullane: Tests Are Grounded" *APA Monitor* (October 1983): 24.
3. Lewis Thomas, *The Fragile Species* (New York: Charles Scribner's Sons, 1992), 182.
4. P. A. Santy, ed., "The NASA In-House Working Group on the Psychiatric and Psychological Selection of Astronauts: Summary and Transcripts" (unpublished paper, 1988), 23.
5. Ibid., 23–24.

# Choosing THE RIGHT STUFF

# 1

# The Beginning of the Future—Project *Mercury*

> How could anyone turn down a chance to be part of something like this?
>
> —Anonymous candidate, 1959

## THE RIGHT PLACE AT THE RIGHT TIME

Dr. George Ruff was a twenty-nine-year-old research psychiatrist when he first came to the Aeromedical Research Laboratory (AMRL) in Dayton, Ohio, in the summer of 1957 to take over as chief of the Stress and Fatigue Section in the Human Factors Division. Ruff, born in Wilkes-Barre, Pennsylvania, had received his A.B. in 1948 from Haverford College, a small liberal arts college in Pennsylvania, and his M.D. in 1952 from the University of Pennsylvania. He finished his psychiatric residency in 1956 at the Neuropsychiatric Institute of the University of Michigan and accepted a U.S. Public Health Service Postdoctoral Fellowship at the Institute of Neurological Sciences at the University of Pennsylvania for one year. His research interest was stress, from both biological and psychological perspectives.

Ruff, like many of his contemporaries who had to deal with the military draft, decided to join the Air Force and negotiate for a position compatible with his research objectives, rather than taking his chances in the draft. He accepted a position at the AMRL, which had been very active for years in studying what was called "the premium man for the premium mission." These studies included many missions that at the time were "classified" and are now known to be the U-2 surveillance flights.

Of interest to behavioral scientists then were the psychophysiological response patterns of the "premium men" who flew high-performance, often top-secret, airplanes. The scientists' interest was a practical one: How do you go about choosing the best man to fly these very dangerous missions?

In 1952, the Human Factors Division of the Air Force Research and Development Command had initiated a program at the AMRL designed to select aviators for special high-altitude research flights. Over a period of years, this test program was extended, using both aviator and university populations. Colonel Don Flickenger had designated the AMRL in 1957 as the premiere center for all aspects of medical research on the new USAF initiative, "Man in Space."

Screening pilots for medical and psychological problems, as well as technical aptitude, had been shown to be essential. The air casualties of the First World War made it evident that not every individual was suited to be an aviator. By World War II, it was noted that a large percentage of the men who went into flight training did not successfully complete it, and Army Air Force psychology personnel developed a screening system called the *stanine* (for "standard nine," where all test scores were adjusted to a standard score between 1 and 9). Every pilot candidate was given a stanine rating on the basis of his scores on the screening tests. The standardized tests that comprised the stanine rating were based on a very practical and common-sense idea. For example, one of the tests asked potential pilot candidates to look at an entire map, then at smaller segments, and to decide which one of the segments was represented in the larger map. Basically, candidates were asked to do tasks that would be a routine part of their job as a pilot. Other tests included special visualization and psychomotor activities that were inherent in the job of an aviator. If a candidate had a stanine score of 7, 8, or 9, the chances were high that he would be able to complete flight training successfully. The stanine rating resulted in a dramatic reduction of the attrition.

Interestingly, it was also true that of candidates with stanines of only 1 or 2, a significant number also were able to complete the training—20 to 30 percent, in fact. As the war progressed the Army Air Force was forced to use lower and lower stanines in order to recruit the necessary number of aviators for the war effort.

Nevertheless, the idea that simple tests could predict who was more likely to do well and more likely to do poorly in flight training was an important concept that has since dominated pilot selection. Considering the enormous cost incurred by the military for training a single pilot, it is not surprising that any system that decreased that expense would be attractive.

About the same time during World War II, psychiatrists were being asked to predict how individual soldiers would adjust to combat. While they proved to be correct about 20 percent of the time, in most cases, they failed dismally. This is not unexpected in light of the fact that being exposed to combat is much less specific than piloting an aircraft (or, to put it another way, the tasks of piloting are fairly well defined and are therefore more amenable to testing). But most psychiatric predictions that an individual will not succeed in a given task had the same result because psychiatrists tend to overpredict failure. This becomes a significant issue, since many potentially competent individuals are therefore excluded. Another way to look at the problem is that psychiatrists tend to underestimate the ability of many individuals to successfully adapt to stressful situations.

Researchers, like Ruff, who specialized in understanding stress and its effects sought to improve psychiatrists' predictions by understanding all aspects of individuals in stressful situations. For this purpose, they intensively studied as many psychological and physiological variables as possible during simulated and real flights.

Using a B-47 aircraft with a three-man crew, extended flight experiments on air-crew interaction and fatigue were conducted. The interior of the B-47 was rigged so that crew members could be instrumented and monitored for several hours while flying at high altitudes. The instruments usually measured heart rate, blood pressure, temperature, respirations, and skin resistance. Often a continuous electrocardiograph could be obtained to measure the electrical activity of the heart. These data were then transmitted to receiving equipment on the ground for analysis.

In addition to studying individual adaptation, Ruff's section was also interested in how *groups* of men performed under stress. For this purpose, the scientists designed an isolation chamber where they could study up to five men under isolated conditions for five days at a time. The research followed a paradigm that had come largely from Ruff's prior experience at the University of Michigan. Subjects were evaluated at every level of experience: from individual biochemical changes to perception, cognition, and group interactions. At every level, measurements were obtained. Group measures evaluated interpersonal interaction. Individual performance was assessed by studying the number of errors made over a period of time. Measures of mood and attitude, as well as individual physiological and biochemical measures, were taken concurrently from each subject.

Although section scientists used isolation as a stressor, many of the pilots they tested, including those who flew the X-15, were not going to be exposed to prolonged periods of isolation. Isolation was easily simulated, however, and it appeared to be stressful to the individual in the same way that the high-altitude surveillance missions were. In fact, prolonged exposure to the chamber induced many of the same psychological effects that were seen on real missions. Many of the pilots who later became NASA astronaut candidates, such as Neil Armstrong, were involved in some of the AMRL research experiments. What is interesting is that no one at the time ever imagined that these studies would someday be applicable to space flight.

On October 4, 1957, an event occurred that changed the course of history forever and initiated the Space Age. The national hysteria that followed the Soviet launches of Sputnik I and later Sputnik II, in November of the same year, is difficult to imagine now. Americans, used to being world leaders in science, quite abruptly had slipped into second place behind their arch rival, the Soviet Union. The normally sedate Eisenhower administration was galvanized into action by American public reaction. The United States must not and would not be left behind. Thus, the *Mercury* program was born out of national anxiety and paranoia about communism. It was logical that under these circumstances the government would turn to the military to help the fledgling civilian space agency select the first candidates for space flight. George Ruff happened to be in the right place at the right time.

## PROJECT *MERCURY*

In November 1957, President Dwight D. Eisenhower established the President's Scientific Advisory Committee. This committee's initial recommendation was to establish a civilian space agency, and the National Aeronautics and Space Act of 1958 was enacted by Congress in July 1958.

A Space Task Group (STG), chaired by Robert R. Gilruth, was established at Langley Field, Virginia, to determine what type of individual would function most effectively in space. These individuals were to be called astronauts, and although no one knew what flying in the vacuum of space would do to the human body, the STG believed that the scientific work going on at the AMRL might be applicable to the space environment. A combination of pure conjecture and real experience with high-performance aircraft operations was to become the basis of the characteristics determined. The primary task for an astronaut, as the STG saw it, was to survive, or to demonstrate the ability of man to fly in space and to return safely. Secondarily, the astronaut would have to demonstrate man's capacity to act usefully under conditions of space flight. The astronaut could also serve as a backup system for all the automatic controls, and this would increase the reliability of spacecraft systems. Finally, the STG concluded, the astronaut would function as a scientific observer and go beyond what the instrumentation and unmanned satellites were able to observe and report.[1]

What kind of person could do these tasks best? The STG considered a number of occupations from which they might draw their first space candidates. The list of possible occupations included aircraft pilots, but it also included balloonists, submariners, deep-sea divers, mountain climbers, explorers, flight surgeons, and scientists. In fact, the original civil service requirements that the STG drew up for the position expected that, since the space agency was civilian and not military, various industries, as well as the Department of Defense, would nominate candidates for consideration as astronauts. It was because of this assumption that the White House forcefully intervened. Eisenhower, it seemed, wanted only active military test pilots to be used as the source of candidates for the astronaut selection.

The cold war was in full swing, and the president was justifiably concerned about national security issues. As a former Army general, he instinctively believed that military personnel could best manage and implement a national space program. So, by presidential decree, all astronaut candidates were to be graduates of military test pilot school and have at least 1500 hours of flying time in high-performance jets. This considerably decreased the number and heterogeneity of possible applicants. Looking back, it is hard to imagine how different the U.S. Space Program would be today if the White House had not made this decision.

## THE PHILOSOPHY OF PSYCHOLOGICAL SELECTION – CLINICAL VERSUS STATISTICAL METHODS

Research in Ruff's section at the AMRL was proceeding along at its usual slow, but steady, pace when, in 1958, the lab personnel were given an unusual request. The managers of a new organization, called the National Aeronautics and Space Administration (NASA), had asked the military services to assist them in screening 110 test pilots who met the initial criteria of 1500 hours in jet aircraft. The purpose of the screening was to decide which of these test pilots would be first to fly in space in Project *Mercury.*

The Air Force Command handed down NASA's request for assistance in the selection of astronauts to the AMRL, and the task of doing the psychological evaluations fell to Ruff and his section. Although Ruff was to be in charge, each of the military services was to be involved in all phases of the selection process, including the psychological evaluations. The Navy's representative to the selection process was Dr. Robert Voas, a psychologist who had also worked on the STG, and the Army designated flight surgeon William Augerson to participate. The actual psychiatric and psychological evaluations on the finalists would be done at the AMRL in Ohio.

The idea of participating in the beginning of the space flight era generated an enormous amount of excitement among Ruff and his colleagues. They saw their task as one of great responsibility as well as one of great opportunity: the chance to study men in the most unique and most hostile environment imagined. The work, as they envisioned it, was a natural extension of all the research studies they had done on pilot and crew isolation and adaptation.

Captain Ruff's team included Captain Edwin Z. Levy, also a psychiatrist and the associate investigator for the project. Captain Victor H. Thaler, Captain John K. Jackson, Dr. Mildred B. Mitchell, and Lieutenant Gilbert E. Johnson also participated as psychological investigators.

With Ruff and Levy in charge, the team developed a basic philosophy that would guide it in the psychiatric screening of those remarkable men who wanted to fly in space. Ruff wrote later that the "concept underlying the astronaut selection program deviated little from that stated by Plato over 2000 years ago—'In the first place, no two persons are born exactly alike, but each differs from each in natural endowments, one being suited for one occupation and another for another.' "[2] But before they could implement specific procedures, they first had to answer the question of what psychological characteristics might be desirable for an astronaut and then determine methods to assess for these desirable characteristics.

Ruff today acknowledges that the evaluation of the first astronaut candidates was very much influenced by two major personality researchers dominant in the field at the time. The first was Raymond Cattell, well known for his work on the description and measurement of personality. Cattell, who was at the University of Illinois, became a consultant to the project. The second scientist whose work was influential in the development of the procedures was Paul E. Meehl, who was chairman of the Department of Psychology at the University of Minnesota.

Meehl had written a book in 1954 called *Clinical vs. Statistical Prediction: A Theoretical Analysis and a Review of the Evidence,* wherein he discussed the philosophical issues of the two conflicting methods of predicting behavior.[3] The two methods in question were referred to as the *clinical* method and the *statistical* (or *actuarial* method). A heated debate raged in the academic community between proponents of one or the other method. The clinicians claimed that "one could not predict or manipulate the behavior of a person by filling numbers into a multiple regression equation."[4] Statisticians, on the other hand, argued that this was, in fact, exactly what they could do. Their dissatisfaction with the clinicians' methodology arose from the inability of the clinicians to prove, or to validate, that their predictions were accurate.

> For example, in response to a demand for validation data, clinicians will sometimes state that they "do not work in a mechanical, additive way" and that the usual statistical procedures are therefore not applicable to their clinical behavior. More often than not, this is hokum. . . . [The clinician] has the obligation of showing statistically that his own predictions, on different assumptions, tend to be correct.[5]

The problem, as Meehl saw it, was "to predict how a person is going to behave."[6] Meehl succinctly summarized the arguments of both sides and examined the existing data to support or dispute each view. Numerous studies were reviewed. T. R. Sarbin, for example, found that clinicians systematically overpredicted the positive, what he called the "leniency error."[7] So in addition to overpredicting failure, clinicians also tended to overpredict success. However, Meehl's review noted that "[W]hen he predicts success, each of the psychiatrists is slightly better than the statistical method (85% and 80% versus 76% correct). However, when predicting failure, each psychiatrist is inferior to the statistician (30% and 50% versus 69% accuracy)."[8]

Meehl concluded that for the purposes of selecting individuals for certain jobs, the critical factor was the prediction of failure, and in such a case, one would have to favor a more statistical approach. Indeed, his book placed the burden of proof on the clinicians, since in another series of reviews, in all but one case, "the predictions made [by the statisticians] actually were either approximately equal or superior to those made by clinicians."[9]

Meehl pointed out that much depended on specifically *what* was being predicted. In the case of the studies he reviewed, the predictions were for (1) success or training in school, (2) recidivism, or (3) recovery from major psychosis. Which method could predict other behaviors was still undecided. However, one continuing problem with the clinical method was that there is a wider variability of skill among clinicians. Some may just be worse at predicting than others are. This problem is not dissimilar to why the "house" is the long-run winner at the casino blackjack tables. The casino dealer, like the statistician, is bound by a set of rules, which are applied to every hand ("draw at 16, hold at 17"), while the player (like the clinician) has the flexibility of not following the rules. Sometimes the player wins big, but over the long run, with many players at different levels of skill, the dealer—and the casino—wins. In spite of this, it is important to remember, as G. W. Allport noted, that "psychological causation is always personal and never actuarial."[10]

In real-world practice, those who favored the clinical approach preferred to use the psychiatric interview and projective-type tests (such as the Rorschach Inkblot test) in their evaluations. Those who favored the statistical approach tended to use multiple-question psychometric tests, such as the Minnesota Multiphasic Personality Inventory (MMPI), which could be given to thousands of individuals and normalized.

Ruff and his colleagues took both sides of Meehl's argument to heart. They were primarily clinicians, but they believed that the astronaut selection project was extremely important, and they wanted to make sure that both clinical and statistical approaches were used in their screening. From the beginning, they had every intention of refining the psychological selection battery over a number of years on the basis of the outcome data and thus identifying, without prejudice, which factors were most important in predicting success as an astronaut. The statistical method employs variables that could be measured objectively. By finding how these variables are correlated with the performance of those who have held a given assignment in the past, predictions could be made of how well others will do in the future. Since both the test results and the criteria of success or failure are expressed quantitatively, objective and precise relationships between test data and effectiveness in the assignment can be identified.

To the extent that they were able, Ruff and his colleagues tried to quantify all their clinical data by developing a numerical rating system. In that way they would also be able to analyze (using a statistical method) their clinical information and see what data were predictive.

Both the statistical and the clinical approach have serious shortcomings when applied to the problem of astronaut selection. For one thing, since this was the first time astronauts had ever been selected, there were no data yet available on space-crew performance; no one had flown in space yet. How could psychiatrists and psychologists objectively (rather than arbitrarily) decide which factors might be important? Also, the lack of computer power in those days very much limited the practicality of the statistical approach, particularly since no relevant actuarial tables were even available at the time. While scientists could make intelligent guesses by examining previous results from other demanding missions, projecting from one set of circumstances to another involves the kind of subjective judgments that the statistical method seeks to avoid in the first place.

Another disadvantage was that many psychological variables could not—then or now—be measured objectively (although this situation has markedly improved now, with better test construction, reliability, and validation). Intellectual functions could be assessed quantitatively, but methods for measuring personality characteristics were far from satisfactory. This is true even today, primarily because the dimensions of personality are not well defined. There are no concepts such as "length" or "color" for describing emotions. In the absence of such dimensions, personality tests often give a less-than-complete picture of the person they seek to measure. In recent years, much work has been done to refine the personality constructs used for self-report psychometric instruments and to determine both the reliability (does the test always measure the same thing?) and validity (does the test do what it says it will?) of each test, but at the time of the first astronaut selection, these analyses were obviously not available.

The clinical method also had its drawbacks for astronaut selection. After examination of background information, interview results, and test scores, clinical judgment is used to predict how a candidate will carry out the assignment. Unfortunately, such judgments are often unreliable. Clinicians tend to rely excessively on their own experience, and they often make imperfect decisions based on a limited number of experiences. By evaluating the importance of their data according to personal biases and theories, they may weigh some factors too heavily and others not heavily enough.

The best solution, Ruff and Levy concluded, was to combine as far as possible both clinical and statistical techniques. The computer can weigh and interpret objective data more accurately than can the clinician. But the clinician can use information that cannot be quantified precisely for the computer.

Even today, a comprehensive selection program should employ the best features of each approach. During the early stages of any selection program, where guidelines are not firm, the broadest possible measures must be employed. At the same time, promising objective tests should be administered and data recorded. Results should be interpreted according to the best clinical judgment obtainable. As data accumulate, the best predictive techniques can be identified—regardless of whether they are clinical or statistical.

In the case of astronaut selection, the AMRL team had to contend with a paradoxical situation: Validation could not be ideal, since the most effective validation studies required that *all* candidates evaluated be selected and assigned to space missions. If those for whom success is predicted perform significantly better than do those who are predicted to fail, the usefulness of the measures is established. But this approach, though ideal, obviously could not be applied to astronaut selection. The purpose of the evaluation in the first place was to select only those for whom success was predicted. Ruff had to content himself with the fact that validation of the psychological selection criteria would eventually depend on examination of all data to determine which measures accurately predicted performance and which ones did not.

Validation, it is important to note—particularly with respect to what happened later on—is essential for both the statistical and the clinical approaches. Even where clinical judgment is primarily relied on, consistent revision of criteria based on outcome is necessary to substantiate the selection criteria. Both clinicians and computer must examine the accuracy of their predictions and determine which information has been weighed too heavily and which has not been weighed heavily enough. At the same time, clinicians must be alert for factors which may have influenced the success of the mission, but for which neither they nor the computer had possessed proper data.

After deciding on the underlying philosophy, the psychological team outlined the steps that were necessary for the psychological evaluation. First, the specific job of the astronaut had to be defined. In other words, what was the applicant expected to do? The determination of the personal characteristics required would logically derive from the job description. Once the personal characteristics were determined, reliable methods of assessing the characteristics in applicants had to be found. And, finally, after candidates were selected for the job, it was essential to follow those individuals and their subsequent job performance. Performance data would then feed back to validate the original selection criteria or justify changing the criteria.

The *Mercury* program psychological selection had four basic, but distinct, tasks:

1. *Determination of job requirements:* Selection criteria cannot be established without knowing what an astronaut is chosen to do. Evaluators must have a good idea not only of the crew's duties, but also of the conditions under which they must be carried out.
2. *Determination of personal characteristics requirements:* After the nature of the mission has been defined, it must be determined what capacities a person will need to accomplish it. Decisions must be made on which physical and mental

characteristics are desirable and which are undesirable for the success of the mission.

3. *Determination of assessment methods:* Reliable tests must be devised to determine who has the best combination of desirable and undesirable characteristics.
4. *Validation of the selection criteria:* Once a sufficient number of space missions have been carried out, predictions must be checked against performance. Measures that correlate well with criteria of success are retained and improved. Those with little selection value are discarded. New measures are devised in the light of increased experience.[11]

With the fundamental issues decided, the AMRL psychological team proceeded to develop the specific selection procedures they would use for the *Mercury* selection. They were assisted in this process by the Navy's Robert Voas and the Army's William Augerson. The AMRL also invited Bryce Hartman to participate. Hartman, an Air Force psychologist stationed at Brooks Air Force Base in San Antonio, Texas, was doing research on complex behavior, and one of his tests, the complex behavior simulator, was being considered to measure performance in the evaluation.

## JOB DESCRIPTION

Voas, who was also a member of the STG, later summarized in memoirs written for NASA archives the major tasks that the group decided the astronauts in Project *Mercury* would have to accomplish.[12] These tasks were (1) *sequence monitoring*—monitoring all the critical phases of the space mission, such as the staging of the booster, the separation of the escape tower, the firing of the retrorockets, and the deployment of the parachutes; (2) *systems management*—operating all the onboard systems and managing the critical consumable supplies so as to ensure that any out-of-tolerance condition is recognized and corrected before a critical situation develops; (3) *attitude control*—maneuvering the vehicle to the proper relationship to Earth whenever it is required during the mission; and (4) *research observations*—carrying out the special activities related to research and the evaluation of spacecraft function under flight conditions.

It was generally believed that space flight would subject the astronaut to stressful conditions—not unlike those experienced by the pilots of high-altitude aircraft—such as high acceleration, reduced pressure, heat, noise, vibration, and even weightlessness. All of these factors would increase the difficulty of satisfactory performance on the part of the astronaut.

The most obvious professional requirement was that the astronaut exhibit a high skill-level in the pilot's role. Additionally, each astronaut would need to have "appropriate" personal characteristics, as well as a high level of physical fitness.[13] More specifically, the astronaut would have to have a good knowledge of engineering; a good knowledge of operational procedures typical of aircraft or missile systems; broad general scientific knowledge and research skills; high intelligence; and excellent psychomotor skills, similar to those required to operate aircraft.

In the area of general psychological characteristics, the candidate would have to demonstrate good stress tolerance; an ability to make decisions; an effective ability to work with others; emotional maturity; and a strong motivation for *team*, rather than personal objectives. From a medical perspective, the candidate would have to be free from disease, be able to tolerate the physical stresses of space flight (reduced pressure, weightlessness, high temperature, and other stressors), and have a medium or small build so as to be able to fit into the small *Mercury* capsule.

## PERSONAL REQUIREMENTS – DEVELOPING THE PSYCHOLOGICAL REQUIREMENTS

With the STG guidelines on what was to be expected of the *Mercury* astronaut, Ruff and the psychological team developed the specific psychological requirements:

1. Candidates should have a high level of general intelligence, with abilities to interpret instruments, perceive mathematical relationships, and maintain spatial orientation.
2. There should be sufficient evidence of drive and creativity to ensure positive contributions to the project as a whole.
3. Relative freedom from conflict and anxiety is desirable. Exaggerated and stereotyped defenses should be avoided.
4. Candidates should not be overly dependent on others for the satisfaction of their needs. At the same time, they must be able to accept dependence on others when required for the success of the mission. They must be able to tolerate either close associations or extreme isolation.
5. The astronaut should be able to function when out of familiar surroundings and when usual patterns of behavior are impossible.
6. Candidates must show evidence of ability to respond predictably to foreseeable situations, without losing the capacity to adapt flexibly to circumstances that cannot be foreseen.
7. Motivation should depend primarily on interest in the mission rather than on exaggerated needs for personal accomplishment. Self-destructive wishes and attempts to compensate for identity problems or feelings of inadequacy are undesirable.
8. There should be no evidence of excessive impulsivity. The astronaut must act when action is appropriate, but refrain from action when inactivity is appropriate. He or she must be able to tolerate stress situations positively, without requiring motor activity to dissipate anxiety.[14]

The AMRL behavioral scientists developed a series of seventeen psychological categories on which each candidate would be rated, using a ten-point scale. In this way, they hoped to quantify the clinical impression of each candidate. All interview and test information obtained from candidates would be used to rate each individual. The psychiatrists rated the candidates on the following categories:[15]

1. *Drive:* An estimate of the total quantity of instinctual energy.
2. *Freedom from conflict and anxiety:* A clinical evaluation of the number and severity of unresolved problem areas and of the extent to which they interfere with the candidate's functioning.
3. *Effectiveness of defenses:* How efficient are the ego defenses? Are they flexible and adaptive or rigid and inappropriate? Will the mission deprive the candidate of elements necessary for the integrity of his defensive system?
4. *Free energy:* What is the quantity of the neutral energy? Are defenses so expensive to maintain that nothing is left for creative activity? How large is the "conflict-free" sphere of the ego?
5. *Identity:* How well has the candidate established a concept of himself and his relationship to the rest of the world?
6. *Object relationships:* Does he have the capacity to form genuine object relationships? Can he withdraw object cathexes when necessary? To what extent is he involved in his relationships with others?
7. *Reality testing:* Does the subject have a relatively undistorted view of his environment? Have his life experiences been broad enough to allow a sophisticated appraisal of the world?
8. *Dependency:* How much must the candidate rely on others? How well does he accept dependency needs? Is separation anxiety likely to interfere with his conduct of the mission?
9. *Adaptability:* How well does he adapt to changing circumstances? What is the range of conditions under which he can function? What are the adjustments he can make? Can he compromise flexibly?
10. *Freedom from impulsivity:* How well can the candidate delay gratification of his needs? Has his behavior in the past been consistent and predictable?
11. *Need for activity:* What is the minimum degree of motor activity required? Can he tolerate enforced passivity?
12. *Somatization:* Can the candidate be expected to develop physical symptoms while under stress? How aware is he of his own body?
13. *Quantity of motivation:* How strongly does he want to participate in the mission? Are there conflicts between motives—whether conscious or unconscious? Will his motivation remain at a high level?
14. *Quality of motivation:* Is the subject motivated by a desire for narcissistic gratification? Does he show evidence of self-destructive wishes? Is he attempting to test adolescent fantasies of invulnerability?
15. *Frustration tolerance:* What will be the result of failure to reach established goals? What behavior can be expected in the face of annoyances, delays, or disappointments?

16. *Social relationships:* How well does the subject work with a group? Does he have significant authority problems? Will he contribute to the success of missions for which he is not chosen as pilot? How well do other candidates like him?
17. *Overall rating:* An estimate of the subject's suitability for the mission. This is based on interviews, test results, and other information considered relevant.

These categories are not independent dimensions. They represent several different levels of abstraction. They were primarily developed for organizing and interpreting data, not for quantifying data. The psychiatrists believed that this would be a useful way to compare one subject with another. Many of the categories were included because of their importance to the specific *Mercury* mission (e.g., 8, 9, and 11–16).[16]

The psychiatric evaluators were instructed by NASA to rate the candidates as to whether they were "well qualified," "qualified," or "not qualified." After the evaluations were completed, they were told to rank all the candidates. In fact, the seven who were chosen came from the top nine psychologically ranked candidates.

## THE MEASURES

The next step was to determine how to actually measure the desired psychological qualities in the candidates. Many of the methods and tests chosen for this purpose were already in use at the AMRL and at Brooks Air Force Base and were considered at the time to represent a state-of-the-art psychological evaluation (in fact, several of the tests are still used today, over thirty years later). Other reasons for including a particular test or excluding one related simply to personal preference or were an attempt to be overinclusive (remember that at the beginning of a selection process, it is necessary to include the broadest possible base of experience). When several members of the team disagreed on an approach, generally both approaches would be included in the battery, since the scientists were aware that, in most cases, there were no data to support one approach over another.

### The Psychiatric Interview

Psychiatric interviews of each candidate by each of the two psychiatrists (Ruff and Levy) were planned. One interview was devoted to a review of the candidate's life history and current life adjustment, while the other interview was relatively unstructured and open-ended, giving the candidate a chance to present himself. Ruff and Levy intended to compare notes and pool information, ending up with a combined psychiatric rating of each candidate. Areas of doubt and disagreement would be recorded for subsequent investigation.

### Personality Tests

Two types of personality tests were to be given to candidates. The first type, called *projective personality tests,* were favored by psychologists such as Bryce Hartman and by most of the psychiatrists. Projective tests are, by their very nature, difficult to score quantitatively. They create a situation in which the psychological issues of the person taking the test can be reflected. At the time of the *Mercury* selection, these types of tests were extremely popular and were favored by those psychiatric clinicians who believed that psychiatry was not a "mechanical" science.

The other type of personality test is the objective test. Objective personality tests are generally self-report questionnaires that have been standardized against a normal population. All tests given to the candidates are listed in Table 1.1. A detailed description of each test is included in Appendix 1.

### Performance Tests

A number of tests, some of them quite demanding, were included by the medical team. These tests assessed the ability of the candidate to perform under stressful conditions. Two of these tests—the complex behavior simulator and the isolation test—were part of the psychological evaluation. The former, developed by Bryce Hartman, measured the capacity of the individual to simultaneously perform several complex psychomotor and cognitive tasks. The isolation test was done under the same conditions as the isolation research the section had been involved in, but unfortunately, due to time constraints, candidates only spent two hours in isolation.

### Ability and Intelligence Tests

Many of the tests given to the astronauts were tests of ability in specific areas (e.g., engineering) and tests of intelligence. These tests are described in detail in Appendix 1.

## THE *MERCURY* SELECTION PROCESS

The Project *Mercury* astronaut selection process had four distinct phases in its implementation. The first phase involved the screening of military records for suitable candidates per the president's criteria: a graduate of test pilot school with 1500 hours or more flying jet aircraft. All four military services participated, although the Army did not have any qualified candidates, since none of their personnel had gone to test pilot school. Over 500 service records were screened by a committee consisting of Dr. Stanley C. White, a physician for the National Advisory Committee for Aeronautics (NACA); Dr. Robert B. Voas, Navy psychologist; Dr. William S. Augerson, an Army flight surgeon; Charles J. Donlon, the assistant director for

Table 1.1
*Mercury* Psychometric Tests

**Measures of Motivation and Personality**

1. Rorschach Inkblot Test
2. Thematic Apperception Test
3. Draw-A-Person Test
4. Sentence Completion Test
5. Minnesota Multiphasic Personality Inventory (MMPI)
6. Who Am I?
7. Gordon Personal Profile Test
8. Edwards Personal Preference Schedule
9. Shipley Personal Inventory
10. Outer-inner Preferences
11. Pensacola Z-scale
12. Officer Effectiveness Inventory
13. Peer Ratings

**Measures of Intellectual Functions and Special Aptitudes**

1. Wechsler Adult Intelligence Scale (WAIS)
2. Miller Analogies Test
3. Raven Progressive Matrices
4. Doppelt Mathematical Reasoning Test
5. Engineering Analogies
6. Mechanical Comprehension Test
7. Air Force Officer Qualification Test
8. Aviation Qualification Test (U.S. Navy)
9. Space Memory Test
10. Spatial Orientation Test
11. Gottschaldt Hidden Figures Test
12. Guilford-Zimmerman Spatial Visualization Test

**Stress Experiments Simulating Conditions Expected during the Mission**

1. Pressure Suit
2. Isolation
3. Complex Behavior Simulator
4. Acceleration
5. Noise and Vibration
6. Heat

Project *Mercury*; Allen Gamble, a personnel psychologist; and Warren O. North, chief of Space Flight Programs and an engineer test pilot.[17]

Only 110 pilots met the basic criteria. They included 5 Marine pilots, 47

Navy pilots, and 58 Air Force pilots. The 110 potential candidates were divided into three groups and invited to Washington, D.C., for a briefing. In February 1959, the first two groups of sixty-nine pilots arrived at the Pentagon, where they were briefed about the new program and why they were here. The pilots were asked to volunteer to go through the extensive and grueling selection process that was at that moment being set up.

Fifty-five members of the group volunteered, and the process of choosing the right stuff began. The second phase of the process took place immediately, in the Pentagon. All the volunteers submitted a detailed personal history statement, and an initial brief psychiatric interview was conducted by Ruff or Levy. In addition to the brief psychiatric interview, the volunteers in Washington received several of the intelligence and ability tests (including the Wechsler Adult Intelligence Scale, Miller Analogies Test, Doppelt Mathematical Reasoning Test, and Engineering Analogies Test). The remainder of the psychological evaluation was to be completed at the AMRL on those individuals who made it through the Washington screening and an intensive medical exam scheduled next at the Lovelace Clinic in Albuquerque, New Mexico.

George Ruff and Ed Levy each interviewed every candidate in the group of fifty-five volunteers. The seventeen traits listed previously were rated, and the result of the combined rating was reported to the selection committee, consisting of Donlon, North, and Gamble. The only individuals eliminated at this point for psychological reasons were those who did not make the arbitrarily determined minimums for the cognitive tests, including IQ.

Bill Augerson, the Army flight surgeon, carefully evaluated every candidate's medical history and eliminated individuals with any medical problems, as well as six pilots who unfortunately were too tall to fit into the *Mercury* capsule. A technical interview assessing job capabilities was given to every volunteer by Donlan, North, and Gamble. All together, twenty-one of the volunteers were eliminated for a variety of psychological, medical, and technical reasons during these first two phases of the selection in Washington. A total of sixteen candidates declined to participate in or dropped out during the first two phases.

The thirty-two individuals remaining reported to the Lovelace Clinic for the third phase of the selection process—rigorous and comprehensive medical examinations. Each candidate spent seven and a half days at the clinic undergoing medical tests. The responses and attitudes of the applicants were depicted with uncanny accuracy in the book *The Right Stuff.*[18] However, only one candidate was eliminated at this stage, in spite of all the medical screening. These pilots had already undergone considerable medical evaluation in the course of their routine duties, and the results were not surprising.

The fourth phase of the process was scheduled at the AMRL in Dayton,

Ohio. The Air Force Research and Development Command under Brigadier General Don Flickinger—command surgeon and also a member of the NASA Special Committee on Life Sciences—worked closely with the STG to provide the general direction of this phase of the program, which began on February 16, 1959, and was completed on March 27, 1959.[19]

George Ruff remembers:

> We were told by NASA Management that we were not to choose the people who were going to get involved. They would do that on the grounds of the requirements for the mission. We were simply to identify any people we think were risky from a psychiatric standpoint. That is exactly the perspective from which we operated. Originally, NASA intended to choose twelve individuals to become astronauts, eventually they actually chose seven.[20]

The chronology of events and the attrition during the four phases of the *Mercury* selection can be found in Appendix 1.

NASA had asked the behavioral scientists how much time was required per candidate. Ruff, Levy, Voas, and Hartman met to work out the schedule.

> When we put the Navy Program tests and the AMRL program tests together, it came out to 30 hours of testing! I knew that was ridiculous. They were never going to allow 30 hours of psychiatric evaluation and psychological testing! But we put it in, and lo and behold, they said OK, we'll do it. They told us that we could work at night. So after a day of stress testing, the pilots would come back and do the psychological tests with us. To our amazement, they did the whole program as we set it up. I felt it was a little bit like Parkinson's law. The amount of testing will expand to the amount of time psychologists are given to test. That is the reason so much was done.[21]

## RESULTS OF THE PSYCHIATRIC INTERVIEW

Ruff and Levy conducted additional psychiatric interviews to compare to their initial screenings of the candidates in Washington. Today, reading the summaries of these interviews is fascinating, particularly because of the relationships that developed between the psychiatrists and the applicants. It is clear from the reports that the psychiatrists saw these men as somewhat remarkable. Candidates are described as "extremely well-integrated," "exceptionally mature," and "highly intelligent." Every once in a while, some negative comments crept into the evaluations: "loud and obnoxious," "arrogant," and "seems to think very highly of himself and not so well of others." But on the whole, even the irritating candidates were felt to be reasonable candidates for the program.

Candidates were questioned about their adolescence, their families, and their goals in life; they were given the chance to "sell" themselves and discuss what they felt were their strong and weak points. To assess motivation, candidates were queried on their thoughts about the *Mercury* program they had applied for.

"The Program is carefully thought-out, it's not what they're saying in the press; it's not a 'half-cocked bid for publicity.' "

"If I had any doubts, I would have quit two tests ago, on that damned Harvard step," stated one candidate who had developed leg cramps on the test.

"It's a chance to get in on the ground floor of the new era—a chance to do something really important," said another.

The general feeling among the candidates was that this was something exciting and different. While they all stated how much they enjoyed their military flying, the opportunity to get into something new, to "go where no man has gone before," appealed to them in almost a fundamental or essential kind of way.

"I want this very badly," admitted one candidate.

These men were not the sort who thought very introspectively about themselves or their feelings. Very few had any real insight into their own motivations or drive to achieve. They tended to deny or greatly minimize the potential dangers of both their current work as test pilots and space flight. When asked about the drawbacks of becoming a *Mercury* astronaut, one candidate expressed it this way, "Well, I guess flaming out would be a drawback, but that can happen in what I'm doing now."

The psychiatrists were very impressed. In spite of the lack of insight, in spite of the obvious denial of danger, all of the thirty-one men were thought to be extremely high functioning, extraordinarily resourceful, and very sure of themselves. It was also apparent from the personal histories recounted to the psychiatrists that in many ways these men were not very different from most other people. Some had stern, disciplinarian fathers, loving mothers, and the usual conflicts with siblings. What is not so typical is that very few of them had to contend with broken homes, none had significant mental or physical trauma as a child, and none was experienced in dealing with failure in any aspect of their lives. A few were more obviously insecure than were others; some had overt anxiety when questioned about their life, while others were more comfortable with the most intimate of questions and seemed to even enjoy the psychiatric probing. Many were able to hide any anxiety about passing or failing the numerous tests very well. Most of them were exactly where they wanted to be in their life, and though they wanted to be astronauts very badly, on the whole they were pretty much satisfied with their lives at the time.

The mean age of the group was thirty-three, with a range from twenty-seven to thirty-eight. All but one were married. Twenty-seven were from intact families. Twenty were only or eldest children. It is perhaps worth noting that four of the seven men ultimately chosen were named "Junior."[22]

Captains Ruff and Levy found that the 31 men had a wide range of motives, making it difficult to generalize about the dynamics underlying the desire to pursue a career as an astronaut. The reasons they listed why they decided to volunteer for Project *Mercury* were a combination of professionalism and love of adventure. One candidate commented, "How could anyone turn down a chance to be part of something like this?"

The two psychiatrists revised very few of their clinical ratings on the basis of the second psychiatric interviews performed at the AMRL. All of the candidates were felt to be qualified.

## RESULTS OF THE PSYCHOMETRIC TESTS

Psychological tests of thirty-one men indicated a high level of intellectual functioning. The mean full-scale Wechsler Adult Intelligence Scale (WAIS) scores for the seven selected ranged from 130 to 141, with a mean of 135. The verbal and performance subtests were both consistently high, with verbal IQ slightly higher in candidates selected than in those not selected.

Projective measures validated the healthy adaptations to life that were seen in the psychiatric interviews. Responses to the Rorschach were well organized and showed no evidence of any bizarre or psychotic thinking in any of the candidates. While the responses to the Rorschach or the Thematic Apperception Test (TAT) were not particularly rigid, neither did they suggest much imagination and creativity. Aggression and aggressive impulses tended to be expressed in actions rather than fantasies.

Objective personality measures, particularly the MMPI, revealed a personality pattern that was significantly different from that of the general population, but did not differ significantly among candidates. Figure 1.1 displays the MMPI profile of the typical *Mercury* candidate. The most striking aspects of this profile, from a clinical perspective, are the V-shaped L, F, K scale configurations and the low social introversion (Si) scores compared to norms. The LFK configuration is frequently seen in very highly defensive individuals who are trying to project themselves well in the testing. This type of profile is probably not uncommon in job applicants in general and is certainly characteristic of astronaut job applicants, as we shall see.

The Si scale is low compared to a normal population. Many of the candidates were extremely socially extroverted and friendly. All scales were in the normal range, even when corrected for defensiveness (K scale). What was most interesting was how little individual candidates deviated from the general profile. The standard deviations were quite small, considering the number of applicants, and this suggests a very homogeneous population.

## RESULTS OF THE STRESS TESTING

Behavior during the isolation and the complex behavior simulator tests showed evidence of great adaptability. No candidate terminated isolation

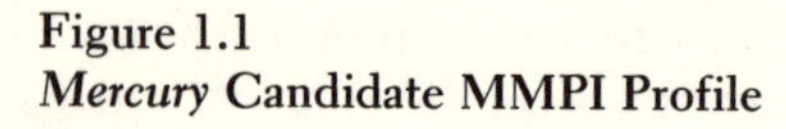

**Figure 1.1**
***Mercury* Candidate MMPI Profile**

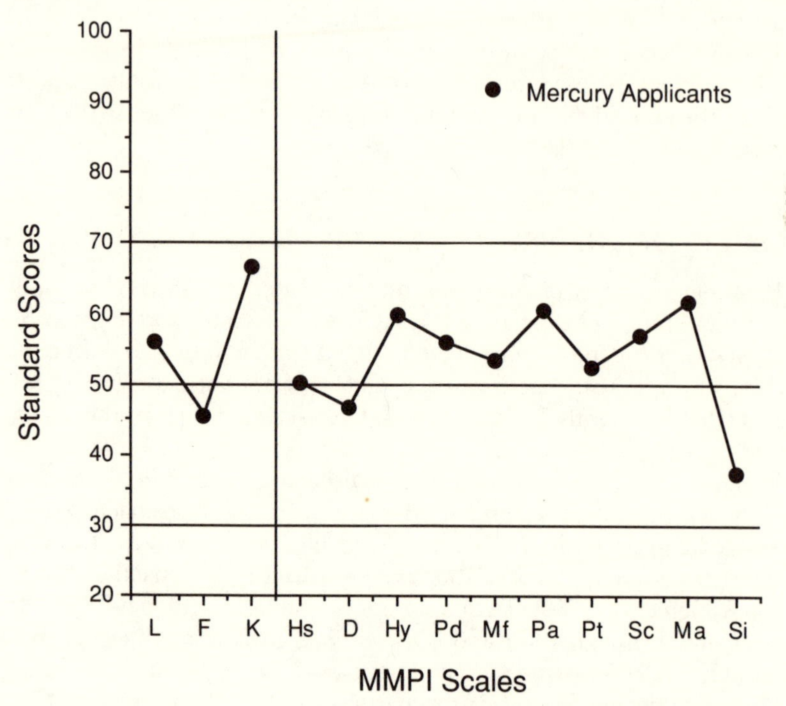


prematurely, and none viewed it as a difficult experience. One candidate even made up a poem about his experience, which he titled "Levy's Box":

> On TV quiz shows, dough is loose
> like water running free,
> And when I had my chance at that
> I took some home with me.
> But let me tell you here and now,
> a pearl of salient truth:
> The pay is pretty lousy here,
> in Levy's isolation booth.

As might be expected for this brief exposure, no perceptual changes were reported. Fifteen subjects "programmed" their thinking (i.e., they structured or forced themselves to think about certain things) during the period of isolation. In five of these men, the attempt to organize thoughts was considered evidence of an overly strong need for structuring. Sixteen permitted random thought, relaxed, and enjoyed the experience. Most slept at least part of the time. Since the amount of time allotted for the isolation test was only two hours, the scientists did not expect to see the kind of behavior that could routinely be elicited when individuals were kept for much longer times.

As for the complex behavior simulator, while some of the candidates did better than others—some required assistance from the scientists to complete the task—no one was eliminated as a result of poor or inferior performance. All subjects met the performance criteria within the time allowed except two who were allowed an extra 1 to 1½ minutes to achieve the criteria.

Performance on the simulator was evaluated in relation to the scores of an "ideal" subject. Measures of proficiency (based on response time) and efficiency (based on the number of signals processed) were derived for each subject. On the average, this special group showed a decrease of 23 percent in efficiency and an increase of 16 percent in proficiency, values generally like those demonstrated by the "ideal" subject.

## THE FINAL SELECTION

In an article written for the *American Journal of Psychiatry* in 1959, Ruff and Levy wrote:

> Impressions from the interviews were that these were comfortable, mature, well-integrated individuals. Ratings in all categories of the system used consistently fell in the top third of the scale. Reality testing, adaptability and drive were particularly high. Little evidence was found of unresolved conflict sufficiently serious to interfere with functioning. Suggestions of overt anxiety were rare. Defenses were effective, tending to be obsessive-compulsive, but not to an exaggerated degree. Most were direct, action-oriented individuals, who spent little time introspecting. Although dependency needs were not over-strong, most showed the capacity to relate effectively to others. Interpersonal activities were characterized by knowledge of techniques for dealing with many kinds of people. They do not become over-involved with others, although relationships with their families are warm and stable.
>
> Because of the possibility that extreme interest in high performance aircraft might be related to feelings of inadequacy in sexual or other areas, particular emphasis was placed on a review of each candidate's adolescence. Little information could be uncovered to justify the conclusion that unconscious problems of this kind were either more or less common than in other occupational groups.
>
> Most of the final thirty-one men had made excellent school and social adjustments. Many had been class presidents or showed other evidence of leadership.
>
> Candidates described their feelings about flying in a variety of terms: "something out of the ordinary," "a challenge," "a chance to get above the hubbub," "a sense of freedom," "an opportunity to take responsibility." A few look upon flying as a means of proving themselves or of building confidence. Others consider it a "way for good men to show what they can do." . . . Danger is admitted, but deemphasized by the conviction that nothing will happen to them. But this seems to be less a wishful fantasy than a conviction that accidents can be avoided by knowledge and caution. They believe that risks are minimized by thorough planning and conservatism. Very few fit the popular concept of the daredevil test pilot.[23]

The NASA Astronaut Selection Board decided that the general appearance and social effectiveness of the individual would be kept in mind; but, in keeping with the need to select the most technically qualified individuals, no one would be eliminated for lack of social effectiveness unless he was grossly inept in this area.[24] Since all of the candidates were reported to be highly personable and were relatively effective in communicating their ideas, this did not prove to be an issue. In the end, the Selection Board chose seven individuals with the following rankings by the psychiatrists: first, second, third, fifth eighth, tenth and fifteenth.[25]

Whether these rankings meant that the NASA Selection Board was actually choosing the people because of the psychiatric recommendation or that the behavioral scientists were simply looking at the same qualities as the Selection Board is still controversial. Ruff believes it was the latter. His group did a factor analysis of all the psychological data afterward, and what really saturated everything was a sort of "social skills" factor. The psychology team had collected peer ratings and was familiar with the personnel records of the candidates. Team members therefore knew which candidates were ranked as superior by their military supervisors and were able to analyze after the selection what factor or factors came into play. The major factor that predicted which individuals were ranked better was how they had done previously. It made sense that a person who had done well in the past was most likely to do well in the future—a simple concept that has been regularly supported by the psychiatric literature.

In the early space flight programs such as *Mercury,* what was expected of the selected astronauts was not fundamentally different from what had been expected of them as pilots. The space missions envisioned at that time were of very short duration—from twenty minutes to about six hours. At the time, no one imagined that astronauts would fly even as long as twenty-four hours. What Ruff and his colleagues were really evaluating the applicants for was tolerance of an *acute* stress environment: "I really felt that we were picking what other people might have picked and never really had any feeling that the psychiatric screening itself had been that important."[26]

On April 2, 1959, NASA announced the names of the seven *Mercury* astronauts, and the future began.

## NOTES

1. Joseph D. Atkinson and Jay M. Shafritz, *The Real Stuff: A History of NASA's Astronaut Recruitment Program* (New York: Praeger, 1985), 31.
2. George E. Ruff, "Selection of Crews for Space Flight" (unpublished paper, 1962).
3. Paul E. Meehl, *Clinical vs. Statistical Prediction: A Theoretical Analysis and a Review of the Evidence* (Minneapolis: University of Minnesota Press, 1954).
4. Ibid., 10–11.

5. Ibid., 14, 28.

6. Ibid., 1.

7. T. R. Sarbin, "A Contribution to the Study of Actuarial and Individual Methods of Predictions," *American Journal of Sociology* 48 (1942): 593–602.

8. Paul E. Meehl, *Clinical vs. Statistical Prediction*, 96.

9. Ibid., 119.

10. G. W. Allport, *The Use of Personal Documents in Psychological Science*, S.S.R.C. Bulletin No. 49 (1942), 146.

11. Ruff, "Selection of Crews for Space Flight."

12. Robert B. Voas, "Operations Part of the *Mercury* Technical History" (preliminary draft, NASA-JSC Archives, 1963), 3.

13. Ibid., 4.

14. George E. Ruff and Edwin Z. Levy, "Psychiatric Evaluation of Candidates for Space Flight," *American Journal of Psychiatry* 116 (1959): 385.

15. Ibid.

16. Ibid.

17. Atkinson and Shafritz, *The Real Stuff*, 37–38.

18. Tom Wolfe, *The Right Stuff* (New York: Farrar, Straus & Giroux, 1979).

19. Charles L. Wilson, ed., *Project Mercury Candidate Evaluation Program*, WADC Technical Report 59-505 (Wright-Patterson Air Force Base, Ohio: Wright Air Development Center, 1959), 3.

20. Patricia A. Santy, ed., "The NASA In-House Working Group on Psychiatric and Psychological Selection of Astronauts: Summary and Transcripts" (unpublished paper, 1988), 18.

21. Ibid., 19.

22. Ruff and Levy, "Psychiatric Evaluation of Candidates for Space Flight."

23. Ibid.

24. Atkinson and Shafritz, *The Real Stuff*, 30.

25. Carlos J. G. Perry, "Psychiatric Support for Man in Space," *International Psychiatry Clinics* 4 (1967): 197–221.

26. Santy, ed., "The NASA In-House Working Group on Psychiatric and Psychological Selection of Astronauts," 20.

# 2

# One of Those Interesting Social Phenomena—*Gemini* and *Apollo*

> It is as if the absence of reports concerning man as a psychological entity were a product of denial in the psychoanalytic sense of the term. If this be true, of course, it would only underscore the importance for the place which psychiatry may eventually have in support of man's progressive exploration of space.
>
> —Carlos Perry[1]

## THE RESEARCH PLAN TO VALIDATE THE *MERCURY* PSYCHOLOGICAL SELECTION CRITERIA

Initially NASA was impressed with the thoroughness of the psychiatric and psychological evaluations during the *Mercury* selection process. George Ruff and Edwin Levy were asked to continue their research on the seven selected astronauts and to participate in studying them during training and actual space flight. This was exactly what Ruff's section had planned for from the beginning.

In order to validate their selection criteria, they believed it would be necessary to follow the selected astronauts for a number of years. While there were some obvious difficulties (in particular, the total number of astronauts was only seven, so the number of subjects would probably be insufficient for some years in order to make any significant conclusions), Ruff and his colleagues developed a research protocol that would follow the astronauts and assess their behavior in both training and flight activities. The scientists had confidence in the fact that, over time, enough data would be obtained on a sufficient number of astronauts to validate their selection criteria or justify changing them.

In general terms, the study they developed involved the observation of psychological and physiological functioning before and during flights in order to assess the degree of stress imposed by suborbital and orbital flight. Ruff hoped that this type of data would elucidate adaptive behavior under stress and that such information could then be used in future selection and training programs. This was an integral part of the original psychological selection strategy.

The selection of the *Mercury* astronauts was made by NASA management on the basis of the professional background and knowledge of the candidates. Astronauts were chosen from a group screened by job interviews, medical examinations, stress testing, and psychological evaluations. As in any assessment program, these evaluations were based on defining the job requirements, deciding the personal qualifications required for the job, and selecting or devising tests to assess these qualities.

Many of the assumptions about what psychological qualifications were needed had been purely speculative. After all, space flight had never been attempted before. It was only because of President Eisenhower's insistence that test pilots made up the only candidate pool evaluated for the position of *Mercury* astronaut. In the beginning, individuals from many different professions were deemed potential candidates for the job. Also, it is worth noting that there was a great controversy among engineers and other experts about whether or not the astronauts would be simple, *passive* monitors of automatic systems or whether they would exercise *active* control over their vehicles.

Ruff and his psychology colleagues freely admitted that it had been extremely difficult to decide which personality qualities might be desirable and which undesirable for space flight. Choosing the method to evaluate those qualities posed similar problems. These were some of the reasons why the group opted for a broad range of tests, preferring to narrow the battery down after outcome data (i.e., data from training and actual space flight) had shown which factors were significant.

The goals of the original study developed by Ruff were to (1) determine whether significant psychophysiological changes are produced by suborbital or orbital flight, (2) assess the degree of stress imposed on the astronauts, (3) investigate mechanisms employed for maintaining adaptive behavior under stress, and (4) provide data for application to future pilot selection and training programs. The study was not dissimilar to the research already being done at the Aeromedical Research Laboratory (AMRL); in other words, it was wide-ranging and looked at astronauts from "whole person" and behavior down to the physiology of a single cell. The approach was research-oriented, rather than clinical, which made sense in view of the fact that there were no data—physiological or psychological—about man in space. It also made sense in the context of NASA's original charter, which focused on "research and development."

Recognizing the importance and complexity of the study, Ruff brought in collaborators from the universities where such research was state of the art. Sheldon Korchin and Kristen Eik-Ness, researchers whose expertise was in stress biochemistry and psychoneuroendocrinology, were included in the research effort to assess functioning on the physiological level. The physiological measure most accessible to the researchers was heart rate. Biochemical studies on urine and blood collected before and after flight, analyzing these fluids for catecholamines and steroids, were also done whenever possible. Experiments were controlled for relaxed situations, for stressful training missions, and for the diurnal cycle.

The researchers intended to comprehensively follow each of the seven astronauts during training flights as well as during the actual suborbital and orbital *Mercury* flights. By 1962, the first suborbital flights and the first two orbital flights had been flown without a hitch. Alan Shepard on May 5, 1961, became the first U.S. man in space, flying in *Freedom* 7 for fifteen minutes and twenty-two seconds. Gus Grissom followed in June of the same year, and Scott Carpenter blasted off in *Aurora* 7 in May 1962. Ruff was able to participate in all the postflight debriefings and was also permitted to interact personally with the astronauts after the flights. The data collected were to be analyzed after all seven astronauts had finally flown.

But the intellectual climate was beginning to change at NASA. Tom Wolfe eloquently described what happened in *The Right Stuff*,[2] which does not exaggerate the developing hostility between engineers (or the "operations" people, including the astronauts) and the scientists. In 1988, George Ruff talked about what happened:

> This is one of those interesting social phenomena, that when the space program began, no one had done anything like it in this country. People were willing to do anything, and if we said we ought to do a study like this, they were inclined to agree. It happened also that the first group of astronauts, the first orbital group was Glenn and then Carpenter. Slayton, too, had been very interested, but then he was disqualified because of an EKG problem. Then you had some flights, and people realized that it could be done . . . and the spirit began to change away from *research* to *operations*.[3]

NASA, after a long search and many refusals from prominent physicians, now had its own medical director, Dr. Charles Berry. Berry came from a background in the Air Force. He had, in fact, been a major in the Department of Aviation Medicine at Randolph Air Force Base in Texas and had participated in the medical evaluations for the Project *Mercury* selection.

By 1962, Ruff had left his military position to return to an academic post at the University of Pennsylvania. At that time, Don Flinn was appointed chief of neuropsychiatry at Brooks Air Force Base School of Aerospace Medicine (SAM) in San Antonio, Texas, close to the Manned Spacecraft Center in Houston. Flinn was scheduled to supervise the psychiatric interviews for the next NASA program, known as *Gemini*. Ruff came down to the center to do his routine preflight interview with Wally Schirra, scheduled to fly in October of 1962. Although he was primarily interested in continuing the stress study, he intended to work closely with Flinn on the selection in order to add data from the *Gemini* psychological evaluations to the study. Berry took Ruff aside and bluntly informed him that the study was over.

"You're out. Don is going to take over and you can go back home."

"What about the rest of the stress study?" Ruff protested.

"It's over," Berry responded.

A confused Ruff returned home. What had happened? Why was the study suddenly terminated? What had the psychiatrists done to irritate NASA? Shortly after being "fired" by Berry, he was instructed to write a progress report on the study for NASA. The report was incomplete. Schirra's and Gordon Cooper's flights could not be included, and none of the biochemical data could be included, since they had not yet been analyzed. The report covered the first four flights and the training period. It came out in late 1962 and was dutifully sent to NASA—which never published it. The report disappeared into a black hole in the spacecraft center's archives. Ruff was explicitly told by NASA management that nothing else was to be published in scientific journals about the study or anything relating to the *Mercury* psychological evaluations except for the information contained in the progress report. All of the original data from the *Mercury* psychiatric and psychological evaluations were confiscated and mysteriously disappeared from not only the medical records of the astronauts, but also the NASA Archives where I looked for them twenty-five years later.

Don Flinn was not unsympathetic to the psychiatrists who had been terminated. He did what he could to continue to get psychological data from the astronauts, who were by now less than eager to cooperate, since NASA management obviously perceived the behavioral scientists as a threat. Ruff remembers:

> That is why there were never a lot of papers, and a lot of them sound sort of the same, because they were somewhat different permutations of that particular progress report. We kept pleading that these data are valuable. Even if we don't finish the entire study, let's put together what we have already collected. Let's put together the performance data, the individual psychological data, the biochemical data.[4]

But no paper was ever published with all the data compiled. The physiological data obtained by Korchin and Eik-Ness were incorporated into their own data bases, but never published separately, since without the availability of the performance and the psychological correlates, the data were basically meaningless. Although the psychological team had summarized the psychometric data collected during the *Mercury* selection in a U.S. Air Force Report,[5] the report was not generally available and, in fact, only included information on the psychological selection process and correlation matrices for the 104 medical variables. It did not contain any raw data. Appendix 2 contains the correlation information from the psychiatric variables and information on the performance test results.

What happened? It is clear from talking to those who were involved at the time that no one knows for sure. For some reason—possibly merely organizational paranoia—NASA became fearful that information on the psychological status and performance of their astronauts would be detrimental to the agency. No longer was there a normal reluctance to participate in such research—there was outright hostility toward the collection of any psychological data. From a psychological perspective, one cannot help but speculate that NASA's behavior strongly suggested that the agency believed that there was something to hide.

The exclusion of behavioral research from NASA's research agenda has continued until very recently. The termination of Ruff's study, as we shall see in the next several sections, was only the beginning of the end.

## PROJECT *GEMINI*

Before all seven of the *Mercury* astronauts had flown space missions, plans to select the next group of astronauts were already under way. *Gemini,* so named because the space capsule would accommodate two astronauts at a time, was a more ambitious and complicated program than *Mercury* had been. NASA planned to shift the aeromedical evaluations, including the psychiatric exams, to the Brooks Air Force Base SAM. The reason was primarily geographical. The AMRL was in Ohio, and the Manned Spacecraft Center (now the Johnson Space Center) was by now, thanks to Vice President Lyndon Johnson, firmly established in Houston, Texas. Logistically, it made more sense to utilize resources nearby, and Brooks was already achieving a reputation as the Air Force center for aeromedical research, education, and clinical support.

Don Flinn was handed the responsibility for the psychiatric evaluations for the *Gemini* program. A graduate of Harvard Medical School, Flinn had completed his psychiatric residency at the Menninger School of Psychiatry in Topeka and at the University of California at San Francisco. He considered himself primarily a clinician and tended toward the clinical approach in selection evaluations. At that time, it was widely and firmly believed by many psychiatrists and psychologists that the statistical method was not particularly useful in evaluating a healthy individual's psychological status and motivation. The other members of the team Flinn worked with at Brooks also used a more clinically oriented methodology. Team members included Carlos Perry, another psychiatrist, and Doug Powell, Earl Cramer, Bernard Flaherty, Richard McKenzie, and Bryce Hartman, all psychologists.

After Berry "fired" him from the research project, George Ruff talked with Flinn and convinced him of the necessity to continue the work begun during *Mercury.* Flinn did not believe in the work as strongly as Ruff did, but he basically had no objection to completing the research. However, he recalls that "[t]he climate at the time at NASA was that it was unacceptable to continue. The data were unimportant, unnecessary and inconvenient as far as the astronauts were concerned. The tests just didn't have any face validity for them."[6]

Although he had promised Ruff to continue to collect pre- and postflight data, Flinn was not allowed to do so by NASA. His approach was to find a compromise between all the psychological and performance testing done during *Mercury* and NASA's obvious desire both to become "operational" in this area and to limit potentially "embarrassing" research. NASA believed that since nothing "bad" (from a psychological perspective) had happened during the first five *Mercury* flights, it would be a waste of time to collect so much data on the next group of applicants. Many of the *Mercury* astronauts were uneasy about the continued psychological evaluations, describing them, according to Flinn, as "annoying and irritating." For them, it made no sense to continue to collect data on their psychological status. After all, hadn't they proved they had the right stuff by the fact that all the flights to date had been successful?

It did not occur to them or to NASA managers that only five data points had been collected, or that the original psychological selection criteria had never been operationally validated, or even that *Gemini* missions, with two crew members, might have different psychological requirements. In fact, they were so eager to conclude that psychological factors in space were not important, it did not matter that no one had a clear understanding of stress adaptation in the space environment. Charles Berry, NASA's chief of medicine, wrote in 1973, ten years later, that "we still know very little about the effects of the space environmental complex on personality and psychic well-being. These aspects of the human could prove to be the factors which limit the duration of space flight."[7] Considering Berry's role in terminating the only psychological research at NASA in 1962, his statement is ironic at best.

NASA has always had an intense political need to publicly present the astronauts as national heroes and therefore without psychological flaws or weaknesses. Even the psychiatrists were willing to admit that the astronauts were in many ways remarkable men. Any pilot who met the basic NASA requirements in terms of professional background and flying experience would have to be both emotionally healthy and reasonably tolerant of aerodynamic stresses. But that does not inevitably lead to the conclusion that psychological research into selection is not worthwhile. On the contrary, research was and is clearly needed to confirm clinical or even operational observations, if only to put an end to speculation about the impact of personality characteristics on mission success. Additionally, research on personality is particularly essential for future space missions where the composition of the crew would clearly be of operational importance. Unfortunately for the behavioral sciences, the overt success of the program became the rationale for discontinuing any research efforts and for making the psychiatric assessment merely a token process by the conclusion of the *Apollo* era. At least through the *Gemini* and *Apollo* programs, however, NASA continued to support—albeit reluctantly—a psychiatric evaluation on all applicants.

The *Mercury* psychological team had been given thirty hours to evaluate each candidate, but Flinn's team was told that the time allotted per candidate for the psychiatric evaluation would be reduced to ten hours. Don Flinn remembers:

> We considered that the original screening for pilot training, the experience that individuals had subsequent to that—the college, the advanced degrees, the test pilot school, and so on—meant that this was a group of individuals who were very qualified, any one of whom was probably qualified for the mission. We had little expectation that we could contribute anything to an opinion about how qualified they were to perform the mission. That was an operational consideration. We felt that our focus should be on identifying people who should be selected out and we expected to find very few people in that category and that was true. We also felt that another focus would be stress tolerance. Remember that this was still in the early days of space flight.[8]

## PSYCHOLOGICAL CRITERIA

Table 2.1 lists the operational psychological criteria used by the Brooks team for both the *Gemini* and the *Apollo* programs.

- "General Emotional Stability" referred to the absence of neurotic or psychotic symptoms and freedom from problems in the social, marital, and financial spheres, as well as the ability to tolerate stress and frustration without emotional symptomatology or impaired performance.
- "High Motivation and Energy Level" took into account the demonstrated ability of the candidate to pursue realistic and mature goals with determination and initiative and the capacity to think in a creative, flexible manner when unforeseen events occur.

**Table 2.1**
***Gemini* and *Apollo* Astronaut Psychological Selection Criteria**

General Emotional Stability

High Motivation and Energy Level

Adequate Self Systems (self concept, emotional control, and adequacy)

Satisfactory and Productive Interpersonal Relationships

- "Adequate Self Systems" was a clinical term suggesting a high self-confidence and the capacity to give opinions and make independent decisions, with the ability to depend on the judgment of others when the mission warrants.
- The ability to form satisfactory and productive relationships with supervisors, peers, and subordinates and to function as a team member in any role without being overly dependent on people for satisfaction encompasses the "Satisfactory and Productive Interpersonal Relationships" criterion.

These psychological criteria were very general, and they reflected the more clinical philosophy of the Brooks team, compared to those at the AMRL. Don Flinn winces slightly now when asked about these psychological criteria. While they reflected a more general philosophy about personality prevalent in psychiatry at the time, now they would be regarded as somewhat vague and unmeasurable. On the other hand, Flinn believes that his team overall had a balanced approach toward evaluating the candidates.

The two psychiatric interviews performed by two different clinicians were similar to the interviews done during the *Mercury* selection. The decision on which psychological tests to retain and which to eliminate was made on the basis of what the Brooks clinicians felt most comfortable with. Psychological tests chosen from the original *Mercury* battery included primarily the projective tests from that battery (e.g., the Rorschach Inkblot and the Thematic Apperception Test). These tests were the ones traditionally favored by the clinical approach. Statistical or actuarial tests such as the MMPl were not officially retained. The cognitive/intelligence tests included were also a subset of those done during the *Mercury* selection, and the only performance test that carried over was the complex behavior simulator, which, in fact, had been developed by a Brooks psychologist—Bryce Hartman. Table 2.2 lists the psychometric and performance tests that made up the final battery.

Three separate rating systems were developed: one each from the two psychiatric interviews and one from the psychological testing and performance on the complex behavior simulator. Each selection variable was superimposed on a semiquantitative six-point scale. The raters assigned an objectified value to "the overt or manifest degree of each of these traits" noted in each candidate. At the completion of the evaluation, the three ratings independently arrived at were compared and combined to yield a unified rating. This final rating served as the basis for psychiatric recommendations on the given case.[9]

Two hundred applications were received by NASA after the first *Gemini* program announcement. Prospective applicants were still required to have a minimum of 1500 hours of flight time in high-performance jet planes, and because of this requirement, all of the applicants were pilots, and male. Of the 200 pilots, 32 were selected to come for an interview and receive the intensive medical and psychological evaluations. Nine pilot as-

Table 2.2
*Gemini* and *Apollo* Psychometric Tests

1. Wechsler Adult Intelligence Scale (WAIS)
2. Miller Analogies Test
3. Doppelt Mathematical Reasoning Test
4. Minnesota Engineering Analogies Test
5. Rorschach Inkblot Test
6. Thematic Apperception Test
7. Draw-A-Person Test
8. Bender Visual Motor Gestalt Test
9. Gordon Personal Profile Test
10. Edwards Personal Preference Schedule
11. Complex Behavior Simulator

* The MMPI was also used (by Bryce Hartman) with the *Gemini* and the early *Apollo* candidates, although this was not an "official" part of the battery and was unfortunately not used with the later *Apollo* candidates.

tronauts were selected in September 1962. Another call for applications went out in June 1963. This time 720 applications were received. In October 1963, fourteen men were selected, eight of whom had advanced degrees (master's degree or higher). Two civilians were included in this group, along with seven Air Force pilots, four Navy pilots, and one Marine pilot.

The psychiatric evaluations of the *Gemini* candidates resulted in no actual disqualifications. Instead, the team rank-ordered the candidates. Those candidates who the clinicians believed had an increased risk of doing poorly were ranked lower. A comparison of the candidates chosen by NASA and the psychiatric rankings demonstrated many similarities. But, as in the *Mercury* selection, the lists were not identical. Sometimes someone ranked highly by the psychiatrists was ignored by NASA, while other times someone ranking very low on the psychiatrists' list was chosen. The psychiatrists were unsure if their rankings had any real meaning to the NASA Selection Board.

We have seen how, as the selection cycles progressed, the psychiatric evaluation was shortened in the interests of time. By the end of the *Apollo* evaluations, each member of the three-man psychiatric team had assigned one holistically determined rating to each candidate. This holistic rating was not further broken down into the subscales for each area. But with the advent of the *Apollo* program, NASA for the first time began to consider selecting *non-pilot* astronauts. As we shall see, the psychiatric evaluation of non-pilot astronauts, or scientist astronauts, brought a new dimension to the psychological assessment.

## THE SCIENTIST ASTRONAUT

Sometime in 1963, NASA decided to include scientists as potential astronaut candidates for its future space programs, since President John F. Kennedy, during an impassioned speech that year, had committed the nation to the goal of landing men on the Moon before the end of the decade. The *Apollo* program was thus born, along with the concept of the scientist astronaut.

Carlos Perry became the chief of psychiatry at the School of Aerospace Medicine in 1964. Perry was a classically trained psychiatrist, having gone to medical school at the University of Kansas and then having done his psychiatric residency training at the Menninger School of Psychiatry. He was also a scholar and contributed several important papers about astronaut selection to the scientific literature.

The 1965 selection of scientist astronauts was the result of extensive discussion and negotiation between NASA and the scientific community. Applicants for this new type of astronaut were required to hold a doctoral degree in medicine, engineering, or one of the natural sciences. The National Academy of Sciences did the preliminary screening of the applications for scientific criteria, and NASA made the final decision on candidates to interview.

The introduction of the scientist-astronaut category stimulated considerable debate among the Air Force behavioral scientists, who were used to dealing only with pilot applicants. What was the appropriate motivation for a scientist desiring to go into space? It was argued that space flight was the natural extension of a pilot's basic professional motivation. But what of a scientist's?

For some years, the aviation psychiatric literature had discussed the ambivalent motivations of some pilots. Douglas D. Bond wrote extensively about the "love and fear of flying" that is part of the pilot's personality.[10] He also explored the psychoanalytic, dynamic basis of the pilot's attraction to flight. But as Perry pointed out, "as the complexity of motivational factors increases, so will increase the chances of their being neurotically overdetermined."[11] The problem was, and still is, that there has never been any scientific relationship (only an intuitive belief) demonstrated between "neurotically overdetermined motives" and pilot performance. Since World War II, the operational demands of military aviation have precluded the selection of individuals who might have a risk of failing from a psychiatric point of view. If they were excluded from flying to begin with, how could the military services or psychiatrists "prove" that such individuals could not be good pilots? No one wanted to risk lives or airplanes for such a study. It seemed to be simple common sense that psychologically impaired individuals would be at risk flying airplanes. So the basis of psychiatric disqualification was experience from other occupations where the results of poor performance are not so deadly.

This question remains: What are the appropriate motivations of scientists who want to go into space? Perry's hypothesis about the more complex motivations of scientists proved prophetic. One of the first scientist astronauts selected by NASA resigned within a short time, ostensibly because his wife disapproved of the hours. Perry pointed out that this individual had been selected on the basis of pilot standards, where it was assumed that the motivation to fly was extremely strong.

After much debate among the psychiatrists and the psychologists, however, the same criteria and the same selection process were used for the scientist candidates. The psychological testing proposed for this group of candidates was cut down from ten hours to about six and a half hours, but still included intensive psychiatric interviews by two psychiatrists. Two selection cycles (1965 and 1967) were devoted exclusively to selecting scientist astronauts. In the first, scientists from both the military and academic environments were considered, resulting in the selection of six scientist astronauts. They included one geologist, two physicians, two physicists, and one electronics engineer. Five civilians and one military person were also selected.[12] In the second scientist astronaut selection in 1967, the National Academy of Sciences recommended 69 out of 600 applicants to NASA for consideration. Eleven were chosen by NASA. This was the first and only NASA selection that did not include military personnel.

## THE USAF MANNED ORBITING LABORATORY

NASA was not the only government agency that was actively recruiting astronauts in 1965. The Air Force also had great ambitions for space exploration and that same year developed a specialized project known as the Manned Orbiting Laboratory (MOL). The MOL mission included an operational medical research component. Responsibility for management of this program rested with the Space Systems Division of the Air Force Systems Command. Candidates for MOL astronauts came from the U.S. Air Force and U.S. Navy test pilot populations. Their preselection evaluations were conducted at the Air Force's SAM in a manner similar to that used with astronaut applicants for NASA.

Carlos Perry replaced Don Flinn as chief when Flinn accepted an assignment in Washington working with the Air Force surgeon general. Perry was then at the beginning of his career. He was asked to coordinate the evaluation of the candidates who might be selected to participate in the MOL. This proved to be his own entry point in the manned space effort.

All of the applicants for the MOL were pilots. Yet the objectives of the MOL program were to conduct scientific experiments in the unique setting of an orbiting space laboratory. A panel of specialists from the Air Force's Aerospace Research Pilots School initially reviewed sixty potential candidates for the MOL program and identified fifteen to be interviewed. Carlos Perry later wrote:

> As a prerequisite, they were all college graduates with engineering degrees. The men were between 26 and 35 years of age. All but one were married and had children. The only item which was striking was that 12 of the 15 men were the oldest or only male child in the family constellation. Only 5 alluded to the possibility of adjustment difficulty during their formative years. Parental death had occurred in two cases; parental separation in another two; and the fifth man had been a disciplinary problem in early adolescence. Only two of these five candidates mentioned subjective awareness of discomfort connected to such classic types of life-stress. . . . In comparison to other pilots, these candidates had long maintained a higher degree of investment in their work as fliers. In this group, by virtue of their professions, every man had experienced immediate life-threatening stress. None was inclined to talk about it. In one instance, successful ejection from a high-performance craft had been made in the split second before impact. Another subject had flown 100 combat missions in Korea. Each man reported his ability to handle such stress in a cool, detached manner. And this was so because each instance discussed was amenable to corrective, life-saving measures under control of the individual himself. Each man had been able to override the classic flight-fight response until after the emergency had passed. All reported awareness of anxiety from minor to major degrees, but never its presence during the time of actual emergency.[13]

Perry's approach to the evaluations was perhaps the most clinical to date. A three-man team composed of two psychiatrists and one psychologist assigned an arbitrary score from a predetermined scale to each candidate. This score represented the degree of confidence in recommending the candidate for a mission. Results were initially presented and then analyzed for factors that determined the selection. The degree of interevaluator agreement was found to be only fair.

A composite, self-determined ranking of the MOL candidates was then developed. In comparing this ranking to the evaluator composite, only one of the examiners' top five candidates had been similarly selected in by the candidates themselves. Their top man ranked twelfth in the former composite. On the other hand, as it is usually easier to identify potentially poor performers than to identify the "best," the evaluator composite agreed with the peer composite on three of the five in the lowest ranks. Perry argued that these differing results did not necessarily detract from the validity of psychiatric impressions. The psychiatric team used psychiatric criteria, and the candidates used test-pilot criteria. But as Perry pointed out, who, at that point, was to say that the pilots' selection criteria were less valid than the psychiatrically determined criteria were?

The top five selected by the psychiatrists included both of the men whose parents had marital strife leading to separation during the subjects' formative years. One factor seemed to predominate as setting this top group apart from the rest. "They were able to relate comfortably and affectively in the interview situation despite the examiner's intrusion into emotionally loaded areas," and "they could handle aggressive feeling openly without discomfort."[14] Less distinct, but equally characteristic, was the ability of each man to relate to others with sensitivity and empathy. For this top group of candidates, the psychiatrists characterized the fundamental determining factor from their perspective as "character strength achieved in spite of adversity." In other words, the psychiatrists were impressed with those candidates who had achieved success despite negative factors in their life history. Such factors generally do not impress most test pilots.

Peer ratings were constructed and given to each candidate after all the candidates lived together in the competitive situation of selection for ten days. Most were already well acquainted with each other from prior military assignments. There was reportedly better agreement between the fifteen candidates than there was between the three behavioral clinicians on who were the "best" and who were the "worst."

Eight men were eventually selected from the group of fifteen candidates. The psychiatric rankings of the eight were first, second, third, fourth, seventh, ninth, twelfth, and fourteenth. As a group, the MOL candidates were psychologically similar to any other group of astronaut candidates.

The MOL program existed for a few short years. Eventually it was grounded (after no time in space) when funding for it virtually disappeared overnight. NASA apparently used its political clout with Congress to discourage the development of any competing space programs. As a compromise with the Air Force, NASA's seventh recruitment of astronauts turned exclusively to the MOL program. Seven military pilot astronauts from MOL were transferred to NASA. Only one MOL astronaut was not transferred because NASA wanted only those individuals who were under thirty-six years of age.

## AIR FORCE SPACE-CABIN SIMULATOR RESEARCH

Another piece of Air Force history not well known is that numerous isolation studies were done from 1953 to 1963. Years before the *Mercury* program was initiated, and throughout the *Mercury, Gemini,* and *Apollo* programs of NASA, the Air Force conducted hundreds of studies researching the implications—from medical, psychological, and performance perspectives—of prolonged isolation and confinement. The studies done in the period from the early 1950s to the mid-1960s are of considerable relevance to the topic of astronaut psychological selection. These investigations were aimed primarily at determining the effects of isolation, confinement, fatigue, and altered work-rest cycles on proficiency, interpersonal communication, and crew performance capabilities.[15]

Although the isolation and the confinement studies had been initiated before the possibility of space travel was seriously considered by the Air Force (in the early 1950s), by the time NASA had completed its first astronaut selection in 1959, the relevance of such research to space flight had become obvious. Space-cabin simulators were designed for one- and two-man investigations. These simulators were essentially small altitude chambers, just large enough for crew, life-support equipment, and equipment necessary to accomplish the performance tasks that measured operator efficiency. They were designed to permit the manipulation of environmental variables in order to study the effects of such variables (e.g., cabin pressure, oxygen partial pressure, carbon dioxide partial pressure, temperature, and relative humidity) on the ability of the crew to perform essential tasks.

Initial studies were conducted at an altitude of 18,000 feet with a 40 percent oxygen environment, while some later flights were conducted at an altitude of 33,500 feet with essentially 100 percent oxygen atmosphere. Duration of the simulated flights ranged from a few days to approximately two and a half weeks. Later in the program, studies were performed for longer periods. The research was focused on assessing the "logistic, behavioral and physiologic parameters" of a closed ecological life support system"—which was exactly what the space-cabin simulators were.[16]

A comprehensive summary of the simulations done during this period was completed by Bryce Hartman and delivered in a key presentation to the National Academy of Sciences Committee on Manned Space Flight in April 1967.[17] The results of these investigations are summarized chronologically in Appendix 2.

The conclusions reached by the scientists who conducted these studies have important implications for those of us interested in developing scientific psychological selection criteria for space flight. Unfortunately, Hartman's summary was never published, and his conclusions were lost until recently, when I interviewed Hartman (who stubbornly declines to retire and still works in the same office at the SAM that he had in the 1960s) in February 1992. Combining Hartman's data with the few published reports that exist, the following points become obvious.

First, during the one-man isolation studies, it became clear that perceptual aberrations were frequently experienced by lone subjects. Almost any individual, no matter how highly motivated, may develop perceptual illusions when he is alone and isolated from others and when the work is monotonous. During the twenty-first hour of a simulated mission, one subject became convinced that an instrument he was monitoring was on fire and about to explode. He believed he could feel the heat and smell the smoke of the impending disaster. Despite reassurances from the support crew that the instrument was functioning perfectly, the study had to be terminated because of the subject's increasing anxiety and paranoia. While this was the only one-man study that was aborted for this reason, many of the individual subjects in the one-man studies reported similar perceptual problems. After about eighteen hours and during early morning hours, under the combined conditions of fatigue, confinement, and restriction of the perceptual field imposed by the tasks, performance decrements were regularly seen, along with a variety of visual, auditory, and proprioceptive illusions.

> Several reported that the dials on the task panel sometimes looked like faces or other figures. Other subjects heard music or voices, had the sensation that their arms and legs were enlarged, felt that someone else was sitting beside them in the chamber, or believed that a large hole had opened up in the floor of the chamber. Usually these illusions were momentary, although in the last mentioned case the subject reached down and felt the floor to convince himself the chamber was still intact.[18]

In general, older, brighter, and more mature subjects were relatively immune from such aberrations. But no subject, whatever his age, became comfortable performing the tedious and monotonous tasks in isolation. All subjects had some degree of negative subjective response to the experience.

The multi-man studies demonstrated effectively that the more demanding work-rest schedules depleted "psychological reserves" and altered the learning curves of the subjects, as well as lowering the tolerance for additional stresses. An individual's tolerance was found to be modified by the level of his motivation—particularly when the motivation level is low. In spite of the stresses, psychomotor performance was maintained across a variety of workload conditions for extended periods. Difficult work schedules reduced efficiency only to the extent that they compromised physical reserves (for example, by interfering with sleep). Subjective tolerance for the mission tended to remain high, and subjective changes were minimal and without operational significance, except for conditions involving substantial physical and physiological insult—such as continuous performance for twenty-four hours without sleep.

As far as specific behavioral findings in the multi-man studies, interpersonal interaction between crew members was found to structure itself in a task-oriented fashion, which resulted in an avoidance of conflict. While mild negative subjective changes, such as hostility directed outward toward the crewmate, were common, tolerance for confinement was maintained for extended periods. Hartman suggested that such tolerance for isolation and confinement may last indefinitely, particularly when subjects are given meaningful tasks and responsibilities.[19]

What is the significance of these studies for astronaut psychological selection? For one thing, these studies would have provided an ideal environment in which to evaluate the effect of personality characteristics on behavior and performance. That is, they might have provided a formal validation for astronaut psychological selection criteria. Unfortunately, they were not used for this purpose, and much of the data collected were never published or disseminated.

Hartman's 1967 presentation was handwritten. Yet it is apparent from reviewing his notes that a great deal of information useful for the development of formal psychological selection criteria is contained in the data. For example, the studies clearly give some scientific credence to the use of pilots as potential astronauts. Successful pilot behavior requires purposeful, goal-oriented activity; is primarily focused on performance; and is well organized and somewhat compulsive. These same traits were found to be effective in coping with the stresses of living in the isolated, confined mission environment.

In the experiments where pilots and non-pilots were compared under the same conditions, the non-pilots routinely had more difficulty maintaining their performance. Maturity (i.e., emotional stability) and high motivation for the mission were found more often in the pilot group, and pilots were able to overcome a considerable amount of environmental, physiological, and psychological stress. Eisenhower's decision to use only military pilots turned out to have some scientific basis, since pilots—especially jet pilots—seemed to inherently possess the personal traits needed to be successful in stressful environments. Intuitively appreciating this, NASA management initially required the first group of scientist astronauts to become pilots and receive jet pilot training with the military.

But let us examine this policy more closely. In reality, it is probably fair and accurate to say that, in general, pilots possess a certain personality profile that would be successful in space. But this profile is not unique to pilots. Many successful individuals in a variety of professions share identical traits. For example, many scientists will have goal-oriented, performance-based behavior. They are also frequently meticulous and careful in their scientific pursuit of truth and possess the ability to think and react under conditions of stress. What we have found most interesting in reviewing the psychological evaluations of potential astronaut candidates over the last thirty or more years is the remarkable psychological similarity of the individuals chosen to become astronauts, whether or not they were pilots or scientists, males or females. But we will discuss this further in ensuing chapters. First, it is necessary to summarize all the data collected through the end of the *Apollo* program.

## A COMPARISON OF THE PSYCHOMETRIC DATA FROM *MERCURY, GEMINI,* AND *APOLLO*

In 1976, Bryce O. Hartman and Richard C. McNee presented a paper at the North Atlantic Treaty Organization Advisory Group for Aerospace Research and Development (NATO AGARD) Conference on Recent Advances in Space Medicine.[20] The paper summarizes the psychometric data obtained on *Mercury, Gemini,* and *Apollo* candidates, as well as some candidates evaluated at the SAM for special Air Force programs. All totaled, this included over 350 individuals. The report was subsequently published in the proceedings of the conference, but has not been generally available; the data have not been reported elsewhere.

Hartman and McNee performed both univariate and multivariate analyses on the three sets of data (*Mercury* candidates, *Gemini* and *Apollo* candidates, and a group of Air Force pilot controls). In the univariate analyses, five groups of subjects were compared: the seven selected and twenty-four not selected *Mercury* astronaut candidates, the 9 selected and 23 not selected *Gemini* and *Apollo* astronaut candidates, and a pilot control group of fifty. A total of thirty-six psychological measurements were considered. Not all candidates took all tests. Data sets on all tests were not available across selections. I have chosen to compare the astronaut groups across only those psychological tests that were taken in common across selections. A summary of the psychometric variables and their means and standard deviations can be found in Appendix 2 as Hartman and McNee presented them in their paper.

Separate comparisons were made between the selected and the not selected groups for the astronaut candidates. The only difference in means detected at the .05 level of significance was the overall rating for the *Mercury* candidates ($p < .005$). Variances differed at the .01 level for four of the measures, affecting the type of testing performed on the means. The differences in variances detected were for Rorschach-F % (controlled conventional responses to the form of the inkblot components) for the *Mercury* candidates and for Rorschach-P (popular responses) and -M (movement responses) for the *Apollo* candidates.

Separate comparisons were also made between the *Mercury* and the *Apollo* candidates for the selected and the not selected groups for the sixteen variables. Only two variables—Rating and % Response to Chromatic Cards ($p < .001$ )—differed significantly between the two sets. Also, the sets differed for Rorschach-F % for the selected groups and for Rorschach-P and I-W (whole inkblot responses) for the not selected groups ($p < .005$). None of the tests of variances was significant at the .01 level.

The fourteen variables on all three sets of data were factor analyzed using a multivariate, stepwise procedure. Three final factors were extracted, but when Hartman and McNee analyzed these three factors, they noted many inconsistencies among the three sets of data. Since control subjects were felt to be psychologically different from the astronaut candidates, they then decided to look at only the two sets the candidates—*Mercury* and *Apollo*. Factor loadings were determined, and tests were then categorized into four groups for each factor, as determined by the levels of the two loadings.

These data are presented here primarily for historical interest. Only general conclusions can be made from them, due to a variety of inconsistencies. In the initial multivariate analysis, the best combination of variables was Verbal IQ (from the WAIS) and number of responses (R) from the Rorschach. This finding was consistent with the unpublished factor analysis of the *Mercury* psychometric data. The interpretation of this result was that candidates who feel free to "produce" psychologically in response to test material make a better impression on the examiner.

In subsequent factor analysis, the scientists found further support for this idea. Hartman and McNee summarized:

> Major loadings on factor 1 are R (number of responses on the Rorschach), which demonstrates willingness to "Produce," and FM (movement responses on the Rorschach) which reflects ability to keep the content during high productivity within normal bounds. Moderate loadings on M (human movement responses) and W (whole inkblot responses) demonstrate the tempering of productivity by the high intellectual resources of the candidates. The major loading in factor 2 is the sum of the color responses, demonstrating integrated responsiveness to one emotive aspect of the Rorschach stimulation. Moderate loadings on F % (controlled conventional responses to the form of the inkblot components) coupled with shading (responses to shading variations perceived in the inkblots) which is the second emotive component of the Rorschach stimuli. Therefore, factor 2 indicates that the productivity identified in factor 1 is tempered by a combination of controlled sensitivity and responsiveness in conventional ways. Factor 3 adds nothing to this interpretation.[21]

Since projective tests played such an important role in these early selections, a summary of some of the clinical material collected from the Rorschach is presented in Appendix 2. A variety of typical responses from candidates in the *Mercury, Gemini,* and *Apollo* psychiatric assessments is discussed.

The results obtained were not unexpected. Controls were not as high in intellectual resources, were more dependent, and were more heterogeneous in their performance than were the astronaut candidates. Hartman and McNee observed that, overall, the differences among all groups were "small, scattered, and not very striking." Astronaut candidates were slightly brighter, better psychologically integrated, more independent, and more homogeneous as a group than was a randomly selected subset of Air Force pilots. In addition, the successive groups of *Mercury, Gemini,* and *Apollo* candidates were highly similar, undoubtedly because each group met the same initial screening standards and because NASA managers were very consistent in selecting the individuals they believed were best suited for the job. Hartman and McNee concluded with a brief verbal psychometric description of an astronaut: "[H]e has a good balance between sensitivity/creativity and conventionality."

MMPI studies were done "unofficially" (i.e., not as part of the "official" test battery) on a number of the *Gemini* and *Apollo* candidates, since one of the psychologists was interested in comparing profiles from these candidates with those of the *Mercury* candidates. The MMPI in Figure 2.1 is remarkably similar to the MMPI profile of the *Mercury* candidates.

In addition to the formal psychological testing, clinicians from each program had rated applicants in four categories: (1) best qualified, (2) qualified without reservation, (3) qualified with reservation, and (4) not qualified. As we have seen, very few individuals fell into the "not qualified" category, and those were only in the *Mercury* selection. A few candidates fell into the "qualified with reservation" category. The remainder were in the top two categories. When NASA managers made their final selections, most of those candidates selected also were in the upper categories, more than if chance alone were at work. "We felt it was a result of the fact that in our interviews, we were looking at some of the same data that the selection panel was looking at, and so we corresponded in our classification."[22]

Carlos Perry wrote in 1967:

> Only a few psychiatrists have been directly involved with those men who have experienced orbital flight or those who have been selected for future missions. . . . This is a partial reason for the dearth of clinical reports. Another reason at this early date is the obvious maximal need for confidentiality in disseminating medical

**Figure 2.1**
***Gemini/Apollo* Candidate MMPI Profile**

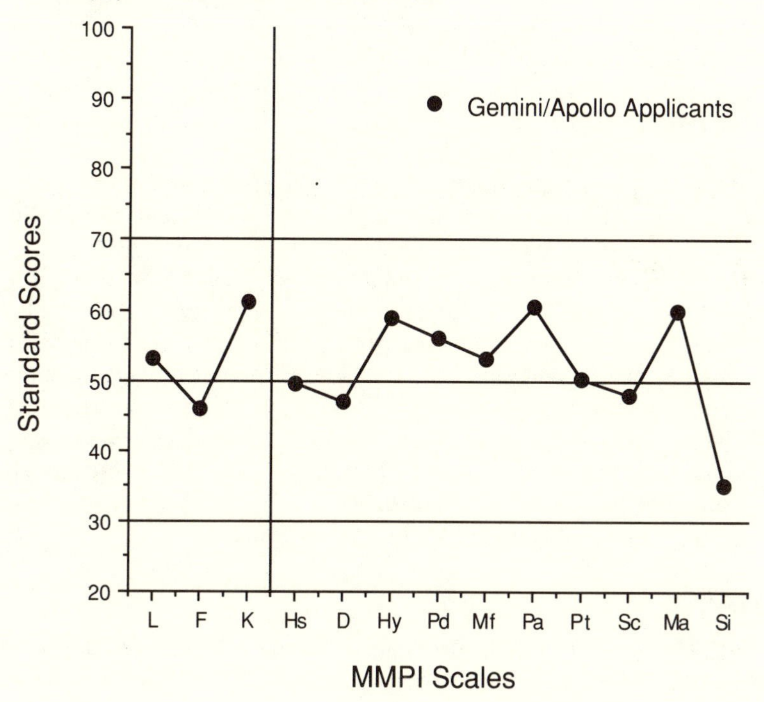

data obtained from prominent public figures. As reported in the popular press, there have been approximately 160 candidates examined in preselection aeromedical evaluations. These evaluations have been required in the course of seven selection increments for the United States space programs to date. Some 70 men have been administratively selected for duties as space crewmembers. This would seem to be a sufficient number of cases on which to discretely base some initial observations without compromising confidentiality. Yet, to date, only one such report has been published in open medical periodicals.[23]

Undoubtedly one of the greatest achievements of the twentieth century is the 1969 lunar landing of *Apollo 11*. I watched, along with the rest of the world, and knew that the image of man taking that "giant" leap indelibly altered the way we viewed our planet and ourselves. The astronaut (who depends on the skills and ingenuity of thousands of other individuals) has become the symbol of that achievement, representing courage, daring, risk, and success.

Even today, the behavioral scientists who were part of astronaut selections use nothing but superlatives in their descriptions of the unusual men who were chosen for the *Mercury, Gemini,* and *Apollo* programs. George Ruff noted, "These were really impressive guys. They were exceptionally stable and integrated men who were intensely dedicated to what they were doing. Even today, thinking back, I have to remark on their intellectual and performance abilities."[24]

But even as the triumph of the *Apollo* Moon landings was waning in the American consciousness, the seeds of psychological denial sown during *Mercury* had by then taken firm root at NASA. The attitude that refused to admit to problems (and that would eventually lead to the *Challenger* disaster in 1986) continued to be particularly hostile toward psychological or behavioral research. As Carlos Perry remarked in 1967, the lack of interest in and the denial that psychological factors played any role in space operations only served to underscore how important those factors actually were.

As we shall see, by the onset of the Shuttle program, NASA had little confidence in or tolerance for psychiatry or psychology. Whether it was official policy or not, the result was that NASA basically ignored all the data and the experience that had been so carefully collected by the military behavioral scientists and clinicians for twenty years. The reasons behind the institutional denial are elusive. Considering the positive statements made by psychiatrists about the men who were selected, as well as the conscientious and thorough job that had been done by the military psychiatrists and psychologists, NASA's attitude was, in retrospect, difficult to defend and ultimately harmful to the agency's image.

## NOTES

1. Carlos J. G. Perry, "Psychiatric Support for Man in Space," *International Psychiatry Clinics* 4 (1967): 197.

2. Tom Wolfe, *The Right Stuff* (New York: Farrar, Straus & Giroux, 1979).

3. Patricia A. Santy, ed., "The NASA In-House Working Group on Psychiatric and Psychological Selection of Astronauts: Summary and Transcripts" (unpublished paper, 1988), 21.

4. Ibid., 21–22.

5. George E. Ruff, "Psychological Tests," in *Project Mercury Candidate Evaluation Program,* edited by Charles L. Wilson, WADC Technical Report 59-505 (Wright-Patterson Air Force Base, Ohio: Wright Aeromedical Development Center, 1959), 81–86.

6. Santy, ed., "The NASA In-House Working Group on Psychiatric and Psychological Selection of Astronauts," 25.

7. Charles A. Berry, "View of Human Problems to Be Addressed for Long-duration Space Flights," *Aerospace Medicine* 44 (1973): 1136–1146.

8. Interview with Don Flinn, M.D. (May 19, 1992).

9. Don E. Flinn, Bryce O. Hartman, D. H. Powell, and R. E. McKenzie, "Psychiatric and Psychological Evaluation," in *Aeromedical Evaluations for Space Pilots,* edited by L. E. Lamb (Brooks Air Force Base, Texas: U.S. Air Force School of Aerospace Medicine, Aerospace Medical Division (AFSC), 1963), 199–230.

10. D. D. Bond, *The Love and Fear of Flying* (New York: International University Press, 1952).

11. Carlos J. G. Perry, "Psychiatric Selection of Candidates for Space Missions," *Journal of the American Medical Association* 194 (1965): 841–844; quotation at 842.

12. Joseph D. Atkinson and Jay M. Shafritz, *The Real Stuff: A History of NASA's Astronaut Recruitment Program* (New York: Praeger, 1985).

13. Perry, "Psychiatric Selection of Candidates for Space Missions," 842.

14. Perry, "Psychiatric Selection of Candidates for Space Missions," 843.

15. B. E. Flaherty, D. E. Flinn, G. T. Hauty, and G. R. Steincamp, *Psychiatry and Space Flight,* Report 60-80 (Brooks Air Force Base, Texas: U.S. Air Force School of Aviation Medicine), 1960. Don E. Flinn, "Psychiatric Factors in Astronaut Selection," in *Psychophysiological Aspects of Space Flight,* edited by B. E. Flaherty (New York: Columbia University Press, 1961), 93. Don E. Flinn, "Behavior and Communication during Space Flights," in *Communication and Social Interaction,* edited by P. F. Ostwald (New York: Grune & Stratton, 1977), 247–261. Don E. Flinn, E. S. Flyer, and F. E. Holdredge, "Behavioral and Psychological Studies in Aerospace Medicine," *Annals of the New York Academy of Science* 107 (1963): 613–634. Bryce O. Hartman and Don E. Flinn, "Crew Structures in Future Space Missions," in *Lectures in Aerospace Medicine* (Brooks Air Force Base, Texas: U.S. Air Force School of Aerospace Medicine, 1963).

16. G. R. Steinkamp, *Human Experimentation in the Space Cabin Simulators,* Report 59-101 (Brooks Air Force Base, Texas: U.S. Air Force School of Aviation Medicine, 1959). Don E. Flinn, J. T. Monroe, E. H. Cramer, and D. H. Hagen, "Observations in the SAM Two-Man Space Cabin Simulator. IV. Behavioral Factors in Selection and Performance," *Aerospace Medicine* 36 (1961): 610–615.

17. Bryce O. Hartman, Presentation to the National Academy of Sciences Committee on Manned Space Flight (unpublished, April 1967).

18. Flinn, "Behavior and Communication during Space Flights," 251.

19. Hartman, Presentation to the National Academy of Sciences.

20. Bryce O. Hartman and Richard C. McNee, "Psychometric Characteristics of Astronauts" (paper presented at the NATO AGARD Conference on Recent Ad-

vances in Space Medicine, Athens, Greece, October 1976).

21. Ibid., C10-7.

22. Interview with Bryce O. Hartman, Ph.D. (February 13, 1992).

23. Perry, "Psychiatric Support for Man in Space," 199–200.

24. Santy, ed., "The NASA In-House Working Group on Psychiatric and Psychological Selection of Astronauts," 22.

# 3

# Opening the New Frontier—Shuttle

The absence of a barrage of psychological tests in the training for potential astronauts is a sign of the maturity of the United States' space program, according to Air Force Lt. Col. Richard Mullane. Through the years, National Aeronautics and Space Administration officials have found that successful candidates for space flight need not demonstrate some superhuman psychological profile, he indicated

—*APA Monitor,* October 1983

An Air Force pilot once asked Carlos Perry if anyone could stand the stress of a flight to Mars, which could take one to two years. Perry responded by noting that

> it is a wonder . . . that Columbus would set out across the Atlantic not even knowing how long it would take, while the scientific authority of that day told him he would drop off the face of the earth. Perhaps it is good that psychiatrists were not around then to screen his crew. Surely many would have been disqualified because they showed a potential for mutiny.[1]

It is perhaps important at this point in our historical review that we pause to consider what role psychological sciences should have in the evaluation of basically healthy individuals for space flight. Is it a waste of time to have potential astronauts undergo an intense psychiatric and psychological evaluation? To answer that question, we must first consider the importance of the human element in space flight engineering systems.

Space flight engineers have adopted a philosophy that all important systems must be fail-safe, meaning that every potential failure in a critical system has to have a backup or a redundant component to prevent a catastrophic failure in flight. There are a number of critical systems in the Shuttle and in important payloads, which have as their backup the utilization of human skills (e.g., extravehicular activity to manually launch a satellite). But there is no effective fail-safe for the human system (except possibly other humans). Humans have been fully integrated into Shuttle systems and are critical for the normal functioning of many Shuttle systems. This was not the case in earlier programs, where the human role was more passive—"spam in a can." This is both good and bad. It is good because, obviously, humans provide a flexibility normally lacking in machines, no matter how complex their design. However, it is bad precisely because success or failure depends mainly on the reliability, performance, and capability of the human element.

All hardware developed to fly in space must pass scrupulously exact tests of strength, endurance, tolerance to a number of environments, and so on before it is deemed safe and "flight-ready." Systems, subsystems, bolts, metal, joints, O-rings—all are studied and tested extensively. It is not good enough that these elements meet standards appropriate to normal functioning on Earth. They must also meet higher standards for the space environment and all the stresses to which they will be subjected by that environment.

If one takes these points into account, it seems obvious that the human element in space flight has had an increasingly important role in every aspect of the mission. How then can it be argued that a less thorough examination of the humans selected to go into space is appropriate? To do so is to deny their importance to mission success. The most minute pieces of hardware on the spacecraft are continually examined and their performance measured to ensure proper functioning. Medical and psychiatric standards for the selection and retention of astronauts provide a similar basis for examining the human "hardware." From the moment they are considered as possible astronauts, an assessment of the potential risks versus benefits of selecting any particular individual is undertaken. And, after an individual is selected, the process continues. Just as a competent engineer obtains objective data to analyze the performance of important flight systems, so must physicians carefully monitor, both medically and psychologically, behavior that may lead to decreased performance, instability, or failure.

Although the psychiatric evaluations of astronaut candidates have only rarely recommended disqualification on a psychiatric basis, it is important to consider the high risk to the space program of selecting an individual who has a significant chance of developing psychiatric illness. Any psychiatric evaluation process may result in the disqualification of some excellent individuals who, if selected, might have performed adequately as astronauts. Sometimes, safety-minded engineers decide to replace a "possibly malfunctioning" component that conceivably could have performed flawlessly, rather than take the chance that it might not. Safety concerns must take precedence over the feelings of any astronaut, manager, or physician. A reasonable approach to minimize the exclusion of potentially good candidates is to collect as much objective data as possible on key factors such as behavior and performance in the space environment, so that psychiatric selection criteria, as well as our understanding of potential risks, will improve over time.

Psychometric data collected in the different programs exhibit striking similarities in the psychological profiles of astronaut candidates, but also demonstrate important differences among individual candidates. Such data might have been used productively by NASA management to assist in assembling crews for Shuttle missions. Crew selection is an aspect of selection that has great relevance for individual selection procedures, but has not yet been studied. NASA managers have never used the psychiatric evaluations (done at the time they select individual astronauts) to help them assemble space crews. Nor has there ever been any psychiatric input to crew selection decisions.

Astronaut Mike Mullane pointed out in a 1983 interview that most potential astronauts are military test pilots and that they often perceive their behavioral observers as "pseudo-scientific."[2] As we have seen, the behavioral scientists involved in the *Mercury, Gemini,* and *Apollo* psychiatric evaluations reflected their times, as did the engineering technology. The controversies in psychiatric and personality theory, as well as fundamental disagreements about the philosophy of evaluation (clinical versus statistical), all contributed to the perception of behavioral sciences, in general, as being less than scientific. The fledgling space program had been fortunate indeed to have a scientist of George Ruff's stature in charge of the initial psychiatric evaluations. In Ruff they found not only an excellent clinician, but also someone to whom the scientific pursuit of truth was of paramount importance.

In retrospect, Ruff's plan to use a broad-based methodology (i.e., one encompassing both clinical and statistical philosophies), coupled with the intent to validate the psychological selection criteria over a period of time, was probably the most reasonable approach that could have been developed. Ruff's firing, with the concomitant termination of his research program, heralded the beginning of the end for behavioral sciences in NASA. Although his successors were willing to continue the research, they were prevented from doing so by NASA, which preferred to believe that it had already been "proved" that astronauts did not have to have some "superhuman psychological profile" to be successful.

But is that what the selection psychiatrists and psychologists had been trying to prove? Ruff had set out to understand psychological *adaptation* to the space environment, not to ensure that astronauts were psychologically superior to anyone else. In fact, they are not. A review of astronaut candidate psychiatric evaluations over the various programs shows the real human side of these high-functioning individuals. In spite of the fact that they possessed many remarkable attributes and even though mental health experts were impressed by these men, these experts did not view them as immune to ordinary human foibles, mistakes, and even errors in judgment.

In many ways, NASA itself required that the public image of astronauts be that of heroes and actively encouraged that public perception. Heroes many of them were, but definitely not superhuman. This perceived need to place the selected candidates unsullied before the public was an important reason why psychiatric, psychological, and behavioral issues were not addressed in a straightforward or scientific manner within the space agency. NASA was so concerned with maintaining and protecting this image that all psychiatric data collected during selection were expunged from the official NASA medical records of astronauts.

NASA currently uses the excuse of medical confidentiality to prevent anyone from accessing or analyzing psychiatric data. But treating psychiatric or psychological data in a confidential manner is not the same thing as destroying or eliminating such data. Yet, in the name of confidentiality, this is what NASA has done.

The absence of overt psychopathology on the first *Mercury* flights does not in any way preclude the possibility that psychopathology may not eventually develop during space missions; nor does it address the problem that George Ruff initially tried to address—the issue of *psychological adaptation of individuals and crews,* particularly to longer-duration space flight. It is important to remember that the Shuttle *Challenger* flew successfully several times with unsound solid rocket booster joint seals until environmental conditions (i.e., extreme cold) facilitated the catastrophic failure of those same seals, resulting in the loss of seven lives.

## THE SHUTTLE PROGRAM

The advent of the Space Shuttle program in the late 1970s presented examiners in astronaut selection with several completely new considerations. First, the position of scientist astronaut was now called mission specialist astronaut. High-performance jet experience was no longer going to be required for the mission specialists, only for the pilot astronauts. Second, an entirely new space crew position was developed—the payload specialist, who would not be a professional astronaut at all, but would join professional space crews to work on very specific aspects of the payload.

In 1964, Bryce Hartman and Don Flinn[3] reviewed the pertinent literature on the interactions among crew members in restricted environments and concluded that, despite minor irritations, well-motivated crew members could suppress antagonistic feelings in the interest of mission completion, especially as members of small crews on short-duration flights. They pointed out that a major difference between the non-pilot, scientist, or engineer candidates and the space pilot candidates was the difference in previous life experiences.

> Whereas the pilot applicants had already been exposed to sudden life and death situations, the scientist and engineer applicants probably had not been so exposed. This latter group could not be presumed to deal with imminent catastrophe as well as experienced pilots, and might react to inflight emergencies in maladaptive ways. . . . Clearly a careful assessment of the nonpilot applicants' performance in any previously encountered dangerous situations would be valuable.[4]

Another new factor to be considered in astronaut selection was that, for the first time in U.S. space history, women could apply to become astronauts. Although Valentina Tereshekova had flown briefly some twenty years before on an early Soviet space mission, this had been primarily considered a public relations ploy on the part of the Soviet Union, rather than any real attempt to include women in space exploration efforts.[5]

NASA believed that the time was right to recruit female astronauts. For some years the agency had been under pressure to hire women, as well as blacks and other minorities, for all agency positions, including astronaut. With the passage of several federal laws prohibiting discrimination on the basis of race or gender, NASA managers realized that they would have to comply to maintain any credibility.[6]

A less obvious reason was that engineers had completed a design for a private toilet that would fly with the Shuttle. The issue of privacy, linked as it was to sexuality and personal hygiene, had been a big factor in NASA's reluctance to include women as astronauts, and the development of the private toilet—probably more than any other reason—encouraged NASA to believe that females could finally (and without embarrassment to the agency) be integrated into Shuttle missions in a way impossible during earlier missions. In *Gemini* and *Apollo* missions, the spacecraft required that crew members live side by side without even the most minimal capability of providing privacy for personal activities.

Whatever the reason to finally include women, it was a huge psychological step for NASA. The decision opened up the "new frontier" of space for many. But there were a considerable number of unanswered medical questions. How might the stresses of flying and weightlessness impact the female physiology? How would the inclusion of women in space crews affect the functioning of those crews? These questions are important—not because women should not be part of space exploration, but because the answers to such questions must be found if the human species is to live and work safely and productively for long periods of time in the space environment. Not surprisingly, most medical and psychological research at the time, especially in relevant aerospace areas, had been done exclusively on male subjects. So, inadvertently, NASA had also opened up a "new frontier" for medical and behavioral research.[7]

All of the above-mentioned factors taken together should have raised a red flag in the mind of competent managers that NASA's psychological evaluation procedures, if not inadequate, might need to be reviewed or revised to take the new aspects of Shuttle missions into account. By now, twenty years into space flight, there should have been sufficient objective psychological, behavioral, and performance data to support a management decision that no further psychological testing was needed in the U.S. Space Program. Instead, there were no data at all on which to base such a conclusion. Psychological selection data had never been correlated with later behavior or with performance. And no astronaut performance data were ever recorded by the agency from 1959 to 1990.

During the course of their careers, military psychiatrists and psychologists involved in previous selections had developed a sense of humility about their ability to predict how a given individual would behave in stressful circumstances. Yet they never abandoned a scientific perspective toward their work, and they believed that over time their hypotheses about the behavior of individuals or groups of individuals would be either proved or disproved by subsequent data on interpersonal interactions and performance.

By the time the Shuttle program began its search for a new group of astronauts, NASA management had made it clear to the military that their services would no longer be needed to perform the psychiatric evaluations of potential candidates. For some time, NASA medical management had been uneasy about the fact that the military (who from the NASA perspective were "outsiders") had medical and psychiatric information about astronauts. From their point of view, medical, and to an even greater extent psychiatric, data represented a potential "smoking gun" that if not carefully controlled might damage an already fragile relationship with Congress and the American public. (Remember the Eagleton affair? Thomas Eagleton was the vice presidential candidate who was dropped by George McGovern from the Democratic tickets in 1972 when it was discovered that he had undergone treatment for depression.) NASA preferred that all psychiatric screening in the future be done by individual consultants on its own payroll. This switch from the use of relatively compulsive and documentation-oriented military psychiatric personnel (who were accountable to their own chain of command) to the use of psychiatric consultants from Houston and the nearby vicinity was to set the stage for the following events.

NASA hired various local psychiatric consultants for the Shuttle selections, which occurred between 1978 and 1985. During that time, no operational review of psychiatric selection data was done. The time allotted for the psychiatric evaluation was limited to two clinical interviews without any psychological testing.

The first Shuttle program recruitment began in 1977, nine years after the *Apollo* selection. On January 16, 1978, 35 astronaut candidates, including the first female and black candidates, were selected. Three subsequent selections were completed in 1980 (nineteen astronauts were selected, including two women), 1984 (seventeen astronauts were selected, including three women), and 1985 (thirteen astronauts were selected, including two women).

Psychiatric assessments by two separate psychiatrists were conducted on all candidates who interviewed for the Shuttle program. The clinical emphasis was supposedly on identifying psychiatric pathology, but psychiatrists during this period ignored the advances in psychiatric diagnostic procedures that were revolutionizing the profession and instead focused on performing an evaluation to determine the candidates "best suited" to become astronauts. Like their predecessors, in earlier selections, they rated candidates as "Exceptionally Well-Qualified" (EWQ), "Qualified" (Q), "Qualified with Reservations" (QR), and "Disqualified" (DQ); but unlike the previous evaluators, there was little, if any, attempt to identify specific psychopathology.

The clinical psychiatric ratings (EWQ, Q, QR, and DQ) initiated during the *Mercury* evaluations actually are representative of the confusion about what the role of psychiatry in selecting astronauts had been from the beginning. According to NASA management, it was primarily a medical role, that is, the identification of psychopathology. *Mercury, Gemini,* and *Apollo* evaluators accepted this task as their primary one and added the clinical ratings to try to rank the candidates according to their predetermined psychological criteria, in order to use such rankings to correlate with future performance. Shuttle psychiatric consultants, on the other hand, appeared to believe that their primary role should be that of determining who, in their judgment, was best suited to be an astronaut. None of the psychiatric consultants during this period bothered to document the psychological criteria they used as the basis of their recommendation. And, even more important, these consultants possessed no data (since they did not exist) to support their belief that their judgment in this particular matter was any better than anyone else's.

The two Shuttle psychiatric consultants hired for each selection did not collaborate or share information. No documentation or standardization of the psychiatric interview process was done, and each examiner probably used his or her own idiosyncratic set of psychological criteria to determine which candidates would make "good" or "bad" astronauts. This led to a situation where a candidate was rated as EWQ by one psychiatrist, and DQ by the other. Candidates were not given specific psychiatric diagnoses when disqualified; nor were they formally cleared of having a diagnosis when found qualified.[8] It is not even certain that all the consultants even evaluated the candidates for current or previous psychopathology. NASA medical/psychiatric standards were vague and generally left it to "the opinion of the examiner" as to whether an individual was a threat to flying safety or not.[9]

Just as in previous selections, it remained unclear how significant the consultants' recommendations were in the final NASA selection process. Operationally, the NASA Space Medicine Board (SMB) (to whom the consultants presented their recommendations) simply reported an individual as either qualified or not qualified to the Selection Board, without commenting on the psychiatric ratings (EWQ, Q, QR, or DQ). This latter board, which is primarily administrative and not medical, has the responsibility of deciding which of the candidates to select as astronauts.

Table 3.1 summarizes the evaluation process from *Mercury* through the Shuttle program (specifically through the 1987 selection cycle).

Table 3.1
**Summary of Psychiatric and Psychological Selection Procedures in the U.S. Space Program (1959–1985)**

| | **Mercury** | **Gemini/Apollo** | **Shuttle** |
|---|---|---|---|
| # Hours for the Psychiatric Eval. | 30 | 10 | 3 |
| Written Medical/ Psychiatric Standards | NO | NO | YES |
| "Select-in" Criteria Used by Psychiatrists | 1. Intelligence<br>2. Drive & creativity<br>3. Independence<br>4. Adaptive motivation<br>5. Flexible<br>6. Motivation<br>7. Lack of impulsivity | 1. General emotional stability<br>2. High motivation<br>3. Adequate "self" concept<br>4. Quality of interpersonal relationships | None documented |
| Screening Method | 2 psych interviews<br>25 psych tests<br>1 stress test | Same<br>10 psych tests<br>1 stress test | Same<br>None<br>None |
| Outcome | One consensus recommendation to SMB | Same | Two independent recommendat. |
| Validation of Criteria? | Data not available | Not done | Not done |

*Source:* Patricia A. Santy, Al W. Holland, and Dean M. Faulk, "Psychiatric Diagnoses in a Group of Astronauts," *Aviation, Space, and Environmental Medicine* 62 (1991): 969–973.

By the late 1970s a revolution was taking place in the field of psychiatric diagnostic and interviewing procedures. The American Psychiatric Association (APA) had adopted clear and objective criteria, which were designed to be used by clinicians as the basis of making psychiatric diagnoses. The third edition of the APA's *Diagnostic and Statistical Manual of Mental Disorders* (DSM-III)[10] was considered the standard of psychiatric care, not only in the United States, but also in many parts of the world. Behavioral researchers were developing specific and reliable structured psychiatric interviews that took the listed criteria as their basis.

This new technology, though available, was not adopted by any psychiatric consultants hired by NASA. Instead, each appeared to have developed his own "system" for evaluating personnel. To be fair, any one of those systems might have been adequate, but since none of the consultants ever documented his methodology, it is impossible to rationally critique them. It is possible that NASA management at that time was more comfortable with the vagueness of these evaluations. When there is nothing specific to point to, it is hard to criticize.

How specific psychiatric consultants were chosen to participate in astronaut selections during the Shuttle period was also rather vague. As an example, one NASA manager informed me in 1990 that a particular psychiatric consultant would have to stay on the NASA payroll because that consultant was a personal friend of the then NASA administrator. This is probably not the best way to determine the credentials of consultants for such an important task.

Since few records exist, it is not possible to directly evaluate the methodologies of the psychiatric consultants employed by NASA in this period. However, one consultant who had been involved in all the Shuttle selection cycles had written summary reports, and these were collected and reviewed. The results of that review will be discussed in the next section.

## THE STATUS OF NASA PSYCHIATRIC SELECTION IN 1987: THE AUTHOR JOINS NASA

In 1984, I applied for and was hired by NASA as a flight surgeon/medical officer in the Flight Medicine Clinic. The clinic is responsible for the medical care of all astronauts and their families. Although I was a board-certified psychiatrist, I had another quality that made NASA medical sciences interested in hiring me: I was female. I only discovered at a later date that I had been hired to replace the only female flight surgeon in the clinic, when she was selected to become an astronaut in the 1984 class. The fact that I was a psychiatrist was overshadowed by the fact that I was the only female applicant for the flight surgeon position.

I remember that several of the interviewers queried me about how important psychiatry was to me or if I had "any axes to grind" about psychiatry. Even then I was clearly aware of how my psychiatric background put me at a disadvantage in getting hired at NASA. Since I desperately wanted to be hired and to participate in the Space Program, I minimized my psychiatric qualifications and emphasized my general medical background and my interest in flying. I was hired and reported to the Johnson Space Center in mid-1984.

Although my duties initially did not include much psychiatry, my psychiatric background naturally led me to be very interested in NASA's procedures for the psychological evaluation of astronaut candidates. I was aware of the literature on the subject from the earlier NASA missions, and I was also aware that no scientific papers had been published on this topic—or any other behaviorally related topic—until my own review article appeared in the *American Journal of Psychiatry* in 1983. This did not unduly concern me, however, since I was certain, at least in the area of selection, that NASA was continuing the important work begun with George Ruff. In the 1983 review article, I wrote that "[t]he bulk of clinical activity for psychiatry in the U.S. space program has been in the area of preselection aeromedical evaluations. As part of a team of physicians, psychiatrists have had considerable input into the astronaut selection process from the beginning."[11]

I was considerably surprised to then discover that NASA had no records of the *Mercury, Gemini,* or *Apollo* psychiatric evaluations and that not only had psychological testing been eliminated from the Shuttle assessment process, but also the two clinicians who did the evaluations did not even discuss the applicants with each other. Nor were they expected to document their methodologies. I was also surprised to find out that there was no record of any behavioral or performance data ever being collected on any U.S. space flight.

One particular psychiatrist (non-board-certified) had been a consultant for every Shuttle selection between 1978 and 1985, but nowhere was there any documentation about what he was doing or how he was doing it. No one at that time had reviewed his records or his recommendations about candidates. Attempts on my part to find out directly from him what he was doing led only to vague comments about utilizing a "process" model and an unwillingness on his part to discuss the details of his interview and evaluation methods. NASA life sciences management did not appear to be interested in accountability or documentation (which is surprising because they paid him over $100,000 a year). Consultants in psychiatry were clearly not expected to document activities in the same manner as consultants in other medical disciplines were.

By late 1985, I was concerned that many of the astronaut selection recommendations made by this psychiatrist were not appropriate, particularly since they disagreed so sharply with some of those made by the other consultants, as well as with my own experience with the same candidates. From a variety of sources, I was able to compile most of this consultant's final psychiatric recommendations on candidates from the 1978–1985 selections. A total of 500 candidates (61 female and 439 male) had been evaluated during selections in 1978, 1980, 1984, and 1985, as summarized in Table 3.2. All available reports for this consultant during that time were then reviewed and compared to SMB records, where a number of numerical discrepancies were discovered (see Table 3.3). There were a total of 7 missing reports on female candidates and 174 missing reports on the male candidates. The fact that so many reports were missing clearly demonstrates the poor record keeping that NASA management permitted during this time. But, that fact aside, an unexpected finding is documented in Table 3.4.

A breakdown by gender of qualified versus not qualified candidates was done, and, astonishingly, over 40 percent of the female candidates had been disqualified psychiatrically by this consultant, compared to 7.5 percent of the male candidates. It is possible because of the discrepancies in the total number of reports that these numbers are not completely accurate. Nevertheless, even if all reports were accounted for, these data are impossible to understand. As we have seen, during all previous selections, psychiatrists and psychologists had actually disqualified only a handful of individuals. Are we to believe that nearly half of all the professional women

Table 3.2
Candidates Evaluated in the Shuttle Program (1977–1985)

| Selection Year | Total Candidates | # Female Candidates | # Male Candidates |
|---|---|---|---|
| 1977 | 208 | 15 | 193 |
| 1980 | 121 | 21 | 100 |
| 1984 | 106 | 22 | 84 |
| 1985 | 65 | 3 | 62 |
| Totals | 500 | 61 | 439 |

Table 3.3
Psychiatric Evaluation Reports Available (1977–1985)

| Total | # Female Reports | # Reports Missing * | # Male Reports | # Reports Missing * |
|---|---|---|---|---|
| 319 | 54 | 7 | 265 | 174 |

* Most of the missing evaluations are from the 1977 and 1985 groups.

Table 3.4
Breakdown of Psychiatric Disqualifications by Gender (1977–1985)

| | Total | Qualified | Unqualified | % Not Qualified |
|---|---|---|---|---|
| MALES | 265 | 245 | 20 | 7.5 % |
| FEMALES | 54 | 32 | 22 | 40.7 % |

interviewed by NASA were considered psychiatrically unfit to become astronauts? That there had been blatant sexual discrimination in the evaluations seemed very likely. But, again, the lack of specific documentation by this consultant did not permit the analysis of any rationale he may or may not have had for his recommendation.

Armed with what I felt was strong evidence of a serious failure of the psychiatric evaluation methods adopted by NASA at the start-up of the Shuttle program, I decided to use the upcoming selection to demonstrate how recent advances in psychiatric diagnosis and evaluation might be applied to the medical/psychiatric portion of the astronaut candidate evaluation. I did not want to present my findings about possible discrimination to NASA management until I had specific recommendations for changing the procedure. With data from this next selection, I believed my recommendation to review the procedures in some detail would have a better chance of being accepted. My fellow flight surgeons in the Johnson Space Center Flight Medicine Clinic were supportive of the idea, particularly since they had frequently dealt with this consultant by simply ignoring his recommendations.

The first step in developing a rational procedure was to detail an appropriate psychiatric interview. Time constraints made it impractical to utilize standard "off-the-shelf" structured psychiatric interviews [such as the Structured Clinical Interview for DSM-III (SCID), the Schedule for Affective Disorders and Schizophrenia–Lifetime (or SADS-L), or the Personality Diagnostic Questionnaire (PDQ)] to evaluate for both Axis I and Axis II disorders. Since my goal was primarily to identify past and present psychopathology in candidates, I settled on a semi-structured interview format. The specific interview is documented in Appendix 3. I adhered fairly rigidly to the interview format with all candidates. DSM-III criteria were used to make specific psychiatric diagnoses. Applicants were also carefully questioned about major psychiatric illnesses in family members, since one of the written NASA psychiatric standards was that an individual could be disqualified psychiatrically if two or more members of his or her immediate family had a major psychiatric illness.

Although I had wanted to include several psychometric tests (specifically, the Minnesota Multiphasic Personality Inventory (MMPI) for Axis I disorders and the Millon Clinical Multiaxial Inventory for Axis II personality disorders), I was not permitted to do this.

## RESULTS FROM THE FIRST PSYCHIATRIC ASSESSMENT OF ASTRONAUT CANDIDATES USING DSM-III CRITERIA

An article reporting my findings in the 1987 selection was published in October 1991 in *Aviation, Space and Environmental Medicine*.[12] Although the data reported in this article were analyzed much earlier, I did not submit it for peer review until after I left NASA in early 1991.

I evaluated 117 astronaut candidates (94 males and 23 females) at the Johnson Space Center during the 1987 selection. Each applicant underwent a two-hour, semi-structured clinical interview at a rate of twenty to twenty-three individuals per week over a six-week period. Fifteen individuals from this group (thirteen males and two females) were eventually selected as astronauts by NASA.

Table 3.5 summarizes the range of Axis I and II diagnoses made in this group of applicants. Of the 117 applicants, 7.7 percent (9) met DSM-III criteria for specific psychiatric diagnoses. Two individuals had more than one diagnosis (one each on Axis I and Axis II). The majority of diagnoses fell in the category of V-codes, or conditions that are not attributable to a mental disorder, but that are a focus of attention or treatment. Such diagnoses include interpersonal problems in marriage, family, and work.

A review of the candidates' family histories indicated that 27 of 117 applicants (23.1%) had at least one first-degree relative with a probable psychiatric illness. The most common disorder they reported in close relatives was alcohol abuse (twenty applicants), followed by substance abuse (three applicants), bipolar disorder (two applicants), and schizophrenia (two applicants). No applicant had two immediate family members with these disorders.

One of fifteen (6.7%) individuals eventually selected by NASA to become astronauts met criteria for a DSM-III diagnosis, compared to 8 of 102 (7.8%) who were not selected. The one candidate selected was diagnosed as having a resolving adjustment disorder and was not medically disqualified. Only two of fifteen (13.3%) candidates selected had close family members with psychiatric diagnoses, compared with 25 of 102 (24.5%) who were not selected.

The absence of major psychiatric disorders (such as schizophrenia and affective illness) in this population is not surprising, considering the academic and professional qualifications necessary in order to be asked to interview for the job of astronaut. V-code problems, on the other hand, are common, and it would be very unlikely if such problems did not also ap-

Table 3.5
DSM-III Psychiatric Disorders in Astronaut Applicants

| Patient Code | Diagnosis | Number of Applicants | Prevalence (from DSM-III-R)[1] | % Psychiatric Patients with Diagnosis[2] |
|---|---|---|---|---|
| *AXIS I:* | | | | |
| V61.10 | Marital Problem | 2 | Common, incidence unknown | |
| V61.10 | Parent-Child Problem | 1 | Common, incidence unknown | 2.8% (psychiatric patients with any V-code diagnosis) |
| V62.82 | Phase of Life Problem | 1 | Unknown | |
| V62.82 | Uncomplicated Bereavement | 1 | Extremely common | |
| 309.28 | Adjustment Disorder | 2 | Common | 9.9% |
| 300.02 | Anxiety Disorder | 1 | Not commonly diagnosed | 6.8% |
| *AXIS II:* | | | | |
| 301.40 | Compulsive Personality Disorder | 1 | Common; males | Not Available |
| 301.81 | Narcissistic Personality Disorder | 1 | Common | Not Available |
| 301.22 | Schizotypal Personality Disorder | 1 | 3% of population | Not Available |

NOTE: Two individuals had two diagnoses, one each on Axis I and Axis II.

[1] Exact data on prevalence and incidence are not available for many psychiatric disorders.

[2] From J.E. Mezzich, H. Fabrega, G.A. Coffman, and R. Haley. "DSM–III Disorders in a Large Sample of Psychiatric Patients: Frequency and Specificity of Diagnoses," *American Journal of Psychiatry* 146 (1989): 212–219.

pear in this group. The fact that there were so few identified instances of marital, family, or work difficulties is probably due to the tendency of the applicants to underreport. Spouses or other family members are not interviewed to confirm clinical material, so clinicians must rely heavily on the candidates' willingness to discuss these types of problems. V-code diagnoses and adjustment disorder diagnoses would not generally lead to disqualification, unless the disorder was considered to impair an applicant's ability to function as an astronaut. In such a case, a V-code problem would relate to an individual's future performance. Again, NASA medical standards at the time were fairly vague, and, thus, the judgment to qualify or disqualify is left to the evaluating psychiatrist.

The individuals in this applicant group for whom psychiatric disqualification was recommended included (1) an individual with both an anxiety disorder (Axis I) and a schizotypal personality disorder (Axis II), (2) an individual with a narcissistic personality disorder (no Axis I disorder), and (3) an individual with an adjustment disorder (Axis I) and a compulsive personality disorder (Axis II).

The high number of applicants having first-degree relatives with alcohol problems is consistent with the prevalence of alcohol abuse in the general population. A community study in the United States from 1981 to 1983 using DSM-III criteria indicated that approximately 13 percent of the adult population had had alcohol abuse or dependence at some time in their lives.[13] These diagnoses could not be confirmed by interviewing the family members.

## DEVELOPMENT OF PROCEDURES BY THE NASA IN-HOUSE WORKING GROUP ON PSYCHIATRIC AND PSYCHOLOGICAL SELECTION OF ASTRONAUTS

After collecting the data discussed in the previous section, I presented the results of the new procedures to NASA life sciences management. I also presented the results of the review of recommendations, which suggested the possibility of sexual discrimination, to the SMB. Management's major concern on viewing the data was that the implication of sexual discrimination not become public, since that would reflect badly on the agency. The psychiatric consultant involved would be dealt with, I was assured.[14] In the meantime, the selection data I had recently collected convinced them that a thorough review of psychiatric selection procedures was probably necessary. This change in attitude suggests that the NASA managers were probably as dismayed as I was about the findings. For the first time, it would clearly be to the agency's benefit to have carefully documented, objective, fair, and reliable methods of psychiatric evaluation. I was given permission to bring together any behavioral scientists I felt appropriate to discuss NASA's current psychiatric procedures and to make recommendations about future psychiatric standards, as well as methodology. The group thus assembled became known as the NASA In-House Working Group on Psychiatric and Psychological Selection of Astronauts.

The Working Group met at the Lunar and Planetary Institute in Houston on April 7–8, 1988. The participants in the initial two-day workshop and their affiliations are listed in Appendix 3. Subsequent meetings of the

Working Group over the next year and a half involved additional experts. Among the distinguished scientists who participated were George Ruff, Don Flinn, and Carlos Perry, leaders of the *Mercury, Gemini,* and *Apollo* psychiatric evaluation teams. Dr. Perry died unexpectedly in 1989. He had been a personal mentor and provided much support for me and my work during my years at NASA. The NASA psychiatric consultant whose work I had critically reviewed was formally invited to participate and asked to present to the Working Group the details of his selection methodology. Although he spoke to the group for over two hours, his talk did not address any specifics of his method, nor did he present any data that might support his approach. This consultant did not participate in the proceedings after the initial workshop. All presentations made during the workshop were recorded and later transcribed as part of the proceedings of the Working Group.[15]

Two astronauts were also asked to participate as members of the Working Group. Dr. Ellen Baker, a civilian physician astronaut selected in 1984, and Dr. Sonny Carter, a military pilot/physician astronaut, both agreed to participate, and their assistance to the Working Group proved to be invaluable. Dr. Baker has subsequently flown on two Shuttle missions, on STS-34 in October 1989 and on STS-50 in June 1992. Dr. Carter flew on Shuttle Transportation System-33 (STS-33) in November 1989. He died in 1991 in a commercial airline crash while traveling as part of his official duties at NASA. His contributions to the process of the Working Group were invaluable, and all of the participants felt a deep personal loss.

The first item of business for the Working Group was to review the Psychiatry Medical Standards for Selection and Retention of Astronauts. Members of the Working Group agreed that psychiatry standards should be based on the most recent version of the diagnostic and classification system described in the American Psychiatric Association's *Diagnostic and Statistical Manual.* NASA's standards were revised to include specific disqualifying psychiatric diagnoses (as is done in all other medical areas of the document). The psychiatry standards, as revised by the Working Group, can be found in Appendix 3.

The next item addressed was the issue of "selecting out" versus "selecting-in." Selecting-out is primarily a medical/psychiatric decision. The candidate cannot previously have had or currently have any of the specified psychiatric conditions listed in the standards, or he or she will be disqualified (i.e., selected out). The medical selection process at NASA has always culminated in a "qualified" or a "not qualified" recommendation to the Astronaut Selection Board. The Selection Board is not informed as to the reason for the disqualification (i.e., whether it is medical or psychiatric, or what specific condition the candidate is disqualified for), since this information is considered confidential.

Selecting-in is a process frequently used in personnel selection, and it does not have specific medical or psychiatric implications. The goal of this process is to identify the "best person for the job." The development of specific, job-related psychological selection criteria is mandatory for select-in evaluations. Experts in the field of personnel selection are generally asked to identify those psychological characteristics believed to be important for performing well on the specific job. A more detailed presentation of the concepts of selecting-out and selecting-in will be undertaken in Chapter 4.

In an effort to clarify whether NASA management wished psychiatrists and psychologists to select out, select in, or do both, the Working Group members requested management to meet with them. Dr. Sam L. Pool, a physician and chief of the Medical Sciences Division at the Johnson Space Center, presented an unequivocal response: The only job the behavioral scientists were being asked to do was to identify psychopathology, or the risk of psychopathology, in potential astronaut candidates. Management did not seek our advice concerning selecting-in.

Over the next year, various members of the Working Group met and divided into two smaller groups. The first subgroup had the task of reviewing the literature and identifying the most reasonable operational plan for selecting out, or psychiatrically evaluating, astronaut candidates. A second subgroup was formed to develop a research plan that would identify psychological selection criteria (i.e., select-in criteria) for future NASA missions (such as Space Station *Freedom*) and then would scientifically validate those criteria, using astronaut behavior and performance as outcome variables. This latter effort was included for several reasons. First, NASA management had requested our general recommendations about psychological characteristics to search for in future space station astronauts. Second, as a group, we believed the behavioral sciences could bring a degree of objectivity and expertise to the select-in portion of selection. We differed with NASA management about the usefulness of such evaluations, and we wanted a research project to answer the question once and for all. The work of both of these subgroups and the results of the implementation of their recommendations are discussed in the next three chapters of this book.

Whatever psychiatry may or may not be able to contribute to the better selection of individuals and crew members for long-duration space missions (and current data suggest that prediction is not one of our strong points[16]), NASA's reluctance to accept and deal with the importance of psychological issues—despite the accumulated evidence in every other hazardous environment and despite numerous efforts on the part of many scientists to enlighten them—is reminiscent of the denial and arrogance manifested by NASA managers prior to the launch of *Challenger* in 1986. Fortunately, at this point in time, the former situation has not yet resulted in the same disastrous consequences as the latter.

Flinn in 1961 wrote:

> It is important that a longitudinal study of these [psychological] factors be carried out in connection with the present assessment program. At present we can afford the luxury of choosing astronauts from a select group with demonstrated adaptability, but eventually larger numbers will be required, and they will need to be chosen before an extensive training program is undertaken. If this is to be done reliably, we will be dependent upon information that can be gained only from a thorough study of the current astronauts, together with observations about future performance.[17]

Over thirty years had passed, and we had succeeded only in coming full circle.

## NOTES

1. Carlos J. G. Perry, "Psychiatric Selection of Candidates for Space Missions," *Journal of the American Medical Association* 194 (1965): 841–844; quotation at 844.
2. C. Cordes, "Mullane: Tests Are Grounded," *APA Monitor* (October 1983): 24.
3. Bryce O. Hartman and Don E. Flinn, "Crew Structures in Future Space Missions," in *Lectures in Aerospace Medicine* (Brooks Air Force Base, Texas: U.S. Air Force School of Aerospace Medicine, 1963).
4. D. R. Jones and C. A. Annes, "The Evolution and Present Status of Mental Health Standards for Selection of USAF Candidates for Space Missions," *Aviation, Space and Environmental Medicine* 54 (1983): 730–734; quotation at 731.
5. Patricia A. Santy, "Women in Space: A Medical Perspective," *Journal of the American Medical Women's Association* 39 (1984): 13–16.
6. Joseph D. Atkinson and Jay M. Shafritz, *The Real Stuff: A History of NASA's Astronaut Recruitment Program* (New York: Praeger, 1985).
7. Ibid. Richard T. Jennings and Patricia A. Santy, "Reproduction in the Space Environment: Part II—Concerns for Human Reproduction," *Obstetrical and Gynecological Survey* 45 (1990): 7–17.
8. Patricia A. Santy, Al W. Holland, and Dean M. Faulk, "Psychiatric Diagnoses in a Group of Astronaut Applicants," *Aviation, Space and Environmental Medicine* 62 (1991): 969–973.
9. NASA *Medical Standards: NASA Class I Pilot Astronaut Selection and Annual Medical Certification*, JSC-11569 (Houston: Johnson Space Center, 1984).
10. American Psychiatric Association, *Diagnostic and Statistical Manual of Mental Disorders*, 3d ed. (Washington, D.C.: American Psychiatric Association Press, 1984).
11. Patricia A. Santy, "The Journey Out and In: Psychiatry and Space Exploration," *American Journal of Psychiatry* 140 (1983): 519–527; quotation at 519.
12. Santy, Holland, and Faulk, "Psychiatric Diagnoses in a Group of Astronaut Applicants."
13. J. E. Mezzich, H. Fabrega, G. A. Coffman, and R. Haley. "DSM-III Disorders in a Large Sample of Psychiatric Patients: Frequency and Specificity of Diagnoses," *American Journal of Psychiatry* 146 (1989): 212–219.
14. In 1992, as I write this book, the individual in question still receives money

for consultation from NASA. While he is no longer involved in selection, he now calls himself an "expert" in crew interaction. When asked why this situation still existed, Dr. Carolyn Huntoon, director of space and life sciences at the Johnson Space Center, stated that it was for the purpose of "damage control."

15. Patricia A. Santy, ed., "The NASA In-House Working Group on Psychiatric and Psychological Selection of Astronauts: Summary and Transcripts" (unpublished paper, 1988).

16. C. K. Aldrich, "The Clouded Crystal Ball: A 35-year Follow-up of Psychiatrists' Predictions," *American Journal of Psychiatry* 143 (1986): 45–49.

17. Don E. Flinn, "Psychiatric Factors in Astronaut Selection," in *Psychophysiological Aspects of Space Flight*, edited by B. E. Flaherty (New York: Columbia University Press, 1961), 91.

# 4

# The Dual Role of Behavioral Sciences—Selecting-In and -Out

> [T]he role of the psychiatrist [in aerospace operations] is ambiguous because the psychiatrist must simultaneously serve multiple tasks, some of which may be in conflict.
>
> —William H. Sledge and James A. Boydstun[1]

## THE ASTRONAUT PROFESSION'S "DEMANDING PARADOXES"

Many observers have noted that the aerospace environment is especially challenging for the psychiatrist, since in that environment even "benign medical disorders may have quite serious occupational implications."[2] Over a relatively short span of years, the tremendous advances in technology that have been a hallmark of the Space Age have substantially expanded the professional requirements of pilots and, by extension, astronauts. Program requirements that depend on maintaining maximal job performance and motivation and on minimizing errors that might have catastrophic consequences have necessitated even more accurate identification and selection of suitable individuals for the profession.

William H. Sledge and James A. Boydstun, in a perceptive article in the *American Journal of Psychiatry*, commented:

> Psychosocial factors of crew interaction, separation from family, fatigue-rest cycles, and time-zone travel are potential stressors. Consequently, professional aviators are caught in a series of demanding paradoxes; they must be able and willing to submit to numerous regulations, procedures, and rules in their professional lives, yet they must maintain a sense of responsibility and initiative in the execution of

their work. They must be calm, thoughtful, and free of impulsive tendencies; however, they must be quick to react adaptively to emergent situations. . . . They must be able to make quick, effective decisions when given incomplete information but they must always be looking for more facts about their present situation. They must also have a healthy, enduring love of flying tempered by a realistic respect for its dangers. They must be self-confident and convinced of their ability to survive dangerous situations; yet they must not be so grandiose that they take foolish chances or so unsure of themselves that they constantly must prove their mastery and invulnerability.[3]

Astronauts must also live with similar paradoxical limitations that continually stress their adaptational abilities. They must be sure of themselves, but not too sure—competent, but not arrogant; they must love to fly, but appropriately fear flying; they must be confident about their abilities, but not grandiose; they must react quickly to danger, but not be impulsive; they must obey authority, and yet be independent; and so on. How many mortals could stand up to these seemingly endless injunctions? The delicate balancing of these paradoxical expectations results in an internal tension experienced by every pilot and astronaut. Frequently that tension manifests itself in the well-described conflict between aviators and flight surgeons—physicians whose job is to observe health and performance in flying personnel. Every pilot and astronaut understands intuitively that the possible termination of his or her career may rest with factors completely out of his or her control. The physician—and particularly the psychiatrist—stands as the reality check when the desire to manifest the "right stuff" and to be perceived as capable and effective in one's job crashes into the reality of human frailty and imperfections of mind and body.

This age-old conflict places the astronaut's physician in a particularly awkward position. While flight surgeons must keep patient care at the forefront of their activities and in that sense represent the interests of their patients, they also are expected to be agents of the organization and to determine those impairments in their patients that may negatively impact the mission or public safety. This latter role has been referred to as the "social control" function of the physician (particularly the flight surgeon).[4]

To some extent the functional dichotomy imposed by the unstated social control function of medicine confronts every practicing physician. But in the Space Program, as in the military, the issue of social control colors practically all interactions between the flight surgeon and his or her patients.

## THE PROFESSIONAL DILEMMA OF THE PSYCHIATRIST

The psychiatric physician's role in aerospace operations is more ambiguous than that of other physicians because of the nature of psychiatry itself. One important function of a psychiatrist is to understand the potential mental health impact of so-called "psychosocial" factors—that is, how the interpersonal functioning of the patient in a social context interacts with physical and emotional health. Many such factors are not generally thought of as medical in the usual sense of that term. Pilots have traditionally had difficulty in understanding how such "emotional" aspects of their lives may impact their job performance. Many astronauts will report voluntarily to the Johnson Space Center Flight Medicine Clinic with a mild head cold, knowing that even such a minor physical problem could have a negative effect on their flying performance because of changes in sinus pressure at altitude. Frequently these same individuals would not even consider the possibility that marital strife—or any feelings they were having at the moment, for that matter—could potentially have even more of an impact on their ability to fly safely.

I can only remember a few instances over seven years when astronauts voluntarily presented themselves to the clinic to seek help with an emotional problem. A typically held tenet of the pilot and astronaut culture is that flying is an effective way to cope with distress in the mundane world below. "When I fly, all the problems I'm having at home go away," one astronaut told me.

Despite its fascination with the private lives and problems of celebrities, American culture, in general, is not particularly sympathetic to individuals with emotional problems, especially if they are open about them. The expression of emotions such as sadness and fear is considered a weakness. The pilot/astronaut culture is overtly hostile to the expression of such problems—in themselves and in others. Denial is the name of the game, and this can be seen even more clearly in the astronaut selection process.

In performing psychiatric evaluations as part of astronaut or pilot selection, it is not surprising that the psychiatrist has historically perceived his or her mission as that of both a physician and a "job performance evaluator" (read "social controller"). The major difficulty with that perspective is that although it reflects the primary dilemma of the psychiatrist, assumption of this dual role has reinforced the confusion about what psychiatry reliably can—and cannot—provide to the selection process. How should the operational psychiatrist deal with this problem?

One way to shed light on this question is to review scientific studies that have looked at how well psychiatric predictions about individual performance and behavior have turned out. Paul E. Meehl[5] did exactly that in his 1954 book, and his review was not overwhelmingly favorable to the clinical psychiatrist.

H. W. Brosin[6] studied four successive medical school classes from the University of Chicago in the 1940s. Students who participated in the study underwent a battery of psychometric tests and an extensive psychiatric interview lasting from one to four hours after admission to medical school (so that the psychiatric results did not have anything to do with whether a particular individual was selected or not). Psychiatrists ranked the students into one of three categories: those with a "good prognosis" (N = 14), those with a "guarded prognosis" (N = 69), and those felt by the clinicians to have a "poor prognosis" (N = 47). The subjects were then followed throughout their medical school career, and the only outcome variable was whether or not the subject graduated from medical school versus dropping out. The psychiatric classification (i.e., good, guarded, or poor prognosis) was found to be significantly correlated with outcome (p = .000105 by Fisher's exact test). Those rated as having a good prognosis were less likely to drop out of medical school prior to graduation than were those with a guarded or poor prognosis.

In 1981, C. Knight Aldrich[7] followed up on subjects from Brosin's Chicago medical student study. Sixteen of the original participants had died, and twelve could not be located, but 94.6 percent of those remaining chose to respond to his questionnaire on their current medical and psychological health. Aldrich was interested particularly in the question of how well psychiatrists could predict future mental health or illness. He found that while there was a higher incidence of impairment in the groups with guarded or poor prognosis (p = .001 by Fisher's exact test), there were also mildly impaired and impaired physicians in the good-prognosis group. He concluded that "[A]lthough for the group the psychiatrist's predictions were better than chance, they were not good enough to justify taking them seriously in individual cases."[8] He goes on to point out that if the medical school selection committee had had access to the psychiatric evaluations prior to making their decisions, at least "20% of the students would probably not have been accepted."[9]

This study is remarkable in that it is analogous in many ways to the process of astronaut selection. The analogy is not exact because few astronauts drop out of or fail astronaut training. No astronaut has ever been eliminated because of poor performance, and all selected astronauts eventually are assigned to space missions. This may be due to the fact that all astronauts have demonstrated outstanding, or at least adequate performance, but that conclusion is extremely unlikely—as the results of recent research will demonstrate in the next chapter.

## SELECTING-OUT VERSUS SELECTING-IN

One way to clarify the dual role of aerospace psychiatrists is to separate conceptually the two conflicting aspects of the role and evaluate each individually.

Select-out or *psychiatric* criteria are medical criteria specifying those psychiatric disorders whose presence would disqualify an individual from consideration during selection. The determination of psychiatric disorders is based on psychiatric/medical assessment of the applicant and is similar to the process by which disqualifying medical disorders (such as cardiovascular disease and cancer) are assessed by the physician. The purpose of the psychiatric evaluation is to determine the presence of current psychopathology and to identify past psychopathology (usually by history) that might place the candidate at risk of developing problems in the future. It is appropriate, therefore, that this aspect of the psychiatrist's job is part of the medical evaluation of the astronaut candidate.

Select-in or *psychological* selection criteria, on the other hand, are an aspect of personnel selection. These psychological criteria are not based on psychopathology or the presence of mental illness, but instead are the basis for identifying desirable personality traits or characteristics linked to a specific job description (e.g., pilot or astronaut). Select-in criteria are developed only in a job-specific manner and are designed to assist personnel managers in selecting the "best person for the job."

Several problems have been associated with the utilization of various personality models for select-in purposes by NASA psychiatrists over the years. First, there is no available accurate or even detailed description of the "job" of astronaut per se. And, second, whatever the job involves, the details have changed considerably since the *Mercury* selection; yet there has not been a noticeable change in the personality models used by evaluating psychiatrists. In fact, during Shuttle selections prior to 1987 the personality models had not even been documented or described and can only be inferred from the results of their application.

Another major problem has been that NASA managers have consistently stated since *Mercury* that they are not interested in the psychiatrist's recommendations about possible select-in characteristics in astronaut applicants. From their perspective, the psychiatrist's job has always been to select out. The Astronaut Selection Board, composed of NASA managers and astronauts, has some justification in believing that the board's assessment of which applicants are best for the job is superior to anyone else's, since presumably they have a clearer idea of what the job itself entails and what it takes to perform it well.

Would the psychiatric evaluation contribute anything of substance to the Selection Board's deliberations? As we have seen, there is no compelling evidence in the psychiatric or psychological literature that would support the proposition that psychiatrists are any better at doing this than anyone else. The psychiatric clinical ratings done on astronaut applicants have frequently been consistent with the decisions of the Selection Board. This suggests, as mentioned before, that whatever qualities the psychiatrists see during their clinical evaluation may also be what members of the Selection Board see when they interview and interact with the applicant.

Are the qualities observed during both the psychiatric and the job interviews really the qualities that will determine job success as an astronaut? And can the behavioral sciences contribute to the process in a productive manner? With the advent of more reliable, theoretically grounded psychometric tests, these questions require a fresh look. While it is true that psychological tests in the past have failed to find significant personality predictors of performance, it does not necessarily mean that such predictors do not exist.

The Brosin study of medical students had a positive correlation between prediction and outcome because their select-in criteria (used to formulate a prognosis) *were not used initially to select the medical students.* This guaranteed that the population studied was heterogeneous with regard to prognostic category, and thus the likelihood of identifying a relationship between the predictor variables and the outcome (i.e., graduation from medical school) was enhanced. This fact is extremely important, and its implications for astronaut selection paved the way to understanding the psychiatric dilemma faced by the Working Group.

If very specific select-in psychological criteria are to be developed for Space Station astronauts, those criteria would then be the predictors of future outcome (i.e., astronaut performance). In order to validate those criteria, *they must not be used to select astronauts initially.* Only after research demonstrates the validity of the criteria would they become part of operations. And, until then, they should have no impact on actual operational decisions, specifically the careers of astronauts or astronaut candidates.

Can anyone, no matter what the psychological status, be selected in that case? Remember that the *operational* task will continue as before. Psychiatrists will still be required as a part of routine operations to select out those applicants with psychopathology or a risk of developing psychopathology. But it is the Astronaut Selection Board—not the psychiatrists—that has been doing the select-in evaluations, and since they are satisfied that their assessment of this is adequate, there is no reason that a research study to formally validate psychological criteria could not be undertaken. In fact, such a study would determine once and for all whether the behavioral sciences could contribute to the selection process. If after all the tests and analysis they do not do any better job of it than the Selection Board does, then there is no reason to include their procedures in selection.

It is absolutely essential to obtain frank and candid answers to psychological questions during the research process, both at the time of selection and during the collection of meaningful performance data in subsequent training and mission activities. Members of the Working Group who had done similar research in other environments—such as Bob Rose (with air traffic controllers),[10] Bob Helmreich (with aviators),[11] and Dave Jones (with Air Force Mission Specialist evaluations)[12]—all pointed out that, as a precondition to obtaining research data, an agreement between the investigators and the subjects was reached that any information collected from them as part of the research would neither help nor harm their careers. For astronauts, this would mean that no data be available to any representative of NASA or other agencies. Such an agreement is both an ethical and a practical necessity to obtain useful data from the subjects participating in the research. Operational data collection, used to make real-time operational decisions (such as who will be selected), must be kept separate from the research activities. Any operational use of such data might negatively impact the individual subjects who participate in the research, would violate the agreement, and would adversely impact the reliability of the information obtained.

> When the psychiatrist represents the needs of the organization, there are various potential problems in the interventions that stem directly from his or her social control task. These problems are most obviously clear in the doctor-patient relationship, in which the usual potential obstacles and resistances may be exacerbated by patient wariness and distrust. For instance, the patient may be reluctant to cooperate and will consciously withhold data that he or she thinks may be condemning in the eye of the psychiatrist. . . . Most aviators fear they will be grounded for an obscure reason that they believe is irrelevant to their safe and reliable function.[13]

Thus, the dilemma challenging the operational psychiatrist during astronaut selection is of such a magnitude that it might impact on research data collection when subjects fear that information freely given might destroy their chances of being selected. While the physician may on occasion be asked to perform a social control function or even a job performance evaluation, it must be remembered that no data exist to suggest that psychiatrists are anything but arbitrary in their decisions.

The psychiatrists' dilemma has had a tremendous impact on behavioral sciences (both in operations and in research) in the U.S. Space Program. Although never explicitly mentioned by NASA managers, it probably played an unconscious role in their decision to remove George Ruff's research project during *Mercury*. The imposition of two conflicting expectations has been the bane of all psychiatric consultants since then. And NASA administrative ambivalence toward the behavioral sciences, combined with the conviction of some psychiatrists that they possess some special or "magical" method to screen astronaut candidates and be able to accurately predict future performance, has consequently never been resolved. The credibility of the behavioral sciences in the U.S. Space Program has been seriously undermined by the ensuing conflict.

All of the above-mentioned considerations were examined by the Working Group when it convened in 1988 to review NASA's psychiatric selection procedures and to develop some psychological criteria for astronauts selected for long-duration space missions. Their deliberation led the Working Group to form two teams: one to develop operational psychiatric strategies for implementing select-out criteria and one to come up with a reasonable research plan to validate psychological select-in criteria.

## THE OPERATIONAL AND RESEARCH TEAMS

After the initial meeting in April 1988, the members of the Working Group met a total of seven more times over the next year and a half. We came to know each other very well and to appreciate the varied professional perspectives that each of us brought to the work of the Working Group. We argued philosophy, dissected every available piece of relevant data, and compared experiences in analogous environments. Sometimes we forged new insights into the complicated relationships of psychiatry, psychology, and astronaut selection. Occasionally we identified an area in which all members of the Working Group lacked the necessary knowledge. At such times we brought experts who could educate us into the process. Finally, we came to realize how extremely important it was to include our international colleagues in the process, since the inclusion of international crew members on Space Station *Freedom* missions adds even more complicated multicultural dimensions to the problem of astronaut selection.

The entire Working Group met together seven times, but in addition to these more formal meetings, the two subcommittees each met several times to hammer out the controversial issues in their respective areas. One team worked on developing the best psychiatric screening method possible for selection operations, while the other studied all aspects of a research plan to validate the psychological select-in criteria developed by the entire Working Group.

### The Research Team

Leading the Research Subcommittee was Dr. Robert Rose, who at the time was the chairman of the Department of Psychiatry and Behavioral Sciences at the University of Texas Medical Branch in Galveston, Texas. I had met Rose many years earlier when I was a resident in psychiatry at the UCLA-Harbor Medical Center. At that time he had just taken the job of chairman at Galveston, and my name had been recommended to him as a potential faculty member. A handsome, imposing man, Rose radiated competence and authority. His experience in research methodology and design was his greatest technical asset. Rose was well known in scientific circles for his studies on air traffic controller stress from both a biological and a psychological perspective.[14] Rose is currently at the MacArthur Foundation in Chicago.

Dr. Robert Helmreich, a professor in the Department of Psychology at the University of Texas at Austin, and Clay Foushee, a Ph.D. researcher at NASA Ames Research Center and a collaborator of Helmreich's, were best known for their work on commercial pilots and cockpit crew interaction.[15] Helmreich's extensive experience in psychometrics and psychological testing was a tremendous asset to the team. He and Janet Spence from the University of Texas had developed a theoretical model of personality that seemed applicable to astronauts. This model came to dominate the thinking of the Working Group because of the compelling data they had collected on high-functioning individuals—including pilots and scientists. Helmreich and Spence are still at the University in Austin. Foushee has since accepted a position as director of a new human factors division at the Federal Aviation Administration in Washington, D.C.

Dr. Harry Holloway, another member of the research subcommittee, was the chairman of the Department of Psychiatry at the Uniformed Health Sciences University in Bethesda, Maryland. A man of incomparable depth and learning, Holloway had an encyclopedic mind and was able to bring his vast knowledge of the psychiatric and psychological literature to the problems at hand. Dr. Holloway has since become NASA's Associate Administrator for Microgravity and Life Sciences.

Sonny Carter was the astronaut representative to the subcommittee. He was a physician and a test pilot, but probably most important of all, he was a sensitive, practical, and warm individual. He was always able to bring the topic back to reality when it had slipped into speculation and realms of psychological fantasy. Carter's "down-home, country boy" approach could not hide the keen intellectual mind he possessed. His frequently humorous observations about psychiatry and psychiatrists were a welcome addition to the discussions. Carter flew on his first Shuttle mission during the two years that the Working Group met. His perspective on space operations and his and fellow astronaut Ellen Baker's ability to convince their astronaut colleagues of the rationality of the research plan we ultimately proposed were major factors in its acceptance among the astronaut corps. As noted earlier, Dr. Carter died in a commercial plane crash in 1990.

Also participating on the research team were the two international representatives to the Working Group. Dr. Chiharu Sekiguchi, a flight surgeon from the National Space Development Agency of Japan (NASDA), had done his residency at the Wright State University School of Aerospace Medicine under Stan Mohler. Sekiguchi was interested in both the operational and the research aspects of the Working Group's activities and enthusiastically participated in the deliberations of both subcommittees. His observations were supplemented by those of several of NASDA's psychological and psychiatric consultants—Dr. Minoru Kume, a psychologist from Waseda University in Tokyo, and Dr. Shigenobu Kanba, a psychiatric consultant from Keio University. Both of these individuals were members of a Japanese working group on psychological issues, which was dealing with many of the same problems our group was confronting.

The European Space Agency (ESA) representative was Dr. Klaus-Martin Goeters, a psychologist from the German Aerospace Research Establishment (DLR) Institute for Flight Medicine in Hamburg, Germany, and an experienced psychological researcher on pilot selection. Goeters and his colleagues had done an impressive amount of work utilizing his own psychological test batteries for European pilot selection. During the team's meetings, he often took a great deal of heat about his group's "secret" methodology and the lack of validating data, but he always remained professional and cooperative. He was eventually willing to incorporate aspects of the research team's methods into his own research, while we incorporated several psychometric scales from his test battery into ours.

### The Operations Team

The Operations Subcommittee was headed by Dr. David Jones, then the chief of the Neuropsychiatry Branch at the Brooks Air Force Base School of Aerospace Medicine. Jones, who has since become the editor-in-chief of the major journal in aerospace medicine—*Aviation, Space and Environmental Medicine*—is a no-nonsense, direct individual with a wry sense of humor and a lot of common sense. He had been the lead Air Force psychiatrist in the evaluation of astronaut candidates for the Air Force's space program (which never came about), and the selection of astronauts had been a matter of interest to him for some time. He shared this interest with one of his psychologist colleagues from the School of Aerospace Medicine, Dr. John Patterson. Jones and Patterson were a formidable team who had years of experience evaluating pilots with emotional impairments. Together they had developed a thorough and coordinated approach to assessing and treating the psychological and psychiatric problems of military pilots and had established a private consulting firm known as Aeropsych Associates.

The Operations Subcommittee was also fortunate to receive input from Dr. George Ruff, now a professor of psychiatry at the University of Pennsylvania. Ruff's career had not been hurt to any extent by his dismissal from NASA as a consultant. In fact, he was known for the excellent scientific papers he had written on the *Mercury* selection process in the 1960s. Because of his historical perspective, Ruff could provide a context for the entire proceedings of the Working Group. It greatly amused him that we were, in some ways, "reinventing the wheel," since our work was similar to what he and his colleagues had initiated thirty years earlier. I can vividly remember the exact moment I made the decision to write this book. Ruff was commenting on his own sense of deja vu about the Working Group's efforts, and with an unusually intense affect he said:

> We kept pleading that these data were valuable. Even if we don't finish it, let's put together what we already have. Let's put together the performance data, the individual psychological data, the physiological data, and the biochemical data. We were told that is just what they [NASA] wanted, but it never really happened. . . . So my plea now is that it is really not too late. You are going to be testing an awful lot more people in the future than we ever did during *Mercury*. There should be some continuity on the follow-up. By now, we would have had a lot of data on outcome, as well as clinical and anecdotal data, you could have used.[16]

It was at that instant that I made a personal commitment to write this history, so that no future group would also have to "reinvent the wheel."

Ellen Baker, another astronaut physician, also participated in the proceedings of the Operations Subcommittee. Baker, a tall and serious individual, worked conscientiously at educating the members of the Working Group on the operational realities of life as an astronaut. Less outgoing than fellow astronaut Carter, she nevertheless developed strong ties with other Working Group members and contributed to the successful implementation of the recommendations that came from the two teams.

When the Operations Subcommittee decided to use a structured psychiatric interview in selection, the members requested consultation from one of the world's leading authorities in that area, Dr. Jean Endicott, professor of psychiatry and psychology at Columbia College of Physicians and Surgeons and the chief of the psychological evaluation section at New York State Psychiatric Institute. Dr. Endicott has an international reputation because of her research on structured interviews, particularly the Schedule for Affective Disorders and Schizophrenia (SADS) and its complement, the Research Diagnostic Criteria (RDC). With Dr. Endicott's help the group was able to review a number of interviews for both Axis I and Axis II psychiatric disorders and select the one that would be most likely to work best in the astronaut candidate population. Dr. Endicott and her colleague Dr. Deborah Hassan were responsible for training the operational psychiatrists and psychologists who implemented the new select-out procedures during the astronaut selection of 1989.

As a NASA employee, I participated on both subcommittees because my responsibility was to coordinate the activities of the Working Group and to report its findings back to NASA life sciences management. I was nominally in charge of the entire Working Group and assisting me in that was Dr. Al Holland, a clinical psychologist and consultant for the Universities Space Research Association (USRA). Holland later came to work with me at the Johnson Space Center and was my deputy in the Biobehavioral Laboratory. When I departed NASA to take a faculty position at the University of Texas Medical Branch in Galveston in 1991, he became chief of that laboratory.

## THE WORKING GROUP COMPLETES ITS ASSIGNMENT

By July 1989, both subcommittees had completed their assigned tasks and had made specific recommendations for me to present to NASA management. Implementation of the recommendations was my responsibility, but first NASA life sciences management had to be convinced of their value. NASA planned to begin an astronaut selection cycle in late 1989, and the Working Group wanted use the new procedures recommended by the Operations Subcommittee then. It was also hoped that at the same time research might begin to validate the psychological selection criteria developed by the Working Group in 1988 (and discussed in detail in the next chapter). The objective was to convince NASA life sciences management to fund the project. Since I was primarily an "operations" person at the agency, I approached Bob Rose and Bob Helmreich (we called them the "two Bobs") and asked them to be the principal investigator and the co-investigator, respectively, on the research proposal, which was aptly titled "Validation of Astronaut Psychological Selection Criteria." The formal White Paper submitted to NASA by the Working Group is reproduced in Appendix 4. In the next two chapters, the research proposal and its results will be outlined, along with the implementation of the new psychiatric evaluation procedures.

One of the more significant outcomes resulting from including representation from the European and Japanese space agencies was that cooperation among the major countries involved in the Space Station project was greatly enhanced. When Japan and Europe selected astronaut candidates in 1991–92, both organizations adopted aspects of the select-out and select-in procedures developed by the Working Group. Chapters 7 through 9 of this book will focus on the unique aspects of astronaut selection in Europe, Japan, and the Soviet Union.

An unofficial reunion of many of the Working Group members took place in May 1992 at the Aerospace Medical Association meeting in Miami, Florida. A scientific panel titled "Psychiatric and Psychological Aspects of Astronaut Selection—An International Perspective," was chaired by Dave Jones and myself. Members of the Working Group or their representatives presented the most recent data reflecting the international activities in the psychological selection of astronauts. Bob Rose and Terry McFadden (one of Helmreich's graduate students) summarized the results of the U.S. validation research begun in 1989. Chi Sekiguchi presented the Japanese psychological selection procedures, which had resulted in the selection of a mission specialist astronaut in May 1992. Dietrich Manzey (a colleague of Klaus-Martin Goeters in Hamburg) discussed the results of the ESA selection of 1991, and Dr. Dean Faulk (from the University of Texas Medical Branch and one of the operational psychiatrists in the NASA selection) presented the psychiatric data collected in the 1989 NASA selection. The panel was both personally and professionally satisfying because it represented the culmination of all the work begun in 1988.

## NOTES

1. W. H. Sledge and J. A. Boydstun, "The Psychiatrist's Role in Aerospace Operations," *American Journal of Psychiatry* 137 (1980): 956–959; quotation at 956.
2. Ibid., 956.
3. Ibid.

4. B. M. Astrachan, D. J. Levinson, and D. A. Adler, "The Impact of National Health Insurance on the Tasks and Practice of Psychiatry," *Archives of General Psychiatry* 33 (1976): 785–794. R. J. Ursano and D. R. Jones, "The Individual's vs. the Organization's Doctor: Value Conflict in Psychiatric Aeromedical Evaluation," *Aviation, Space and Environmental Medicine* 52 (1981): 704–706.

5. Paul E. Meehl, *Clinical vs. Statistical Prediction: A Theoretical Analysis and a Review of the Evidence* (Minneapolis: University of Minnesota Press, 1954).

6. H. W. Brosin, "Psychiatry Experiments with Selection," *Social Service Review* 22 (1948): 461–468.

7. C. K. Aldrich, "The Clouded Crystal Ball: A 35-Year Follow-Up of Psychiatrists' Predictions," *American Journal of Psychiatry* 143 (1986): 45–49.

8. Ibid., 48.

9. Ibid.

10. R. M. Rose, "Predictors of Psychopathology in Air Traffic Controllers," *Psychiatric Annals* 12 (1982): 925–933. R. M. Rose, C. D. Jenkins, and M. W. Hurst, *Air Traffic Controller Health Change Study,* FAA Contract FA73WA-3211 (1978).

11. R. L. Helmreich, T. R. Chidester, H. C. Foushee, S. E. Gregorich, and J. A. Wilhelm, *How Effective Is Cockpit Resource Management Training? Issues in Evaluating the Impact of Programs to Enhance Crew Coordination,* NASA/University of Texas Technical Report 89-2, Austin, Texas, 1989. R. L. Helmreich, H. C. Foushee, R. Benson, and R. Russini, "Cockpit Management Attitudes: Exploring the Attitude Performance Linkage," *Aviation, Space and Environmental Medicine* 57 (1986): 1198–1200.

12. D. R. Jones and C. A. Annes, "The Evolution and Present Status of Mental Health Standards for Selection of USAF Candidates for Space Missions," *Aviation, Space and Environmental Medicine* 54 (1983): 730–734.

13. Sledge and Boydstun, "A Psychiatrist's Role in Aerospace Operations," 957.

14. Rose, "Predictors of Psychopathology in Air Traffic Controllers." Rose, Jenkins, and Hurst, *Air Traffic Controller Health Change Study.*

15. Helmreich, Chidester, Foushee, Gregorich, and Wilhelm, *How Effective Is Cockpit Resource Management Training?* Helmreich, Foushee, Benson, and Russini, "Cockpit Management Attitudes."

16. Patricia A. Santy, ed., "The NASA In-House Working Group on Psychiatric and Psychological Selection of Astronauts: Summary and Transcripts" (unpublished paper, 1988), 22.

5

# Not Everyone Can Fly—The Psychiatric Evaluation

With the present airplane structure, not everyone can fly; high levels of concentration, good motor coordination, resourcefulness and composure are necessary for flying.

—N. E. Zhykovsky (1910)[1]

## PSYCHIATRIC SCREENING

The psychiatric screening of astronaut candidates can be thought of as a process of excluding those individuals who have a personality dysfunction (from either biological or psychological causes) serious enough to represent a major safety risk to flying. As early as 1910, the Russian physician Zhykovsky observed that "not everyone can fly."[2] The profession of the flight surgeon, a physician who specializes in taking care of pilots, came into being because of the extremely high number of fatal accidents due to "pilot error" among World War I aviators. Simply by making the medical qualifications to become a pilot more stringent and assessing the health status of the aviator on a regular basis, military flight surgeons discovered that the number of aircraft fatalities decreased dramatically. Today, it seems obvious that a sick pilot is a poor safety risk in the cockpit.

Psychiatric standards for pilot selection were developed at the same time as the general medical standards, so historically psychiatric and psychological evaluation has always been part of the medical assessment. But, unlike the medical criteria for disqualification, most of the psychiatric standards were not specifically linked to psychiatric disorders or illnesses. Many of the specific psychiatric disqualifications both in the U.S. military and at NASA required more of an evaluation of *psychological* and *personality* status. This is not to say that the job of pilot can be done by anyone with suitable personality features. In fact, intelligence and a certain level of cognitive ability are obviously also required, and most successful pilots have IQs in the high to superior range. But a high IQ does not preclude the possibility of psychiatric or psychological vulnerabilities in an individual.

The psychiatric evaluation used by NASA was based primarily on the experience of the interviewing psychiatrist in dealing with pilots, rather than on any principles of psychiatric medicine. Psychiatrists, like other physicians, are trained to detect disease. For the most part, they understand personality in the context of dysfunction (e.g., in specific personality disorders). The flight surgeon/psychiatrist must think of the consequences of even the most minor psychiatric dysfunction in the aviation or space environment. And, in the context of astronaut selection, the psychiatrist's role has been clearly defined: to identify those psychiatric processes that may make an individual a risk to flying safety.

Unfortunately, as we have seen, this confusion has resulted in many psychiatric physicians making decisions about *what kind of personality* is "best" for a pilot or an astronaut. And when they perform that function, they are no longer acting within the realm of psychiatric medicine. Although this function has traditionally been assigned to the interviewing psychiatrist, there is no scientific evidence to support the proposition that psychiatrists are any better than anyone else at doing such an evaluation on healthy subjects.

Because the psychiatric evaluation of both pilots and astronauts has always been performed as part of the medical examination, the Working Group concluded that the proper function of the psychiatrist—from both management and professional perspectives—was a medical one. That is, the psychiatrist's job is really to determine whether or not the candidate has current psychopathology or a history of psychopathology that puts him or her at risk in the space environment.

By itself, this task is still a difficult one. What psychiatric disorders would make a candidate a high risk in the space environment? How do you reliably assess currently *healthy* individuals for psychopathology, particularly when they are applying for a high-status job like that of astronaut (and have every reason, therefore, to prevaricate about their past histories)? How do you keep the personal prejudices and biases of the evaluator out of the process; that is, how do you make sure that the evaluations are fair? And even if you do the best possible job imaginable, does that mean that someone selected as an astronaut will *never* develop emotional problems? All of these questions require answers. It was also evident from reviewing NASA's psychiatric screening procedures for the Shuttle program that those procedures had not kept pace with psychiatric assessment technology.

## DEFINING PSYCHOPATHOLOGY IN THE CONTEXT OF THE SPACE PROGRAM

One of the first actions of the 1988 Working Group was to re-assess the NASA psychiatric standards, which were formally delineated in an internal NASA document.[3] The first NASA medical standards were written in 1978 and had been based on similar criteria used by the U.S. Air Force and Navy. Most of the psychiatric terminology used in the document came from the 1940s and 1950s psychiatric lexicon. For example, the term *neurosis,* which has come into widespread use outside of psychiatry, is not used much in psychiatric diagnosis at the current time.

In fact, a diagnostic revolution took place in the field of psychiatry in the 1970s and 1980s. The profession sought to develop a more standardized and atheoretical basis for making psychiatric diagnoses. In order to do that, specific criteria for making the diagnosis had to be identified, and this entailed a thorough analysis of clinical symptomatology currently used by psychiatrists to make those diagnoses. Although much research into psychiatric disorders such as depression and schizophrenia had already been done by the 1970s, clinicians could (and often did) use different diagnostic criteria in their practices and in their research. This resulted in studies that could not be compared, since different standards of diagnosis were used. The American Psychiatric Association formed a task force to develop and standardize the criteria based on the most recent data available. Task force members utilized the input of many distinguished psychiatrists and psychologists. They knew that the process they initiated was a longitudinal one. But it would enable researchers and clinicians to use a common language and eventually would result in the collection of data that might feed back and change the original criteria.

In a way this revolution in the conceptualization of psychiatric disorders began about the same time that computer statistical methods became more available. It represented the swing of the pendulum from the clinical psychiatric perspective (where each clinician operated in the framework of his or her understanding of one or more theories of psychiatric illness) to the statistical perspective, which could be used with larger numbers of individuals with the same illness to predict treatment and long-term outcome, without regard for the particular theory. The revolution also came at the same time that *psychological* theories about mental illness were being obscured by rapid advances in the understanding of their *biological* foundations.

These changes in the field of psychiatry enable us to put the process of psychiatric selection of astronaut candidates into a proper historical perspective. Psychiatric interview techniques based on standardized diagnostic criteria have become state of the art. To use an analogy, in the 1950s and early 1960s, the large-scale use of computers for many tasks was constrained by the fact that all available computers were very large and very expensive. Today it would be absurd to use an outmoded, expensive, and out-of-date computer when a small personal computer, with larger memory, more efficiency, and a smaller price tag can do the job even better.

In 1980, the third edition of the *Diagnostic and Statistical Manual of Mental Disorders* (DSM-III)[4] represented the first attempt by the American Psychiatric Association to standardize the criteria for making psychiatric diagnoses. It was followed in 1987 by the revised third edition (DSM-III-R)[5] (and at the time this is written, DSM-IV is about to be published). Each revision has integrated the latest research and clinical data into the diagnostic criteria for each psychiatric disorder. The manual has become the standard of psychiatric practice in the United States and in many countries around the world.

## ASSESSING PSYCHOPATHOLOGY IN A FAIR AND RELIABLE MANNER

The metamorphosis of psychiatric diagnosis made possible the development of standardized and reliable structured and semi-structured instruments for use in clinical psychiatric interviews. The structure of the psychiatric interview in astronaut selection was always left to the discretion of the individual psychiatrist. He or she spent varying amounts of time with the applicant and made a judgment about the presence or absence of mental problems. Most reputable psychiatrists agreed on the general material that should be included in any clinical interview, but each interview was unique and individualized to the particular psychiatrist. This meant that the information obtained in the interview was entirely dependent on the skill of the individual psychiatrist and on the openness of the subject being interviewed. It facilitated a problem known as interviewer bias. One way to minimize such a bias in one's data is to establish inter-rater reliability, a method determining how similarly and reliably different examiners use a standardized approach.

By using a structured or semi-structured psychiatric interview, each psychiatrist would ask the same questions in the same order and generally in the same manner. If inter-rater reliability is high, then the different examiners may be considered interchangeable, and one source of bias in the evaluation has been eliminated.

When a semi-structured interview format was implemented in the 1987 selection,[6] it was the first time that such an instrument had been used in astronaut selection since *Mercury*. The results of that interview were presented in an earlier chapter, but by 1989, the use of even more structured clinical diagnostic interviews had become standard practice.

NASA's In-House Working Group on Psychiatric and Psychological Selection of Astronauts was fortunate to have Dr. Jean Endicott as a consultant. She, along with Robert Spitzer, jointly developed many of the well-known structured interviews used today to reliably assess for psychopathology. Dr. Endicott worked closely with the Operations Subcommittee to identify the best interview for astronaut selection. Along with Dr. Alv Dahl, a representative of the European Space Agency, she assisted the Operations Subcommittee in evaluating the different types of structured psychiatric interviews available for use in astronaut selection.

The pros and cons of each major instrument available were discussed. The interview chosen had to counteract the tendency of astronaut candidates to minimize psychological symptoms. In psychiatric parlance, this was known as "staying clean." If obvious questions are asked to an intelligent subject and he or she can clearly identify the information that is wanted, the chances of staying clean are high. On the other hand, if a question is asked less directly, and in such a manner that it is difficult for a subject to answer simply "yes" or "no," then the chances of staying clean are reduced. An example of this is asking the subject "Have you ever been depressed?" Most healthy subjects would realize it is not a good thing to be depressed and would probably answer "no," particularly if they want to be an astronaut. But if you ask the question in the form of a request—"Tell me about the time when you have been most depressed (sad or blue) in your life"—it is very hard for the subject to be able to escape from giving some clinical information on the topic.

The interview selected had to include questions for *all* potentially disqualifying psychiatric disorders. This meant that both Axis I (major and minor psychiatric disorders that are a focus of treatment) and Axis II (personality disorders) disorders needed to be included. And a final constraint for selecting an interview was the amount of time that would be available to conduct the interview with each astronaut applicant. During the *Mercury* evaluations, psychiatrists spent over thirty hours total on each candidate (including time for taking the psychometric and performance tests); in *Gemini* and *Apollo,* that time had been cut to ten hours. Our time was two to four hours (not including the psychometric tests, which we planned to give to all candidates at one sitting early in the selection week).

Endicott and Spitzer's Schedule for Affective Disorders and Schizophrenia–Lifetime version (SADS-L)[7] was eventually chosen when all aspects of the problem were considered. This interview evaluates for all disqualifying Axis I psychiatric disorders, it has good reliability, and the Operations Subcommittee felt that it provided a more clinical type of instrument and minimized the applicants' ability to stay clean [as opposed to the Structured Clinical Interview for DSM-III (SCID)].[8] The other major advantage of the SADS-L was the fact that it had been translated and used widely around the world, including Japan and Europe, so our international partners could also use the instrument in their astronaut psychiatric evaluations. With only minimal revision to include questions (based on DSM-III-R criteria) on post-traumatic stress disorders, sexual disorders, eating disorders, and impulse control disorders, the SADS-L became the first part of the astronaut selection interview.

The second part of the astronaut interview would assess for the presence of DSM-III-R personality disorders. According to the standards, all DSM-III-R personality disorders are considered disqualifying for becoming an astronaut. After much discussion and review, the Operations Subcommittee chose a European questionnaire, the Personality Assessment Schedule (PAS),[9] which had been developed and validated by Dr. Peter Tyrer in Great Britain. Its advantage was that in answering the questions, it would be difficult for subjects to be deceitful, since the algorithm for diagnosing a DSM-III-R personality disorder was not obvious. The PAS divided each DSM-III-R personality diagnosis into a series of traits (e.g., histrionic personality disorder was divided into "lability + dependence + childishness + irresponsibility"). A candidate would be asked to rank himself or herself on these individual traits and give examples in his or her life.

Other advantages of the instrument were that it was used extensively in Europe and that it took only thirty to forty minutes to administer. Again, like the SADS-L, this structured interview was a more clinically oriented one, which, while structured, allowed clinicians to use their own clinical judgment.

The third section of the interview was added at the suggestion of several clinically experienced members of the Operations Subcommittee. They believed that the use of structured interviews does preclude obtaining a general developmental history of the applicant and that this life history, as well as previous responses to problems in life, would be important to know for purposes of determining past or present psychopathology.

There are basically three reasons why an applicant might be psychiatrically disqualified: (1) the history or presence of a disqualifying Axis I disorder (e.g., major depression); (2) the history or presence of an Axis II disorder (e.g., borderline personality disorder); and (3) the presence of "traits, characteristics, or behavior which, in the opinion of the examiner, make the candidate a potential hazard to flying safety." This last condition, though vague, was left in by NASA management against the recommendation of many members of the Working Group, who wanted to exclude all non-psychopathological conditions from the standards. After all, they reasoned, if NASA only wants the psychiatrists to make a medical decision (i.e., to select out), then why should this very non-medical standard remain? But NASA management—particularly life sciences management—was comfortable with using the vagueness of this particular standard as the justification for excluding certain "undesirable" candidates (rather than simply not choosing them). For example, NASA originally asked the Working Group to include homosexuality as a *psychiatrically* disqualifying condition. We pointed out to them that homosexuality per se was *not* a psychiatric disorder and that we did not intend to disqualify anyone because of sexual orientation.

Since NASA insisted that the standard remain, the Working Group decided to comply. However, we also decided that in the first implementation of the standards, we would not disqualify anyone who did not have either an Axis I or an Axis II disorder.

## AUGMENTING CLINICAL JUDGMENT BY OBJECTIVE TESTING: WHAT CLINICAL TESTS TO GIVE?

To assist the clinician in determining whether or not a candidate should be disqualified for the third reason, a series of psychometric tests was added to the psychiatric interview. Since the beginning of the Shuttle program, psychometric testing had been "grounded." But such tests give some objective credibility to an otherwise vague clinical judgment that an applicant was a "risk to flying safety" without having a specific psychiatric diagnosis.

The following psychological tests were used clinically on all applicants: a brief family psychiatric health history questionnaire, the Minnesota Multiphasic Personality Inventory (MMPI),[10] the Multidimensional Assessment Battery (MAB)[11] (a group form of the Wechsler Adult Intelligence Scale–Revised), the Millon Clinical Multiaxial Inventory–II (MCMI-II),[12] and the Forer Structured Sentence Completion Test (FSSCT).[13] The MCMI-II was included as an Axis II clinical instrument, and the FSSCT, a short projective test, was also included because many clinicians were uncomfortable with the high number of strictly statistical psychometric tests and lobbied to include at least one projective test. The FSSCT was used a great deal by U.S. Air Force clinicians John Patterson and David Jones as part of their evaluations of the hundreds of pilots referred to the School of Aerospace Medicine. As it turned out, the FSSCT was probably the most clinically useful test given, since it elicited a considerable amount of interesting clinical information that could be followed up with the applicants when they were later interviewed by the psychiatrist or the psychologist.

These psychological tests were not considered in and of themselves a basis for a psychiatric diagnosis. In the real world, many clinicians use such tests to give support to their diagnostic formulation. They provide a somewhat more objective way to collect data compared to the psychiatric interview, where a candidate is likely to be less open. Of course, applicants who wish to conceal pathology may fake responses to test questions, as well as to interviews, but most tests have built-in scales that detect such proclivities and at least make the examiner aware that a particular individual is very defensive and may be less than completely honest in answering personal questions.

Applicants were scheduled to take the battery of psychological tests before any interviews, so examiners had the results of the tests in front of them when they began the structured interview. This tactic proved very useful, as it helped focus the third part of the interview, and it also gave the clinicians some basic psychological information about an applicant beforehand.

## THE 1989 SELECTION: IMPLEMENTING THE WORKING GROUP RECOMMENDATIONS

In 1989, the Medical Sciences Division at the Johnson Space Center (JSC) in Houston, Texas, formally revised its methods for the psychiatric evaluation of astronaut candidates, based on the recommendations of the Working Group. That year, 106 astronaut applicants (90 male and 16 female) were asked to come to JSC to have a job interview and to be medically and psychiatrically evaluated to determine their qualifications for the position of pilot astronaut or mission specialist astronaut.

By the time they arrived at JSC, these candidates had already been highly selected from the general population, since they had to meet rigid educational qualifications (bachelor's degree or higher for the mission specialist astronaut candidates) and/or demonstrate exceptional proficiency in operating high-performance aircraft (for the pilot astronaut candidates). In addition, NASA flight surgeons had thoroughly reviewed each candidate's prior medical history for the presence of any disqualifying medical or psychiatric disorders. More than 1,945 individuals had sent in applications for the 1989 selection.[14]

Many clinicians participated in the implementation of the new procedures. Included were Al Holland and I from NASA; Dave Jones, John Patterson, Walter Sipes, Frank Carpenter, and Roy Marsh from the Brooks Air Force Base Department of Neuropsychiatry; and Dean Faulk from the University of Texas Medical Branch.

Each clinician (M.D. or Ph.D.) interviewer received training in the use of each part of the interview instrument and participated in several videotape interview training sessions. Some of these sessions were led by Dr. Endicott. To maintain high inter-rater reliability, a colleague observer was included randomly in some interviews, and ratings were compared. The results of this exercise indicated that the inter-rater reliability was excellent. All of the actual interviews were audiotaped and then later evaluated to ensure that all clinicians were applying the interview in a reasonably uniform manner.

The psychiatric interview was conducted on all 106 astronaut applicants. After all data were collected on a specific applicant, the entire clinical staff met to discuss diagnosis and qualification. A consensus was reached on any Axis I or II diagnosis, as well as on the final recommendation to the NASA Space Medicine Board (either "qualified" or "not qualified").

Of the 106 applicants, 9 (8.5%) met DSM-III-R criteria for either Axis I or Axis II disorders One candidate met criteria for both a dream anxiety disorder (recurrent nightmares that cause the individual to awaken from sleep screaming) and a single episode of major depression. This candidate was disqualified. Also disqualified was an individual who met the criteria for both avoidant and dependent personality disorders. Other disorders that were found, but that did not disqualify the candidate, were bereavement (grief reaction over the loss of a loved one) and marital and life circumstance problems (both are termed V-code diagnoses—i.e., diagnoses that do not represent psychopathology per se, but that are a focus of treatment). None of the nine individuals with these Axis I or Axis II disorders was selected by NASA, although only two of them were formally psychiatrically disqualified.

One criticism of the general approach used in the psychiatric evaluation is that the applicants will not willingly admit to any symptoms suggestive of mental disorders. As an exercise to assess whether or not the clinicians were able with the interview instruments to obtain significant clinical information from their reluctant interviewees, all the interviews were reviewed for the presence of "near" diagnoses (or diagnoses in which 50% or more of the DSM-III-R criteria listed are met, but not enough criteria are met to officially make the diagnoses). An additional twelve individuals (11.3%) met at least half the criteria for one of a number of diagnoses including major depression, obsessive-compulsive disorder, hypomanic episodes, and a variety of personality disorders. The majority of the diagnoses in this "near" category were personality disorders, but they also included two individuals who lacked only one criterion each for having had a major depressive episode. Research criteria for diagnosing hypomania (not currently a specific psychiatric diagnosis, but only a symptom) were included in the SADS-L portion of the interview, since it was theorized that hypomanic episodes without other psychiatric impairment might be a *positive* predictor for selection in this group. Two individuals met criteria for having numerous hypomanic episodes, and one of these individuals was selected as an astronaut by NASA. It is not known at this time if a history of hypomanic episodes is predictive of later psychopathology (e.g., bipolar disorder), but this would be something to watch for in making future selections and in following up selected candidates over the years.

## PSYCHOMETRIC TESTING RESULTS

It is of interest to compare the results of testing on this group of astronaut applicants to results collected over thirty years during the *Mercury*, *Gemini*, and *Apollo* candidate evaluations. Unfortunately, the only tests in common are the Minnesota Multiphasic Personality Inventory (MMPI) and the intelligence tests (the Wechsler Adult Intelligence Scale and the Multidimensional Aptitude Battery). Figure 5.1 compares the MMPI pro-

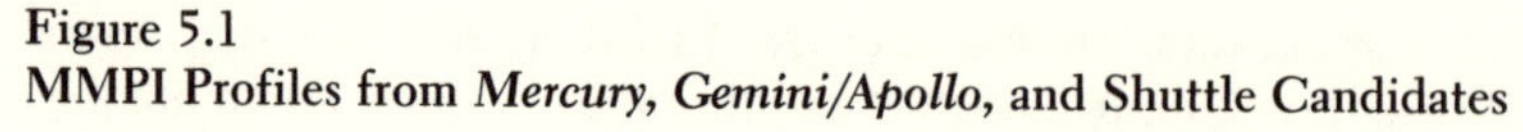

**Figure 5.1**
**MMPI Profiles from *Mercury, Gemini/Apollo,* and Shuttle Candidates**

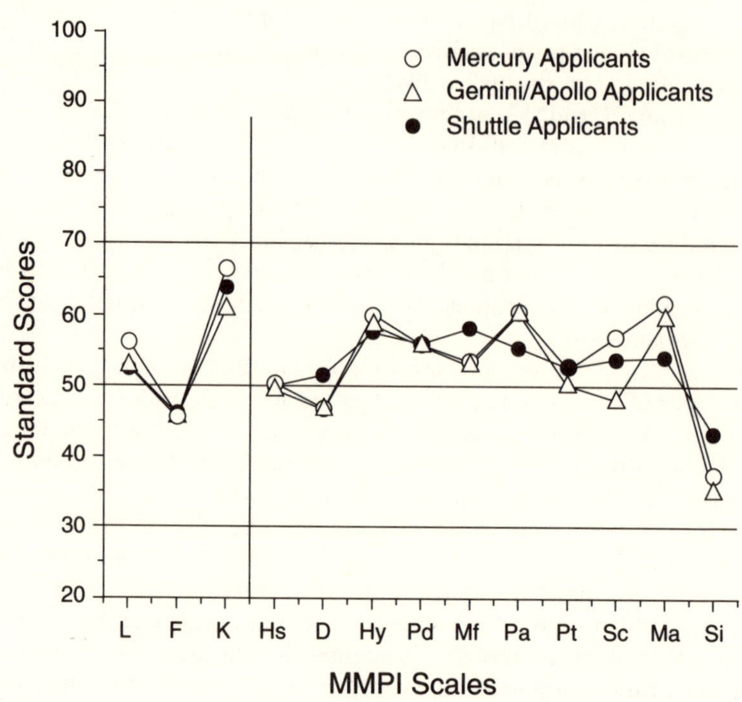


files of the four programs. What is remarkable is the similarity of all four groups of candidates over the thirty years.

All groups, including the Shuttle applicants, were extremely defensive and presented themselves in the best possible light (LFK scales). There were no statistically significant differences in T-scores among the four groups of candidates. As can be seen from the profiles, the Si (social introversion) scale is low in all groups, suggesting a socially outgoing pattern of behavior. If we examine the Shuttle group profile in comparison with the general population (Figures 5.2 and 5.3), we first have to split the Shuttle group into males and females, since the MMPI group norms are different for these two groups. This was the first group of astronaut applicants intensively studied that included female applicants. Comparing male Shuttle applicants to the normal male population, the main differences are in the LFK configuration and in the Si scale, with male Shuttle applicants very much more socially extroverted than the normal male population.

The female Shuttle candidates have these same features, but they also have a significant difference on the Mf (masculinity/femininity) scale. As might be expected, the female Shuttle applicants were significantly more "male" (which merely indicates less adherence to the traditional female role

**Figure 5.2**
**Male Shuttle Candidates versus U.S. Male Norms**

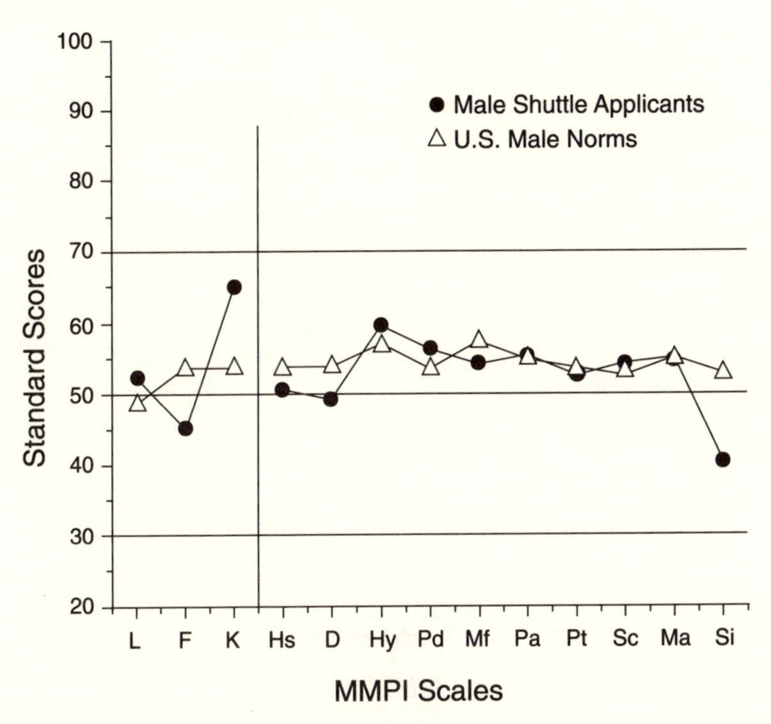


and more interest in traditional male activities). It is important to remember that the MMPI was originally developed in the 1940s and standardized on average subjects (male and female) from that time. The concepts of masculinity and femininity have undergone radical transformations since then. The Mf scale is not considered of any significance in defining sexual orientation. In fact, female Shuttle applicants were much more like their male counterparts than like the normal female population.

A comparison of the intelligence scores is possible, even though a different test was used on the Shuttle group. The MAB is a group form of the Wechsler Adult Intelligence Scale–Revised (WAIS-R) and correlates nicely with it. On the average, all groups, including the Shuttle candidates, were in the very high intelligence range. The only difference is that the range for the Shuttle applicants was larger, with IQs of 90 to 152 (while *Mercury* candidates' range was 130 to 141). It should be remembered, however, that the *Mercury* candidates were selected from a larger pool of homogeneous pilot candidates and that there was an arbitrary decision to select only applicants with IQs above a certain level (see Chapter 1), so it is to be expected that their range would be narrower.

The Astronaut Selection Board during the Shuttle period has not been

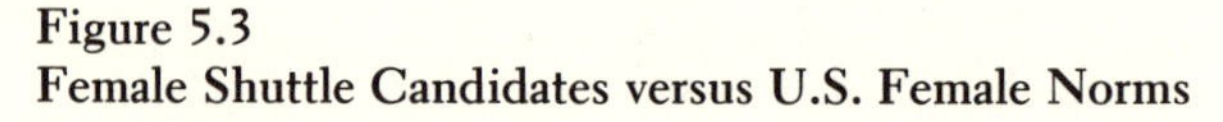

**Figure 5.3**
**Female Shuttle Candidates versus U.S. Female Norms**

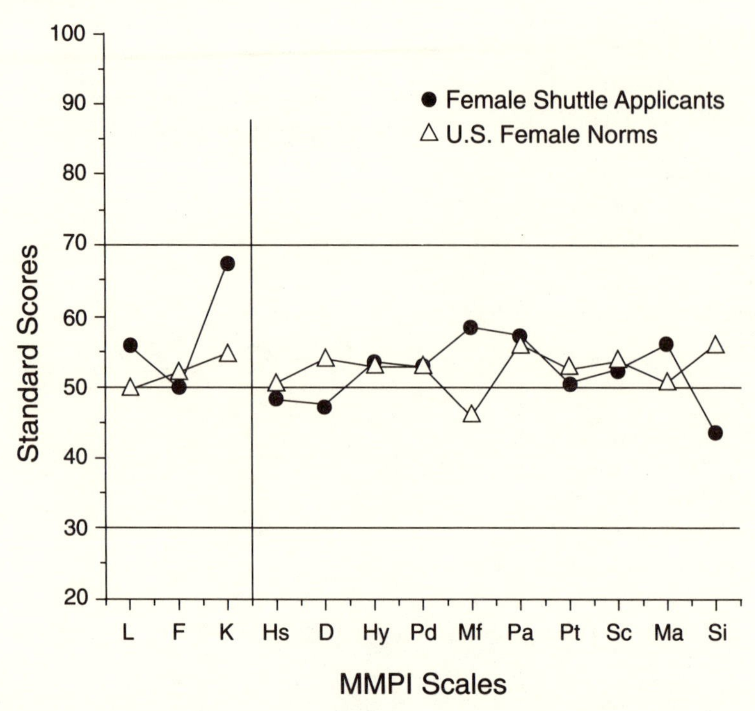


interested in IQ as a determining factor, preferring instead to look at the overall professional performance of the individual candidates as documented in their biographical history. While this approach has merit, it is important to realize that current astronaut applicants come from a much more heterogeneous population and may not have been sufficiently challenged by their environment previously. In one recent selection, for example, the Selection Board chose a candidate with a total IQ score a full twenty points below the average. It will be interesting to see if intelligence is a factor in future performance.

The use of IQ assessment in astronaut screening and evaluation is controversial. It is possible that the use of measures developed and validated in populations that do not share the same characteristics as those individuals who apply for an "elite" position such as astronaut may, at best, be nonapplicable and may, at worst, lead to erroneous results if used in a selection process. In fact, like IQ's relationship with creativity, beyond a certain threshold, there may not be a significant effect of higher IQ on many behavioral and performance domains.[15] Using IQ as a predictor of performance in the *Mercury* astronauts would be useless because there was insufficient variability among the candidates' IQ scores (i.e., they were all in the superior range). Shuttle applicant IQs have a much greater heterogeneity, and it may be worthwhile to pursue the possibility of whether or not a correlation exists.

Psychometric testing was clinically useful in helping the interviewer to probe more thoroughly into potentially sensitive clinical areas. Clinicians found that the FSSCT, a projective test, helped identify sensitive or emotionally significant areas in an individual's history. The MCMI-II, which asked for "yes" or "no" responses to all questions, had a high rate of 100 percent "no" answers. Many applicants who were judged "normal" by the interviewer actually scored high enough on this test to be classified as "obsessive-compulsive." This is primarily because the MCMI-II "assumes" pathology in the examinee. In the astronaut selection, this test did not prove to be very helpful. For the purposes of completeness, Appendix 5 contains a summary of data collected on several of the objective tests not discussed in the text.

In such a high-functioning group of healthy individuals, it is not surprising that there is little evidence of psychopathology. What is surprising is that we were able to elicit significant levels of symptom reporting in a group of individuals expected to be extremely reluctant to admit to any psychological or behavioral problems.

## FINAL THOUGHTS ON SELECTING-OUT

While astronaut applicants are, in general, very healthy, there is no reason to expect that they should be immune to all mental disorders. The most common disorders one might expect to see would be V-code disorders and substance use disorders, since these are ubiquitous to modern life. More severe psychiatric disorders (such as schizophrenia or other psychoses) would be unlikely.[16]

Lee N. Robins and his colleagues[17] in 1984 showed a lifetime prevalence range for a major depressive episode of 3.7–6.7 percent; manic episode, 0.6–1.1 percent; schizophrenia, 1.0–1.9 percent; alcohol abuse/dependence, 11.5–15.7 percent; and drug abuse/dependence, 5.5–5.8 percent. For all mental disorders, the prevalence rates from numerous studies range from 0.4 percent to more than 8 percent. Our results are consistent with these general findings. The lack of alcohol or substance abuse/dependence disorders in our astronaut candidate sample is puzzling. Substance abuse may be fairly easily disguised or minimized on any type of psychiatric interview. Individuals who want to be astronauts are not likely to admit to behavior that involves the use of alcohol or drugs. In fact, only two individuals out of the entire group acknowledged that they had ever even smoked marijuana, and both insisted they had tried it only "once." Alcohol use was frequently acknowledged, but level of intake was clearly minimized.

The psychiatric diagnostic results obtained in this selection cannot be compared with those of previous astronaut selections, since no such information was recorded from previous astronaut candidate groups. Although the likelihood of finding a disqualifying disorder in this type of population is fairly small (only two individuals in this group were disqualified), it is critical that a careful psychiatric screening be done, since the consequences—in terms of mission safety, effectiveness, and success—of selecting an individual with major psychiatric disorder or high risk of developing one are very significant.

The method of psychiatric evaluation reported here to screen individuals according to a set of disqualifying disorders is unique. The current NASA psychiatric interview was developed from existing and reliable structured interviews, modified to make sure that all the disqualifying disorders were screened for and to minimize the likelihood of an applicant staying clean.

As noted previously, in this first application of the new evaluation procedure, clinicians were reluctant to disqualify an individual who did not meet criteria for a specific psychiatric diagnosis unless there was additional evidence of problems indicated on the psychometric tests. No one, therefore, was disqualified because of the presence of a personality "trait," and, as it turned out, the NASA Selection Board did not happen to select any of those individuals with "near" diagnoses. This suggests that the kinds of traits identified by the behavioral clinicians are the kind of traits a perceptive Selection Board might sense—consciously or unconsciously—in its job interview.

The select-out methods described here may be useful in any environment where individuals must function under unusual stress and maintain high levels of performance. Such environments frequently have adopted specific medical/psychiatric standards for job selection (e.g., military aviation, and undersea and nuclear industries). While no psychiatric screening method is foolproof, using a structured, standardized approach to psychiatric selection minimizes the risk of selecting individuals with serious psychopathology or the risk of developing it.

In the United States, it has become "politically correct" to assert that everyone has the right to become or do anything—reality to the contrary. It important to remember that "not all can fly," even though they may want to. And in humanity's expensive and dangerous undertaking to explore the universe (at the public's expense), it is important to remember that those individuals who are disqualified from becoming an astronaut—whether for psychiatric or other medical reasons—are not being judged as worthless. On the contrary, as a group, those who merely apply are among the best and most talented individuals in the country, but for one reason or another, they represent a risk that should not be taken at this stage of space exploration.

## NOTES

1. V. A. Bodrov, V. B. Malkin, B. L Pokrovskiy, and D. I. Shpachenko, *Psychological Selection of Pilots and Cosmonauts* (in Russian), vol. 48 of *Problems in Space Biology*, edited by B. F. Lomov (Moscow: Nauka Press, 1984), 6.

2. Ibid.

3. NASA *Medical Standards: NASA Class I Pilot Astronaut Selection and Annual Medical Certification*, JSC-11569 (Houston: Johnson Space Center, 1988) (published for internal use only).

4. American Psychiatric Association, *Diagnostic and Statistical Manual of Mental Disorders*, 3d ed. (Washington, D.C.: American Psychiatric Association Press, 1980).

5. American Psychiatric Association, *Diagnostic and Statistical Manual of Mental Disorders*, 3d ed., revised (Washington, D.C.: American Psychiatric Association Press, 1987).

6. Patricia A. Santy, Al W. Holland, and Dean M. Faulk, "Psychiatric Diagnoses in a Group of Astronaut Applicants," *Aviation, Space and Environmental Medicine* 62 (1991): 969–973.

7. J. Endicott and R. L. Spitzer, "A Diagnostic Interview: The Schedule for Affective Disorders and Schizophrenia," *Archives of General Psychiatry* 35 (1978): 837–844. R. L. Spitzer and J. Endicott, *Schedule for Affective Disorders and Schizophrenia.* Biometrics Research (New York: New York State Department of Mental Hygiene, 1975).

8. R. L. Spitzer, J. B. W. Williams, M. Gibbon, and M. B. First, *Structured Clinical Interview for DSM-III-R—Patient Version* (SCID-P, 6/1/88) (New York: Biometrics Research Department, New York State Psychiatric Institute, 1988). R. L. Spitzer, J. B. W. Williams, M. Gibbon, and M. B. First, *Instruction Manual for the Structured Clinical Interview for DSM-III-R* (SCID, 5/1/89 Revision) (New York: Biometrics Research Department, New York State Psychiatric Institute, 1989).

9. P. Tyrer, M. S. Alexander, D. Cicchetti, M. S. Cohen, and M. Remington, "Reliability of a Schedule for Rating Personality Disorders," *British Journal of Psychiatry* 135 (1979): 168–174. P. Tyrer, J. Alexander, and B. Ferguson, *Personality Assessment Schedule (PAS)*, 5th revision (London: Author, 1987).

10. S. R. Hathaway and J. C. McKinley, *MMPI-2: Manual for Administration and Scoring* (Minneapolis: National Computer Systems, 1989).

11. D. N. Jackson, *Multidimensional Aptitude Battery (MAB)* (Michigan: Sigma Assessment Systems, Inc., 1984).

12. T. Millon, *Million Clinical Multiaxial Inventory–II* (Minneapolis: National Computer Systems, 1987).

13. B. R. Forer, *The Forer Structured Sentence Completion Test: Manual* (Los Angeles: Western Psychological Services, 1988).

14. P. A. Santy, J. Endicott, D. R. Jones, R. M. Rose, J. Patterson, A. W. Holland, D. M. Faulk, and R. Marsh, "Results of a Structured Psychiatric Interview to Evaluate NASA Astronaut Candidates," *Military Medicine* 158 (1993): 5–9.

15. S. Bishop, D. Faulk, J. Patterson, and P. Santy, "The Use of IQ Assessment in Astronaut Screening and Evaluation" (paper presented at the Aerospace Medical Association Meeting, Toronto, May 23–26, 1993).

16. P. A. Santy, "Psychiatric Support for a Health Maintenance Facility (HMF)

on Space Station," *Aviation, Space and Environmental Medicine* 58 (1987): 1219–1224. Santy, Endicott, Jones, Rose, Patterson, Holland, Faulk, and Marsh, "Results of a Structured Psychiatric Interview to Evaluate NASA Astronaut Candidates."

17. L. N. Robins, J. E. Helzer, M. M. Weissman, H. Orvaschel, E. Gruenberg, J. D. Burke, and D. A. Regier, "Lifetime Prevalence of Specific Psychiatric Disorders in Three Sites," *Archives of General Psychiatry* 41 (1984): 949–958.

The *Mercury* Candidate Selection Board. Drs. Ed Levy and George Ruff are at the 12:00 and 1:00 positions at the table.

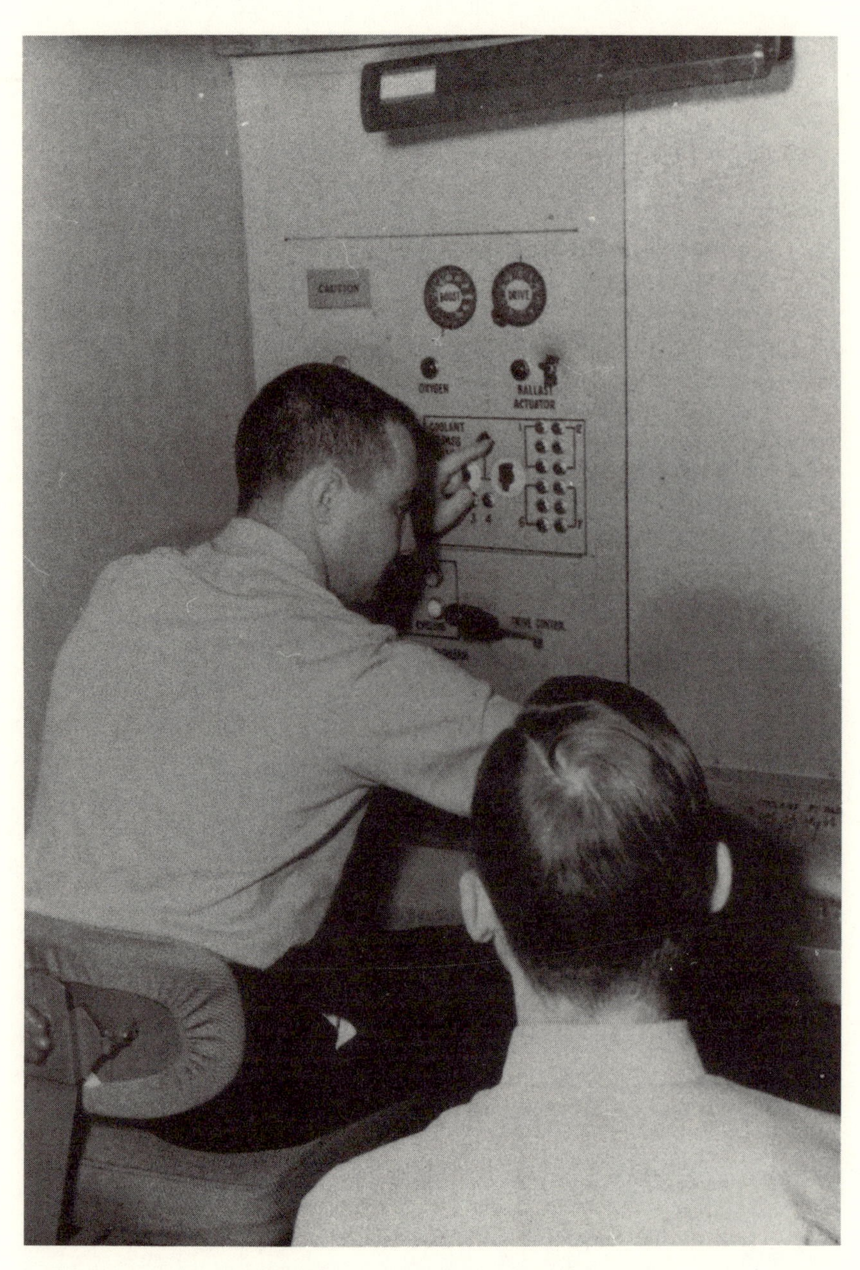


The complex behavior simulator. An observer stands behind the subject.

The isolation test. The light on the subject's face is for photographic purposes only.

Bryce Hartman, Ph.D., in his office at Brooks Air Force Base, Texas, in 1992. Dr. Hartman was involved in the evaluations of *Mercury, Gemini,* and *Apollo* astronaut candidates.

Don Flinn, M.D., currently chief of staff at the Veterans Administration Hospital in Los Angeles.

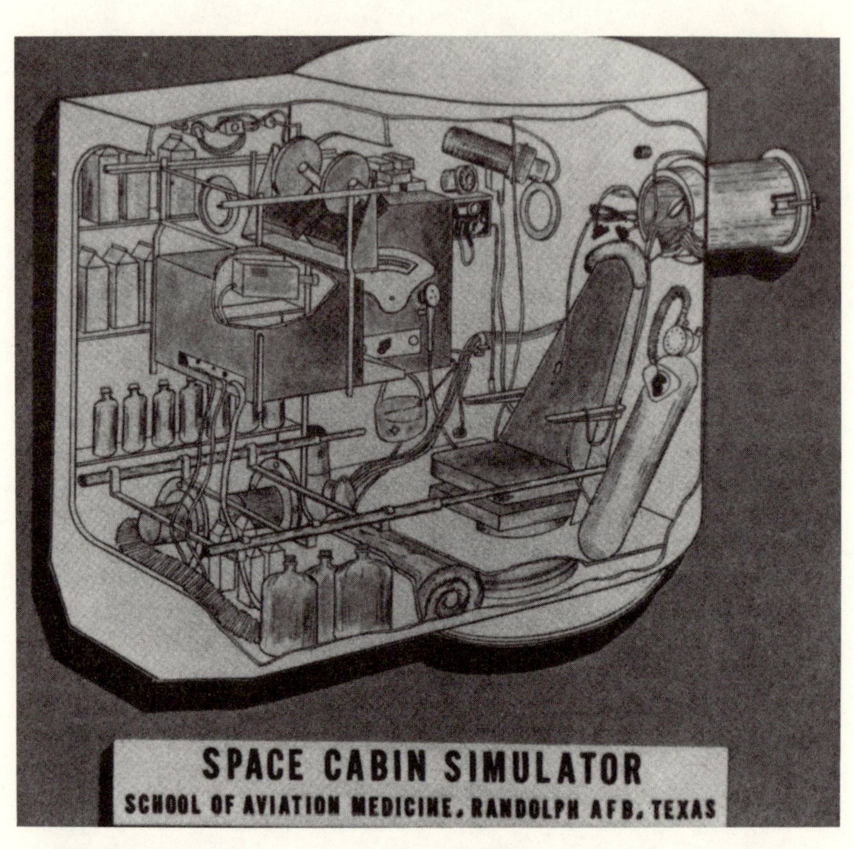


Diagram of the SAM space cabin simulator (1960).

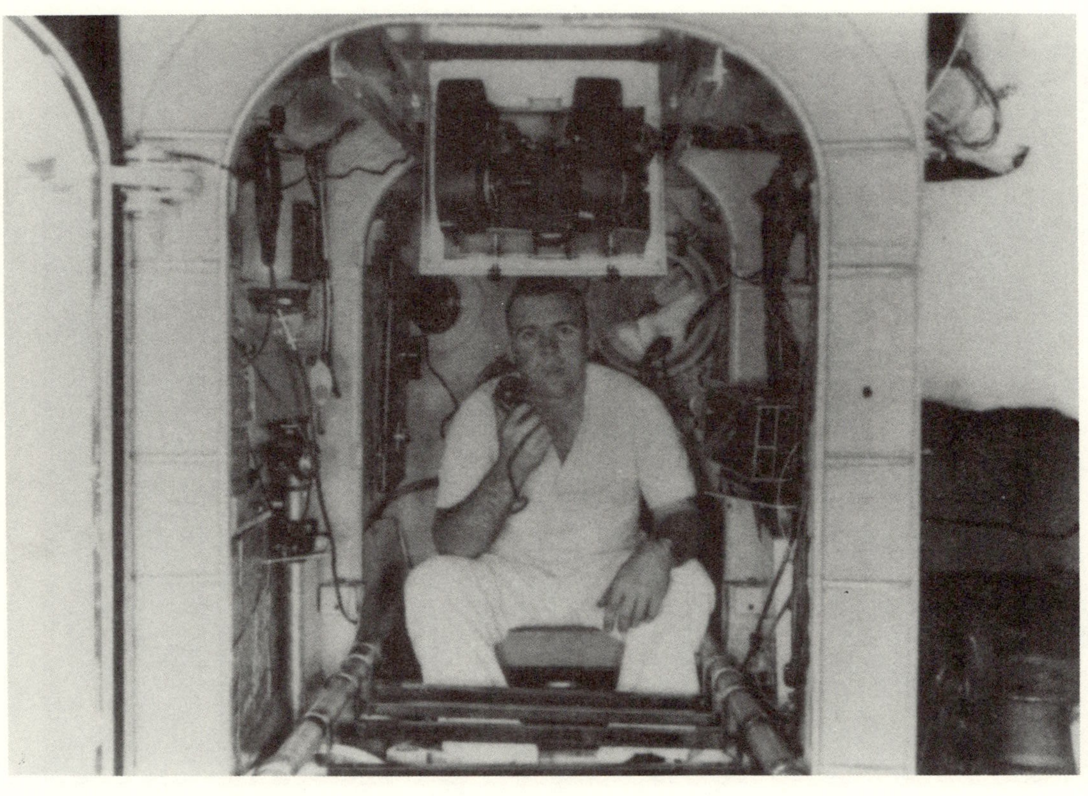

Subject in the one-man space cabin simulator after seven days of isolation.

Carlos Perry, M.D., when he was chief of the Neuropsychiatry Branch at Brooks Air Force Base, Texas (circa 1964).

Drs. Patricia Santy and Anke Putzka in the Johnson Space Center Mission Control Room during the German D-1 Shuttle mission.

David Jones, M.D., M.P.H., when he was chief of the Neuropsychiatry Branch at Brooks Air Force Base, Texas (circa 1984).

Robert M. Rose, M.D., currently at the MacArthur Foundation in Chicago. (Reprinted by permission of Gittings, solely for publication purposes.)

Dr. Robert L. Helmreich in his office at the University of Texas, Austin.

NASDA psychiatrists participating in a training exercise for Japanese astronaut selection, October 1991. Seated from left to right in the front row are Shigeru Ohbayashi, Patricia Santy, Naoki Matsunaga, and Satoru Shima. In the back row from left to right are Hiroshi Kunugi, Takurou Endo, Satoru Okada, Akihisa Takahashi, Kouichi Abe, Genichi Matsuda, Toshio Ohta, and Toshinori Kitamura.

Chamber used for isolation experiments on cosmonauts in Star City, U.S.S.R., in 1985.

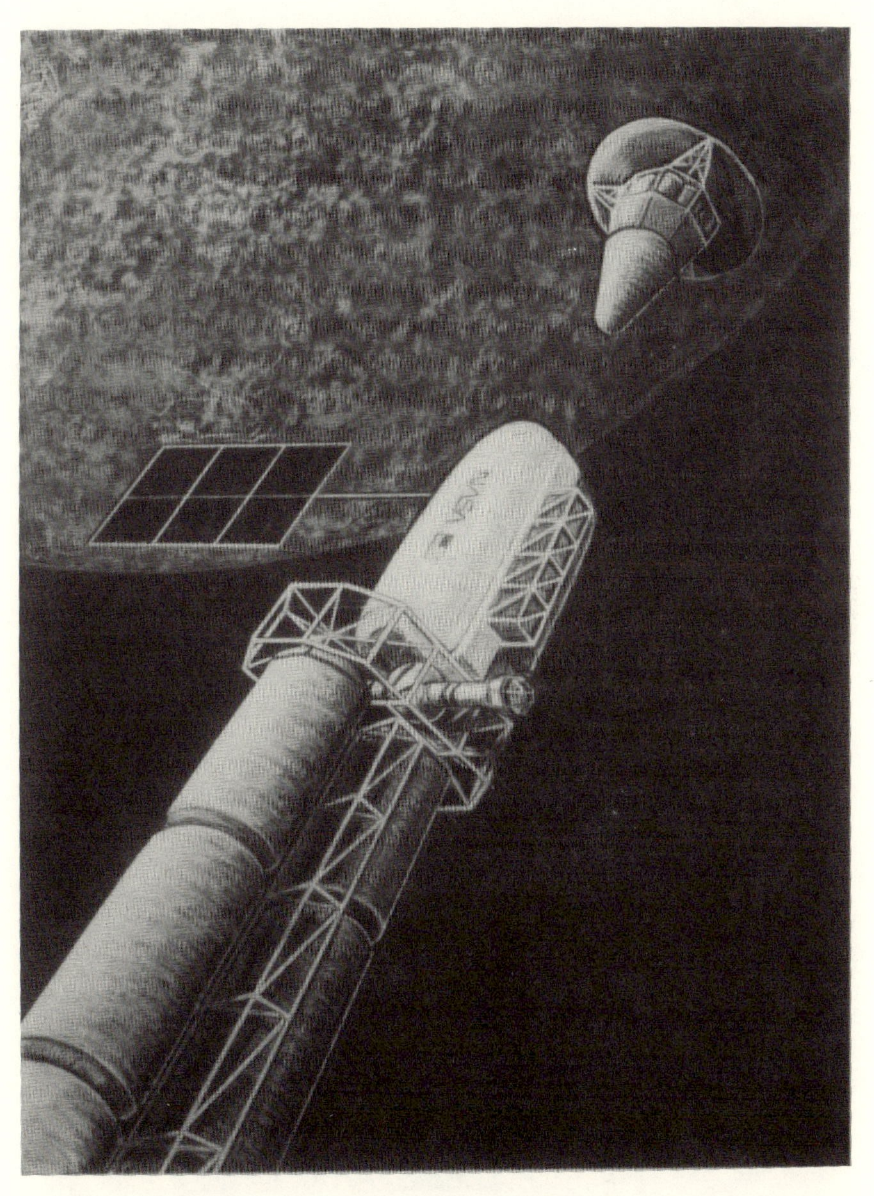


Artist's rendering of an international Mars mission, depicting the Mars transfer vehicle and the Mars excursion vehicle. (Illustration by Renée L. Myers at the Sasekawa Institute of Space Architecture, University of Houston, 1991–1992.)

An isolated science station in the Antarctic. (Photo courtesy of Dr. Regina North.)

# 6

# The "Real" Psychological Stuff for Missions of the Future

> A young man might go into military flight training believing that he was entering some sort of technical school in which he was simply going to acquire a certain set of skills. Instead, he found himself all at once enclosed in a fraternity. And in this fraternity, even though it was military, men were not rated by their outward rank as ensigns, lieutenants, commanders, or whatever. No, herein the world was divided into those who had it and those who did not. This quality, this "it" was never named, however, nor was it talked about in any way.
>
> —Tom Wolfe, *The Right Stuff*[1]

## TALKING ABOUT "IT"

The unnamed quality no one talked about was christened "the right stuff" by author Tom Wolfe in his classic history of the early days of the U.S. Space Program. From a pilot's perspective, "it" represented the essence of manliness and was just as mystical and elusive. Men who had "it" were held in superstitious awe by their comrades. And to achieve "it," one had to embark on a quest for male perfection—with daily displays of bravery, courage, and manhood beyond the capability of most men. To have "it" was to be the best, surpassing all others. You could not have a little bit of "it"—either you had it or you did not. And if you made an embarrassing blunder while flying (like getting yourself killed), or if you did not get that key assignment, after all—well, you probably never had "it" to begin with.

From the point of view of a psychiatrist interested in psychological health, the pilot's conception of the right stuff seems seriously off-key. No one doubts that it takes courage and bravery to endure the dangers and hardships of flying, to put your life on the line for a sense of "duty." But frequently the bottom-line behavior exhibited by those who had the right stuff was narcissism, arrogance, and interpersonal insensitivity. Are those personality traits necessary to be a superb astronaut? And, more important, are those personality traits really necessary to get the job done? If they are not, then what psychological traits comprise the "real" right stuff for future space explorers?

A major task of the NASA In-House Working Group on Psychiatric and Psychological Selection of Astronauts was to re-examine the classical concept of the right stuff and redefine it for space missions of the future. To accomplish that, "it" had to be dissected into essential components—retaining the necessary elements and purging the mystical aspects. The real psychological stuff had to be measurable, and it had to be tested to see if those who possessed it would perform well as astronauts on Space Station *Freedom* or on a journey to Mars.

One of the mystical aspects of the pilot's perception of "it" is the emphasis on *maleness*. In the exclusive fraternity of those with the right stuff, there is no room for women, and the admission of women to the club devalues the membership. This belief accounts for the extreme difficulty women have had in breaking into the ranks of high-performance jet pilots. The 1992 Tailhook Convention scandal, where hundreds of Navy jet pilots were implicated in sexual misconduct, is a case in point. Film clips from that convention showed high-ranking naval officers making negative and hostile statements about women pilots joining the ranks of the "elite." The open resentment of women pilots was obvious. Admitting women to those elite ranks is a frightening prospect for men who seek to "prove" their manliness by becoming a pilot—or an astronaut. Their attitude is very easy to understand—if a woman can be a jet pilot (i.e., have the right stuff), then the right stuff must have nothing to do with manliness, and having "it" does not automatically guarantee virility.

There are, of course, many jet pilots whose masculinity is not threatened by women in the cockpit, just as there are many men whose self-esteem is secure enough to tolerate women as members of a space crew. The exclusively male membership requirements of the club had to be eliminated, and the selection of men whose intense need to prove themselves was, as Carlos Perry so aptly put it, "neurotically overdetermined"[2] had to be reevaluated. Any woman applying to become an astronaut would be expected to have the maturity, motivation, and technical competence expected of the men.

Another uncomfortable characteristic of "it" was the powerful and unemotive "tough guy" image it projected—the image of a man not burdened by emotional sensitivity, either toward himself or toward anyone else. This is all well and good—and may even be necessary—for the lone test pilot of the past. But is emotional insensitivity really relevant for a space crew on the long journey to Mars? Who would be able to live and work comfortably—particularly in isolation and confinement—with a person like that? Imagine yourself locked up for months—even years—in a small room having no exit with an insensitive, albeit technically competent, person. As Jean-Paul Sartre observed, hell is other people.

The key psychological characteristics members of future space crews should possess are psychological select-in traits. These characteristics had yet to be formally identified by the Working Group. The determination of those psychological traits in astronaut applicants would presumably be part of future selections at NASA, initially to choose *individual* astronauts and then to assemble the most compatible individuals for specific *crews* and missions.

To identify these traits was a truly daunting task, and one not likely to be popular. Not only did members of the Working Group have to talk about "it," but also they had to eliminate the most sacred parts of the mystique. Critics pointed out that a man's performance spoke for itself. It did not require any complicated measurement. Either he did the job, or he did not. It was that simple.

But it was *not* that simple. Neither pilots nor astronauts—contrary to the popular myth—are infallible, and the expectation of perfection is a heavy burden that, in itself, is a frequent cause of errors in judgment and performance.

The job of the pilot and the job of the astronaut, which seemed so similar in the days of *Mercury*, had begun to diverge by the time of *Apollo*. No longer were solo astronauts expected to brave the final frontier. No longer was it necessary to go up and risk your life day after day. Embarking on missions to other planets and living and working in the space environment for weeks, months, and years at a time require the long-distance astronaut to possess something *more* (more interpersonal awareness and sensitivity) than the classic right stuff—as well as something *less* (less of a neurotic need to prove one's self).

Based on the combined professional experiences of Working Group members and on all relevant data presented in the psychological and psychiatric literature, a tentative set of psychological select-in traits was fashioned. An operational definition of each specific select-in criterion was then devised, thus linking the desired psychological characteristic directly with a specific facet of space operations. And, finally, a method to reliably measure each criterion was sought. After selection, the performance of selected astronauts would be carefully monitored and correlated with the presence or absence of the psychological criteria. In this way, we would prove that the trait was important to success on the job and, hence, represented the real right stuff.

International members of the Working Group from Japan and Europe pointed out that the *transcultural* validity of each psychological criterion would also have to be determined, since Space Station *Freedom* crews will be culturally and ethnically diverse. Psychological traits that might predict performance in one culture do not necessarily predict them for all cultures.

## THE FOUNDATION OF THE REAL PSYCHOLOGICAL STUFF

Several important assumptions formed the foundation of the Working Group redefinition of the right stuff. Since astronaut selection is assumed to be an *ongoing* process and not a specific occurrence at one point in time, the selection of crews for extended space missions requires continuous evaluation of performance and behavior after the initial selection. NASA has never, until recently, obtained any formal data on astronaut performance. In the true tradition of the right stuff, it just was not talked about. In theory, selected astronauts have to go through a year of "basic training" as astronaut candidates, or ASCANS, before they are promoted to full astronaut and can be assigned to a specific mission.

During the training year, it is significant that no ASCAN has ever "flunked out" or been deemed unacceptable for a mission assignment. It is possible that all astronauts have demonstrated superior performance during the training period, but no data exist to support this assertion. It is much more likely that there have been poor performers (i.e., those without the pilot's classic right stuff) whose liabilities, while not formally documented, are well known to those who make crew assignments. This informal system may be reflected in how long a time elapses before an astronaut is assigned to his or her first flight, or even in the type of mission they are assigned to (while every astronaut wants to be assigned to a mission, some missions are more desirable than others). Since no rationale is documented as to who flies when or even why, the process of crew selection remains mysterious and shrouded in secrecy, and thus contributes to the mythos of the traditional right stuff as it applies to space.

Another facet of the ongoing process of selection as envisioned by the Working Group is that both individual astronaut training and crew training are actually part of selection, since such training provides the key operational settings where performance and interpersonal dynamics may be observed and evaluated.

A second underlying assumption of the Working Group was that, unlike the relatively short-duration Shuttle missions, space missions of the future will last for months or years and will require crews to develop a common group experience during training. The research necessary to define the nature, length, and possible content of that experience has not yet been done, but such research is essential to understand how to maintain individual and crew performance on long-duration missions. No matter how carefully individual candidates are selected, there is no guarantee that those individuals will be able to function flawlessly all the time. Methods for enhancing and maintaining that functioning have to be devised and are an important part of planning for any long space mission. But, in order to enhance or maintain performance, it first has to be *measured*, and, as George Ruff found out, this has always been the sticking point when negotiating with astronauts and NASA management about these matters. Any performance measurements used by behavioral scientists would need to receive the approval of the astronauts and their management.

It was also clear that decisions about the confidentiality and operational use of performance data collected for the purpose of validating psychological selection criteria must be made and agreed on with the participating astronauts prior to initiating the research. While this is a consideration that is routine in all social and behavioral research, it had to be emphasized in this particular instance to reassure both astronauts and NASA that such information would be beneficial only in the long term.

## DETERMINING THE REAL PSYCHOLOGICAL STUFF

The Working Group spent many hours over several months in 1988 and 1989 reviewing the scientific literature on personnel selection in the context of the specific job of astronaut as it was currently understood. Dr. Ellen Baker presented a thorough analysis of the typical tasks expected of astronauts before, during, and after assignments to flights. According to Dr. Baker, most outsiders have little appreciation for how little of the professional astronaut's time is actually spent in the space environment—or even training for a space mission. Astronauts not assigned to a crew participate in a number of ongoing support activities, including assignments at the Kennedy Space Center for one to three days a week to work on vehicle procedures and testing; in mission control as the "capcom" (capsule communicator), who directly interfaces with crews during simulations and real missions; in software verification; and in the development of crew procedures. Other activities for an unassigned astronaut might include making public relations appearances, maintaining skills in flying high-performance jets, and participating in general Shuttle systems training.

Once an astronaut is assigned to a specific space crew and mission, that astronaut then spends most of his or her time over the next year studying specifically for that mission. Although assigned to a mission, the astronaut is also expected to maintain flying proficiency in the NASA T-38 jets. Occasionally the assigned astronaut is involved in other activities, but not many. If one looks at the time breakdown of the astronaut job, only 1 percent of time *or less* is actually spent in the space environment. This will probably not change to a significant degree, even when space missions increase in length, since the training time allotted to them will also be longer, and the down time when an astronaut is not assigned to a crew will also be longer. There is not much glamour—or even risk—in many of the routine, day-to-day tasks expected of the astronaut.

Much speculation ensued about the *future* duties of astronauts, both during a typical space mission and during the down time. Dr. Norm Thagard, an experienced Shuttle astronaut, presented to the Working Group the most recently developed Space Station *Freedom* mission scenarios. But, because *Freedom* and Lunar Base or Mars missions are still on the drawing board and are regularly revised, the actual job specifications of the *Freedom* astronaut remain vague and ill-defined. Chapter 10 discusses the possible activities a typical *Freedom* mission might include.

With the best available information on the nature of the astronaut job, the Working Group entered a period of extensive review, discussion, and debate. Eventually, three major clusters of psychological traits believed to be important for future astronauts were identified: *aptitude for the job, motivation*; and a quality termed *sensitivity to self and others.* These three general areas and the specific psychological criteria encompassed by each are discussed next.

### Aptitude for the Job

An important aspect of the right stuff is ability. By itself, ability will not ensure election to the elite; but without it, making the cut would be impossible. The specific psychological traits comprising aptitude for the job include the following.

*Intelligence and Technical Aptitude.* Candidates must have a high level of general intelligence, with the ability to interpret technical instruments, perceive mathematical relationships, and maintain spatial orientation.

*History of Professional Success.* A candidate must have demonstrated from his or her personal history successful accomplishment of professional goals, as well as outstanding performance of professional activities.

*Adaptability and Flexibility.* Within the context of the job of an astronaut, flexibility is the ability to do or learn to do many different types of jobs, as well as the ability to function when one is in unfamiliar surroundings and when usual patterns of behavior are impossible.

*Team Player.* Being a team player is a highly prized skill at NASA. It is an individual's ability to function effectively as a member of the team, contributing one's individual abilities toward the attainment of team goals, rather than for personal gain.

*Ability to Represent NASA Effectively.* The job of astronaut carries with it many responsibilities related to interfacing with the general public. A candidate must be able and willing to meet with the public with grace and patience. He or she must also be able to represent colleagues within the agency and with outside support contractors.

*Stress or Discomfort Tolerance.* This characteristic has always been an essential component of the traditional right stuff. The candidate must be able to endure (not necessarily enjoy) physical and mental discomfort, deprivation, and hardship. A history of character strength in spite of adversity or stressful life situations was thought to be a good indicator of this trait.

*Ability to Function Despite Imminent Catastrophe or Personal Danger.* The ability to function despite personal danger is an extension of stress tolerance, but includes the ability to function effectively in an environment that is hazardous and potentially life threatening, as well as stressful. Many individuals do fine until their life is threatened; then they panic or become unable to function. Again, this is an integral part of the right stuff tradition.

*Ability to Compartmentalize.* This criterion describes the ability to put aside personal or interpersonal problems and function effectively when one has to. It does not imply the unwillingness to deal with or think about those problems at a later, more appropriate time and/or place. Compartmentalization is a method of psychological *suppression*—a healthy psychological defense—and is not the same as *repression* of emotions or feelings, which is far less psychologically healthy.

*Ability to Tolerate Separation from Loved Ones.* Long space missions may place inordinate stresses on family and personal relationships. Both the astronaut and his or her family need to understand this stress and be able to manage it effectively.

*Ability to Tolerate Isolation.* Future space crews will function in isolated, confined environments. There is nowhere to go to "get away from it all," and those individuals who require—or prefer—privacy from the rest of the group or who have difficulty in handling close, interpersonal relationships will have difficulties adapting to life in space.

## Motivation

Motivation represents a cluster of psychological traits at the core of the definition of the traditional right stuff. Indeed, one cannot imagine a person with ambitions of demonstrating that he or she has "it" who does not have intense desire and drive to be among the "best." Motivation is the degree of interest and enthusiasm an individual exhibits for the job. It assumes that one has high energy levels in pursuing one's goals. Generally, for the job of astronaut, there must be a high degree of mission-oriented motivation and a lesser degree of personal motivational factors. Frequently personal factors may be emotionally overdetermined, meaning that attempts to compensate for identity problems or feelings of inadequacy may interfere with more appropriate behavior. The neurotic aspect of motivation is clearly undesirable, since in the long run it leads to self-destructive behavior and inability to do the job.

If motivation is broken down into measurable components, it has the following psychological characteristics.

*Achievement/Goal Orientation.* All individuals with high motivation are achievement- and goal-oriented. That is, they are able to focus on the accomplishment of a particular goal and achieve a sense of satisfaction from that accomplishment.

*Hard Working/Self-starting.* An individual must be able to work hard for long periods of time without direct supervision. He or she must be able to provide the internal motivation and desire to succeed in order to get the job done, whatever it takes.

*Mastery.* This is a well-studied psychological characteristic that involves the willingness and desire to master challenging environments or tasks.

*Persistence.* A highly motivated person has a willingness and an ability to stick with a goal or task until completion, despite frustration or negative influences from the external environment.

*Optimism.* Without optimism—the psychological trait that enables a person to believe that he or she can accomplish a goal and that success is possible—motivation would be drastically impaired. Individuals must believe that success is possible, but they must not be grandiose or excessively optimistic when the facts suggest the contrary.

*No Excessive Extrinsic or Unhealthy Motivation.* Motivations that are overdetermined by pathological or neurotic needs are contraindicated, since sooner or later the neurotic need will interfere with success.

*A Healthy Sense of Competition.* Competitiveness has some mutually contradicting aspects. Unhealthy competition, or the excessive need to win in spite of any or all consequences, is not desirable, but an astronaut candidate must be able to enjoy the normal level of competitiveness present in most tasks, which contributes to the team's level of motivation and desire for excellence.

*A Capacity to Tolerate Boredom and Low Levels of Stimulation.* Many of the everyday duties of an astronaut, as well as tasks during training and long-duration space missions, are repetitive and potentially boring. A candidate must not have an excessive need to be stimulated and must be able to find some satisfaction even in repetitive or boring tasks.

*Mission Orientation.* A potential astronaut must be willing to subordinate personal goals to mission goals. This does not imply the absence of personal goals, merely a balance between the individual's personal objectives and the mission's.

*Healthy Risk-taking Behavior.* Part of the manliness mystique of the classical right stuff is the willingness (even eagerness) to take risks. But this ability to willingly experience personal danger must not be excessive, since that indicates that the individual is attempting to compensate for feelings of inadequacy. Such compensation might well be detrimental to the mission. Hot-dogging and other flagrantly adolescent behaviors intended to prove one's self are not desirable.

### Sensitivity to Self and Others

A typical right-stuff pilot would become apoplectic at the suggestion that being sensitive is an important aspect of "it." On the contrary, the manly code of conduct proscribes the feeling or expression of emotion. Emotions and the inability to control them mean that you do not have "it."

Yet, in the opinion of most of the experts in the Working Group, sensitivity to one's self and to others is possibly the most important component of the future job of the astronaut, compared to that of the fighter pilot. As the tasks involved in the two jobs—pilot and astronaut—have diverged, it has become evident that, unlike most jet pilots, the astronaut must function in an interdependent group for very long periods of time. The size of future crews may eventually reach ten to twelve individuals or more. Sensitivity reflects the emotional *maturity* and interpersonal awareness of an individual. Getting along with crew mates and being able to develop effective compromises and group problem-solving skills—these are essential for the astronauts assigned to long-duration space missions. Specific psychological traits in this area are the following.

*Overall Emotional Maturity and Stability.* An astronaut must be able to tolerate both success and failure. Maintaining a balanced perspective in life is a sign of maturity, as is the use of humor, sublimation, and altruism as basic defense mechanisms in handling the stresses of everyday life.

*Self-esteem.* A candidate must not have serious conflicts about his or her self-worth. On the other hand, arrogance and an overly elevated estimation of one's worth are traits that can significantly disrupt group functioning. An appropriate self-concept, with knowledge of one's strengths and weaknesses, is what is called for as a long-distance astronaut.

*Ability to Form Stable and Quality Interpersonal Relationships.* A candidate must have demonstrated success not only in his or her professional life, but also in his or her personal life, particularly in being able to establish enduring and satisfying relationships with other people.

"Does this mean if you have a divorce you can't be an astronaut?" I was asked once. The answer is "of course not"—yet the question symbolizes the worst fear of many astronaut candidates. Stability in one's life is reflected by many factors. One important factor is durable and long-lasting relationships with other people. A candidate who has gone through multiple divorces, who has no friends, who is out of touch with any family—such a candidate would be a poor interpersonal risk. He or she might do well as the solitary pilot of a fighter jet, but probably not as a member of a crew of eight on the lunar or Martian surface.

*Expressivity.* Expressivity, or the ability to express emotion appropriately, is an important skill for developing and sustaining interpersonal relationships in one's job and personal life and in group functioning. Self-disclosure and the ability to trust must be present, but not to a degree that they impede effective functioning. The candidate must be able to suppress (not repress) emotions under certain conditions. Expressivity also represents a quality of self-awareness and psychological-mindedness that is frequently absent in the typical fighter pilot. An ability to be introspective when necessary is important in interpersonal relationships. Without that ability, the individual may lose perspective when involved in conflict with others. Insight and self-awareness are traditionally valued by behavioral scientists, but such qualities may also be detrimental to effective functioning in operational environments such as space. Obviously, the astronaut has to be able to balance the two sides of this trait.

*Sense of Humor.* Any individual locked in a remote and dangerous environment with other people begins to value the constructive use of humor, without any sadistic or hostile overtones. Sarcastic wit and humor at the expense of others are frequently counterproductive and disruptive, but the ability to see the humorous in otherwise frightening or embarrassing situations is a valuable asset to any team.

*Insight and Self-awareness.* Not only should the ideal astronaut have insight, but also he or she should relate to others in a way that demonstrates that insight and self-awareness. Sensitivity to the feelings of others and empathy are crucial skills in a group living situation. Disregard for others' feelings can upset a previously well-functioning group. The candidate who is overtly hostile to either authority or subordinates will have a difficult time being integrated into a space crew. This is not to say that individuals cannot experience some negative feelings toward others, but they must have effective ways to redirect the hostility into constructive actions. An individual should not be walking around with a "chip on his shoulder."

*Appropriate Assertiveness.* A candidate must be assertive—that is, be able to assert one's self in the group, without undue aggressiveness or impulsivity. This also implies the ability to handle aggressive feelings without significant discomfort.

*Cultural Sensitivity.* Since Space Station *Freedom* and most other missions of long duration will have international crew members, it is important that crew members be sensitive to cultural and ethnic differences and be able to tolerate those differences.

It would seem that the expectations of future astronauts might require a truly superhuman being. Who could possibly possess all of these traits in the right amount? Can such a person with all the traits as they are listed really exist? No, probably not. But the goal of setting psychological standards and criteria for astronaut selection is not to set superhuman expectations, but to be able to identify potential strengths and weaknesses in each individual character that may prove to be important in designing training programs or in assembling crews for different missions. It is clear that no one individual could possibly have all of the select-in characteristics listed; nor are any two candidates likely to have them in the same degree. In other words, each individual astronaut will have strengths and weaknesses, which will need to be evaluated within the context of the program objectives. Psychological tests can provide relevant data on this issue, but they cannot provide sound judgments that consider the broader context of the particular mission.

## OPERATIONAL DEFINITIONS

After the psychological characteristics important in long-duration space flight were determined, each characteristic was examined in the context of its specific operational definition in the Space Program. In other words, how was a particular characteristic to be measured in any given candidate? A search of the psychological literature revealed a number of psychometric constructs from a variety of tests that might be effective. During this process, we witnessed a telescoping effect, in that several of the characteristics collapsed into one larger construct. The final results of this process of operationalization are listed in Table 6.1 and represent the final psychological selection criteria that were recommended to NASA.

In the summer of 1989, these psychological select-in criteria were formally submitted as a White Paper recommendation by the entire Working Group to NASA life sciences management. At that time, a carefully thought-out research proposal—which had been developed by the select-in subcommittee, composed of myself, Dr. Robert Rose, Dr. Robert Helmreich, Dr. Ellen Baker, Dr. Sonny Carter, and the international members from Japan and Europe—was presented. The goals of this research proposal were very similar in some ways to those of George Ruff's project during the early *Mercury* period, thirty years earlier. The research was designed to answer the question of whether the psychological selection criteria proposed represented the real right stuff—in short, did the presence of these psychological traits in an astronaut predict that this astronaut would function effectively in the space environment?

When the proposal was presented to NASA management, it was argued that it was essential to begin to measure astronaut performance; otherwise, NASA's selection decisions and the psychological criteria would never be formally validated. Behavioral scientists and selection committees may choose candidates on the basis of specific psychological traits they *believe* might be important in future astronauts, but that is not the same thing as *proving* that those traits have any degree of reliability in predicting astronaut performance and behavior. In fact, much of the scientific literature

Table 6.1
Astronaut Psychological Select-in Criteria

| Criteria | Means of Assessing |
|---|---|
| *Aptitude for the job* | |
| Intelligence | MAB |
| Leadership/Followership ("team player") | PCI |
| Ability to Tolerate Stress | PCI—Jenkins Activity Scale |
| Trainability/Flexibility | PCI |
| *Motivation* | |
| Mastery | PCI—Mastery |
| Work Orientation/Energy | PCI—Work and Vitality |
| Optimal Competitiveness | PCI—Competitiveness |
| Optimal Extrinsic Motivation | PCI—Motivation |
| *Sensitivity to self and others* | |
| Emotional Maturity/Stability | PCI |
| Ability to Form Stable, Quality Interpersonal Relationships | PCI—Expressivity and Impatience/Irritability |
| Expressivity | PCI—Verbal Expressivity |
| Appropriate Assertiveness | PCI—Verbal Aggression |
| Sense of Humor/Cultural Sensitivity | PCI—Expressivity |

argues against the concept, as we shall see. It is entirely possible that the best psychological predictions are not any better than current NASA Selection Board procedures.

## PERSONALITY'S DISMAL RECORD IN PREDICTING PILOT PERFORMANCE

The Working Group's position was that personality assessment is underutilized as a resource in astronaut selection, but the empirical record in aviation psychological research of using personality traits as predictors of performance is appalling. This dismal record extends back to World War I and the selection of opponents for the Red Baron.

While aviators intuitively recognize that personality factors are strong determinants of pilot behavior, empirical evidence suggests there is no relationship between personality and aviator performance.[3] Part of the resolution of this paradox can be found by examining three factors: (1) the individuals, specifically the personality models used to evaluate individuals and the dimensions those models assess; (2) what you want the individuals to do and the choice of performance criteria against which the personality constructs are validated; and (3) the environment in which you expect the individuals to perform.

Personality models must be tested in real situations and environments, but the literature on pilot selection deals almost exclusively with initial training. Usually, the defined pilot performance criterion is success or failure in completing initial training or the rated performance during training. But performance training is only a valid criterion if it correlates strongly with desired outcome (i.e., performance in flight operations). Little evidence exists to support performance in training as a strong predictor of subsequent line performance. A phenomenon known as the honeymoon effect suggests that persons selected for desired positions are likely to exert maximal effort during training and initial operational experiences, but are unlikely to sustain performance over time if they are deficient on relevant personality dimensions.[4] These data suggest the honeymoon effect may attenuate true relationships between personality and performance in a number of occupations, including aviation.

In order to test the effects of personality on performance, one needs to use *relevant* personality models. The basic unit of personality is the *trait,* which is defined as a relatively stable, enduring behavioral disposition that an individual exhibits over time. Traits can be reliably measured and may combine to define broader behavioral categories.

Two recently developed and comprehensive models of personality are of interest in the process of astronaut psychological selection: the Spence-Helmreich model and the Big Five model—both of which have been shown to have operational (i.e., real-world) significance.

## The Spence-Helmreich Model

In the Spence-Helmreich model, two broad underlying trait dimensions have been identified: *instrumentality* (I), or goal orientation, and *expressivity* (E), or interpersonal capacities.[5] The central element of instrumentality is *achievement motivation,* which is defined as motivation directed toward the attainment of goals. Three distinct components have been delineated: (1) *mastery needs,* or the desire to undertake new and challenging tasks; (2)

*work needs,* or the realization of satisfaction and pride in working well; and (3) *competitiveness,* or the desire to surpass others in all areas of endeavor. Positive instrumentality (I+) is the orientation and motivation for achievement. Negative aspects of instrumentality (I−) are reflected in an autocratic, dictatorial orientation. Individuals high on this dimension tend to achieve goals at the expense of others and without regard for their sensitivities (part of the constellation of the classic right stuff).

Expressive attributes have been divided into four categories, one positive and three negative: (1) *Expressivity* (E) consists of traits reflecting warmth and sensitivity to others, including what is often called empathy; (2) *verbal aggression* refers to a type of nagging hostility directed toward others; (3) *negative communion* encompasses a constellation of passivity and servility in interpersonal relations; and (4) *impatience/irritability* refers to a characteristic pattern of drive and annoyance in dealing with others.

Helmreich and his colleagues have identified frequently occurring patterns of the above traits. Using large samples of pilots, and other professionals, three major clusters were defined. The first, a combination of favorable characteristics, is defined by high scores on positive I and E attributes (EI+) and low scores on the negative attributes. This constellation has been referred to by the investigators as their version of the "right stuff."

A second cluster is characterized by high I, particularly of the negative form, and low positive E and is termed the "wrong stuff." This cluster is designated as the negative instrumentality (I−) group.

A third prominent cluster is notable for being defined by low scores on both positive I and E dimensions, with elevated scores on verbal aggression. The authors refer to this as the low motivation, or "no stuff," cluster.

A modal cluster is defined by individuals who do not show elevation (positive or negative) on either instrumental or expressive traits.

In addition to the above cluster, a group of individuals characterized by high response desirability (RD) makes up a final commonly seen grouping of personality characteristics. These individuals answer uniformly very positive (above two standard deviations from the mean) on the traits. It is suspected that these individuals are faking positive test responses. As we shall see, the RD group has some interesting behavioral and performance problems.

The use of these clusters to define subgroups has proved valuable in understanding and communicating determinants of performance in operational aviation environments.[6]

## The Big Five Personality Model

The Big Five Personality model represents the independent development of a comprehensive method to characterize the major domains of adult personality structure. This is primarily an atheoretical model, which came into being through the review of personality traits in many personality systems. Paul Costa, Jr., and Robert McCrae[7] found that the traits labeled "Neuroticism" and "Extraversion" appeared in most of them. They later added "Openness to Experience" as a third trait.[8] A competing research group demonstrated good results using a personality concept composed of five traits—neuroticism, extraversion, agreeableness, conscientiousness, and culture.[9] Since the late 1960s, factor analytic techniques have refined the model to include the five domains of personality currently referred to as the "Big Five." While it is possible that other personality domains exist outside the five included, "no other system has a better claim to comprehensiveness."[10]

Each of the three broad domains of the model has six specific facets or subscales: (1) *neuroticism* (N)—anxiety, hostility, depression, self-consciousness, impulsiveness, and vulnerability; (2) *extraversion* (EX)—warmth, gregariousness, assertiveness, activity, excitement seeking, and positive emotions; and (3) *openness* (O)—fantasy, aesthetics, feelings, actions, ideas, and values. In addition, two other major scales—*agreeableness* (A) and *conscientiousness* (C)—are included. A detailed discussion of these traits can be found in the Appendix 6.

The history of the Big Five concept has recently been the subject of several reviews,[11] and investigators have begun to extend its application to new situations and environments.[12]

## RESEARCH SUPPORT FOR USING PERSONALITY MODELS IN ASTRONAUT SELECTION

The Personal Characteristics Inventory (PCI), developed by Helmreich and his colleagues, is composed of scales from three psychometric instruments: the Work and Family Orientation Questionnaire, the Extended Personal Attributes Questionnaire, and the Impatience/Irritability and Achievement Striving subscales of the Type A Questionnaire.[13] The PCI measures instrumentality (I), expressivity (E), and their positive and negative subscales. It is the primary means by which the constructs of the Spence-Helmreich model are assessed.

The PCI has been shown to exhibit adequate reliability and validity in a number of populations, including pilots and scientists.[14] A consistent pattern of results begins to emerge on the various constellations of traits described by the model. The results demonstrate that instrumental and expressive traits relate consistently to crew performance in flight operations and to the acceptance of training by pilots in crew coordination concepts. The same cluster that predicts successful aviator performance also appears to be involved in academic success.

Work, mastery, and competition (three scales of instrumentality) all predicted scientific attainment and academic performance.[15] The three components each acted differently as predictors. As expected, mastery and work were *positive* predictors of achievement (as measured by citations to the scientists' published work), but competitiveness was *negatively* associated with attainment. The pattern associated with highest achievement was high mastery, high work, and low competitiveness. This same pattern was found to predict salary five years after graduation in MBA graduates, as well as academic grades in college students. Achievement motivation also related significantly to job satisfaction in a study of supertanker crews, with the pattern of relationships depending on whether the crew member held a non-demanding job involving routine labor (deck force) or a demanding job involving technical duties (engineering force).[16]

Research employing the more global measures of instrumentality and expressivity also demonstrated a number of significant relationships. Individuals scoring high in both instrumentality and expressivity were found to have higher self-esteem than did those with any other pattern of these attributes and to be most satisfied with their lives and relationships.[17] The negative components of the two dimensions were found to relate to neuroticism and acting-out behavior. A study conducted in Mexico with a Spanish version of the instrument demonstrated that the structure of personality was similar in that culture.[18] Parallel results were found with a German version in a West German sample.[19]

More recently, the battery has been extended to include what has been known as the Type A pattern of behavior.[20] The Type A syndrome is alleged to be a syndrome of time urgency, impatience, and workaholism that resulted in both high achievement and proneness to cardiovascular disease. Helmreich's results, obtained through factor analytic studies of the most widely used objective measure of the Type A construct, suggest that the concept has two active ingredients: (1) achievement striving, which is a particular form of achievement motivation related to superior performance, but not to health; and (2) impatience/irritability, which is related to a number of minor illnesses (perhaps cardiovascular disease), but not to achievement. These patterns were demonstrated with students in predicting college grades and health and with scientists in predicting research attainment.

A program of research by Julian Barling and Stephen Bluen and their colleagues[21] provides further validation of these scales and extends the range of criteria. One of their studies replicated the pattern of achievement striving related to performance, but not to health, and impatience/irritability related to health, but not to performance. The subjects of another of their studies were life insurance salespersons. Job satisfaction and depression were used as outcome criteria. In this group, achievement striving predicted job satisfaction significantly and positively, while impatience/irritability was a significant predictor of elevated depression scores.

In a very different setting, Barling and Bluen administered the scales to a group of medical practitioners and their spouses and found that high impatience/irritability scores were significantly associated with marital dissatisfaction among both the doctors and their spouses.

Another study examined the relationship between the constellations of traits defined by Helmreich's cluster analysis and managerial performance at an airline. Managers' performance was rated by their superiors, peers, and subordinates. The highest performance ratings were achieved by managers in the cluster containing high instrumental and high expressive traits (EI+), while those who were average on both dimensions received average ratings, and those with negative characteristics or lower on both dimensions received below-average evaluations.

With the cooperation of a national airline, pilot applicants were screened using the PCI, and their later flying performance was rated by specially trained check airmen, unaware of the personality research. Results confirm that the constellation of optimal traits is related to superior evaluation as a commercial pilot. A dissertation by Thomas Chidester, using the same clusters, reached similar conclusions.[22]

One distinction that can be drawn from all the research reviewed is between attitudes and personality traits. Attitudes are acquired and modified through experience. Much of our life activity is surrounded by changing attitudes—or by resistance to such change. Personality traits, on the other hand, are acquired quite early in life and are highly resistant to change, even through such Draconian mechanisms as intensive psychotherapy. This distinction relates to what we can and cannot expect to accomplish in training. We can often change attitudes through effective training programs, but personality traits are not likely to be impacted. Thus, to the extent that behaviors are driven by traits, an organization is stuck with what it initially selects.[23]

In a separate series of research experiments, flight crew members were studied in real-time flight operations. Crew member attitudes regarding cockpit management were isolated and validated in terms of line performance. It has subsequently been found that these attitudes can be significantly changed in most pilots by formal training in cockpit resource management (CRM).[24] A recent finding relevant to pilot selection is that the personality constellations isolated by the PCI predict the degree of acceptance of CRM concepts in training. Additional data from a large sample of Air Force pilots confirm the significant relationships between personality clusters and the ability of CRM training to change cockpit management attitudes.

A study conducted in the experimental B727 simulator at NASA's Ames Research Center provides important findings regarding personality and performance and the limitations of training.[25] In this study, volunteer crew members from a major airline were tested using the PCI, and crews were composed on the basis of their personality traits. The crews then flew a two-day, five-leg simulation that involved significant mechanical abnormalities and marginal weather on two of the legs. Extensive performance data on individual and crew functioning were collected.

Results looking only at the *captain's* personality show that the most effective crews were led by captains with personalities from the right stuff cluster (IE+), while the least effective crews had "no stuff" leaders (low motivation cluster). An interesting pattern emerged in crews led by captains with the autocratic, domineering wrong stuff constellation (I–). These crews performed worst during the abnormal situation on the first day of the simulation, but their performance was much improved and almost as good as the best group during the second day's activities. It is hypothesized that subordinate crew members may learn to adapt to the autocratic and domineering, but technically competent, type of leader over time. These findings persisted in spite of having had formal training in CRM and Line-Oriented Flight Training (LOFT).[26]

The follow-up of individuals with high response desirability from the PCI has proved to be extremely valuable. Approximately 10 to 15 percent of pilot applicants will have the RD profile. This made it possible to determine if their performance was in any way different from that of other candidates—particularly if it was as good as their highly positive profiles suggested.

When their subsequent performance is evaluated, the results clearly indicate that the RD employee is a poor job risk. For example, while 13 percent of pilot applicants during one period were classified RD, more than half of those terminated for behavioral causes were in this category, suggesting that it is an excellent *negative* predictor of performance in its own right.

Parallel results were found in a sample of flight attendants. It is particularly noteworthy that RD individuals are generally rated extremely favorably by interviewers and tend to impress others initially with their capabilities. They are often articulate and express enthusiasm and motivation for the position for which they apply. In many ways, their behavior fits the classic profile of the psychopathic personality. This reinforces the utility of objective testing as a companion methodology to the job interview.

A concern with the use of psychological testing is the possibility that it may unfairly disqualify a member of some protected group. Considerable data have been obtained from ethnic minorities and women. The findings show that racial differences are not significant.[27] There are some significant sex differences, but these are smaller when the comparisons are between men and women seeking demanding positions (such as pilot or astronaut) rather than between men and women in the general population. Generally, the pattern of residual sex differences would tend to favor women as candidates (as they exhibit higher expressivity and lower competitiveness as a group).[28]

### The NEO Inventory

The NEO–Personality Inventory (NEO-PI) is a 181-item inventory used to assess the five major dimensions or domains of normal adult personality [that is, the Big Five—neuroticism (N), extraversion (EX), openness (O), agreeableness (A), and conscientiousness (C)]. Respondents indicate the degree to which they agree with scale items on a five-point scale ranging from "strongly disagree" to "strongly agree." The reliability and validity of this instrument have been demonstrated.[29]

Many members of the Working Group were impressed with the thoroughness of the Big Five model of personality and strongly believed that it should be tested in the context of astronaut selection. If the five traits and subtraits are characteristics that appear to cross all theoretical models, then it is possible that the evaluation of these traits in astronaut candidates may prove useful. Additionally, Helmreich's group has data showing a parallel between the personality constructs from his theory and the personality constructs of the Big Five. Since it provides a slightly different perspective on those constructs, the NEO-PI was included in the select-in test battery.

## CRITERION MEASURES EVALUATED: HOW TO ASSESS ASTRONAUT PERFORMANCE

After a series of discussions with the JSC Astronaut Office about possible astronaut performance measures for the psychological validation project, a decision was made to use peer and supervisory ratings. There were two major reasons for selecting these particular criterion measures: (1) Such ratings were acceptable to the Astronaut Office, while other performance measures were felt to be too intrusive and time demanding; and (2) obtaining peer and supervisor ratings would be relatively simple and inexpensive, compared to any other method.

Peer evaluations have been used successfully in the Antarctic,[30] as well as in other isolated, confined environments.[31] The rating system developed by Robert Rose and Helmreich asked astronaut subjects to rate each other on two broad dimensions: technical skill and knowledge (job performance), and the ability to live and work with others (group living). From a list of those peers with whom the astronaut has trained or flown, each astronaut was asked to nominate two to five individuals for the "top" or excellent performer group and two to five individuals for the "bottom" or poor performer group. Scores for each individual were then tabulated and ranked by adding the number of nominations received for each group.

The specifics of the peer evaluation form were determined by the Research Subcommittee, working closely with the astronauts on the Working Group. Table 6.2 summarizes the general and specific categories on which astronauts were asked to rate each other. Separately, the chief astronaut

Table 6.2
Astronaut Peer Rating System

Job Competence (Knowledge)
- complete fund of information
- absorbs new information quickly
- reduces complex issues into essential elements
- "I value this person's opinions on technical matters"

Job Competence (Performance)
- accomplishes tasks thoroughly and efficiently
- develops innovative solutions to difficult problems
- consistent performer under stress/discomfort
- excellent job performance
- endures mental and physical discomfort well
- avoids unnecessary risk

External Relations
- presents him/herself well
- speaks clearly and effectively
- is able to represent the Agency well

Leadership
- motivates others to complete tasks
- is not authoritarian
- is decisive when required; flexible when required
- "I would work in a group with this person as leader"

Teamwork
- puts group goals ahead of individual goals
- works effectively with many different kinds of people
- pulls his/her own weight
- shares credit; accepts blame
- "I would choose this person for my team"

Personal/Group Living
- easy to get along with; "I am comfortable with this person"
- good listener
- absence of irritating qualities
- considerate of others
- wears well over time
- helpful
- tolerant of individual/cultural differences
- has a sense of humor

and his deputy were asked to rank members of the corps according to the same performance categories.

## OUTLINE OF THE RESEARCH PROPOSAL

The three hypotheses of the research proposal to validate the psychological select-in criteria were as follows: (1) Current selection screening procedures may emphasize operational concerns and downplay interpersonal skills, (2) clinical psychiatric (select-out) screening may not reliably identify performance-related personality dimensions, and (3) personality correlations measured by the select-in tests will predict astronaut performance. It is entirely possible that, given a rigorous selection and a large pool of applicants, there may actually be no relationship between personality and performance. This possibility was the null hypothesis.

## HOW THE RESEARCH PROPOSAL WAS PRESENTED TO THE ASTRONAUT OFFICE

The first step in initiating the research plan was to get the cooperation of the Astronaut Office. Rose and Helmreich proposed that the "select-in" tests be given first to the group of astronaut applicants scheduled to be interviewed in the fall of 1989. This group would then be followed over time and their performance assessed through peer evaluations.

A second and even more important phase of the research was to assess *current* astronauts on the select-in personality dimensions and then to collect peer and supervisory evaluations as outcome criteria. This second data collection with current and active astronauts was considered important because the number of astronaut applicants who are chosen to be astronauts is small (usually only ten to twenty from each selection). With such a small sample size, it would take many years before the project could accumulate a significant N. By using current astronauts, the process could be shortened considerably.

## RESULTS OF THE SELECT-IN RESEARCH

### Astronaut Applicants

Of the 106 astronaut applicants, 104 volunteered to take the select-in tests at the time they came to JSC for the job interview and the medical and psychiatric evaluation. However, none of the data collected as part of the select-in research could be used by the operational psychiatrists in determining a candidate's psychiatric qualifications. In fact, to ensure that this would be so, the research and operations data were collected and maintained separately by different individuals.

The relationships between the psychological select-in traits and the psychiatric traits measured by the select-out tests and the clinical psychiatric ratings were determined.[32] Two scales were found to be significantly correlated with the psychiatrists' clinical ratings: positive instrumentality (I+) and vigor (this trait was borrowed from the European psychological selection battery and included in the U.S. research; it will be discussed in more detail in Chapter 7). This relationship makes sense, since the psychiatrists were actually making an assessment of their overall impression of each candidate's technical capability and competence.

Several correlations were found between the select-in battery and the MMPI (one of the select-out tests). Specifically, achievement striving (from the PCI) was *negatively* correlated with hypochondriasis (Hs), depression (D), and psychopathic deviation (Pd). Impatience/irritability (from the PCI) was *negatively* correlated with Hs, D, Pd, and hysteria (Hy). Vigor (the scale added to the PCI from the European Space Agency's selection test battery) was *negatively* correlated with mania (Ma) and positively correlated with social introversion (Si). These correlations were all modest (.3 to .5) and intuitively made sense. But, despite these small connections, overall the select-in battery measured constructs that differed fairly significantly from the psychiatric clinical ratings and the MMPI—suggesting that the select-in and select-out processes are quite different, as expected. There was a much greater agreement on whom to reject, but the two methods identified different individuals who should be selected.

## Current Astronauts

The extensive work to collect data from current astronauts was performed by Robert Rose, Robert Helmreich, Lou Fogg, and Terry McFadden and is recounted in several scientific articles that have been recently submitted for publication.[33] These data represent the first time in over thirty years that psychological data were obtained from astronauts.

Eighty-four current astronauts were asked to complete peer ratings and take the psychometric tests. A total of 65 astronauts actually completed the ratings and tests, for 77 percent participation, including thirteen female mission specialists, thirty-four male mission specialists, and thirty-seven male pilots. Considering the historically negative attitude of the astronauts toward psychiatry and psychology (documented in the first three chapters), this is an amazingly high participation rate. It is due primarily to the efforts of Drs. Sonny Carter and Ellen Baker, both of whom worked very hard to convince their peers of the worth of the project.

The chief astronaut and his deputy completed supervisory ratings on all astronauts. Peer and supervisory ratings were obtained on all astronauts, even those who elected not to participate in collecting the psychological data. Data were collected in a highly confidential manner.

Probit analysis of the peer ratings was done to transform the peer and supervisory data into nearly normally distributed continuous dimension scores. Supervisor ratings and peer ratings were found to be very highly correlated (.81), which suggests that there is some agreement among astronauts and their supervisors on which astronauts perform well and which do not.

A factor analysis with oblique rotation was performed to reduce the number of peer rating dimensions to only two: group living/personality (or interpersonal effectiveness) and job performance (or technical effectiveness). The first factor is related to interpersonal skills and the second to technical competence. These two factors were found to be correlated at the .29 level, suggesting that they are measuring two very different aspects of performance. The group living/personality dimension explained 52 percent of the total variance, while the job performance factor explained 36 percent of the variance. The quality of leadership loaded almost equally on both factors, but all other dimensions of the peer and supervisory ratings loaded on the expected factor. Rose and his co-authors concluded:

> Results indicate that peer ratings and supervisory ratings show reasonable convergence and that peer ratings on the *Performance/Competence* dimension were related to flight assignments of mission specialists. . . .
>
> It is not surprising that *Performance/Competence* was associated with who got flight assignments (at least among the mission specialists), while *Group Living/Personality* was less related, although trending towards significance in the expected direction. Decision-makers often place more value on technical abilities than on interpersonal ones. . . . These results . . . [and those obtained by Helmreich's group] . . . suggest that it might be possible to improve decision-making by incorporating measures of interpersonal ability into the decision-making criteria for selecting astronauts from the pool of candidates and for missions.[34]

McFadden and Helmreich[35] performed factor analysis on the personality test data collected from current astronauts to obtain theoretically meaningful clusters. The astronauts fell into the following clusters: (1) 13.6 percent of the population was classified in the positive instrumental/expressive cluster (IE+), (2) 11.1 percent were in the negative instrumental cluster (I−), (3) 16.7 percent were in the "low motivation" cluster (i.e., they had low levels of instrumentality, expressivity, work, mastery, and competitiveness), (4) 31.8 percent of the population were classified as "modal," and (5) approximately 12.1 percent of the astronaut population did not fall cleanly into any cluster.[36]

Both psychological factors associated with high performance (nominated in the "top" category) and those associated with low performance (nominated in the "bottom" category) were analyzed.

Psychological factors associated with high performance ratings in the group living/personality (interpersonal effectiveness) category and in the job performance (technical effectiveness) category were positive instrumentality and positive expressivity (the IE+ cluster). Interestingly, the "low motivation" cluster was predictive of the job performance factor, but not of the group living factor. This finding is interesting, since the "low motivation" individual has not usually been studied, and it is not understood why they should be ranked high by peers. However, while subjects with this personality type may actually perform the technical aspects of the job fairly well, they are not, in general, individuals who do well in interpersonal interactions with co-workers. These results are shown graphically in Figures 6.1 through 6.3. The trends are apparent in the data and strongly suggest that the EI+ personality cluster is predictive of astronaut performance (at least as measured by peer and supervisory evaluations). The relationship between peer evaluations and real-time performance in the space environment has yet to be explored.

Rose and his colleagues have concluded that an astronaut's personality characteristics appear to be most likely related to interpersonal effectiveness (this conclusion is nearly identical to George Ruff's comments that the factor analysis of the *Mercury* psychological data was saturated with a "social skills" factor). The most effective astronauts had high scores for agreeableness and negative instrumentality–negative communion (to score high on this scale, the subject describes himself as "servile," "gullible," and "subordinate"; although the latter scale is described by these negatively

**Figure 6.1**
**Group Living Performance Factor**

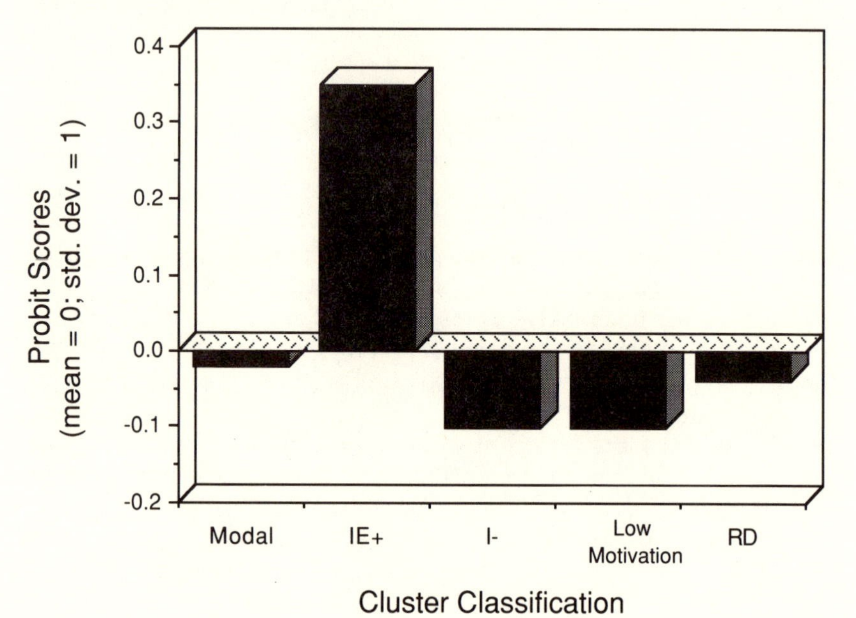

Figure 6.2
Job Performance Factor

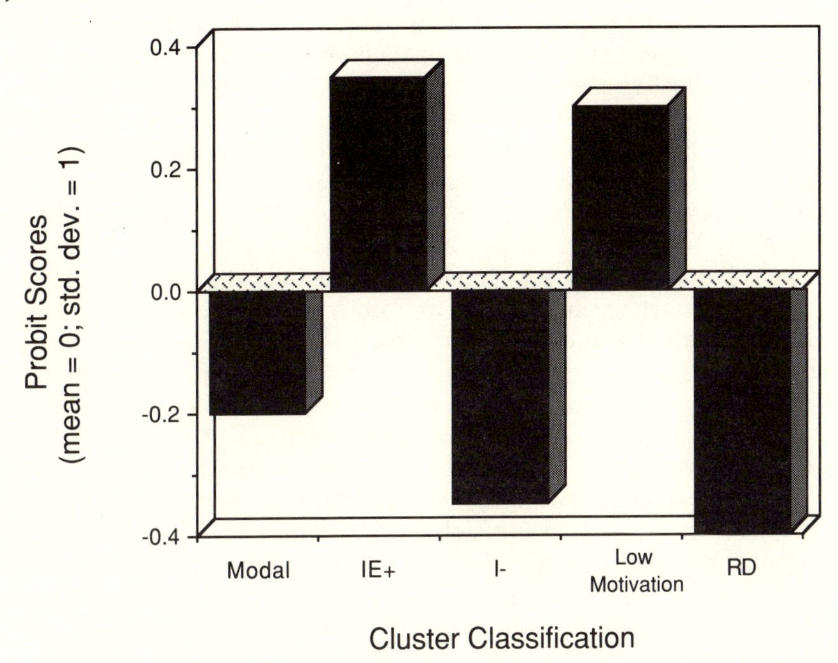


Figure 6.3
Combined Group Living and Job Performance Factors

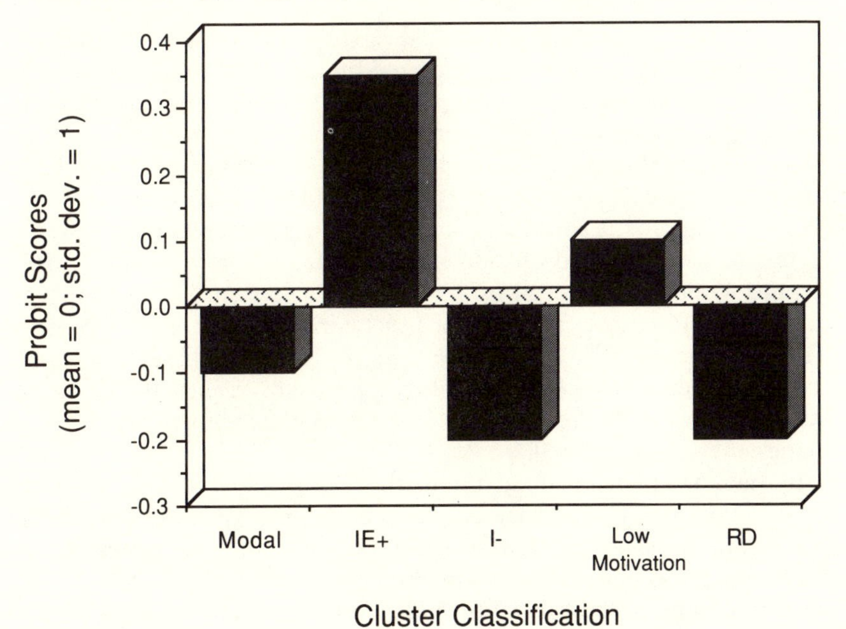

perceived traits, it is plausible that astronauts who describe themselves in that manner are more interested in their co-worker's needs and desires). The most effective astronauts also had low scores on the impatience/irritability scale and on negative instrumentality and openness (Rose and his colleagues describe openness as a sort of "freefloating intellectual curiosity"; apparently the job of astronaut does not tap this characteristic). The researchers conclude that the most effective astronaut can best be described by their data as a "hard-headed humanist"—someone concerned about other people, but who is not a daydreamer.[37]

The results of the research to validate psychological selection criteria suggest that there was a substantial variance in the current astronaut corps on the personality dimensions that were measured (i.e., astronauts have a more heterogeneous personality pattern than the MMPI data would indicate) and that these personality dimensions relate to effectiveness as an astronaut (as measured by supervisory and peer ratings).

While based on a small sample size, these data support the idea that psychological assessment of astronaut candidates may provide valuable information for astronaut and crew selection. Further work is under way to refine the select-in battery, using a large sample from analog environments with available behavioral criteria. A study of the relationships between the peer and supervisory ratings and the actual inflight performance over the next two to three years will also be undertaken, as will a prospective study of selected candidates (if NASA continues to fund the project). The last two studies mentioned are crucial steps in validating the data collected on current astronauts and in ultimately determining the "real" psychological stuff for future space missions.

The Working Group formed in 1988 was extraordinarily successful in accomplishing its goals. The data presented in the previous chapter on the implementation of state-of-the-art psychiatric screening and in this chapter on psychological criteria validation demonstrate that the behavioral sciences might contribute a great deal to the process of individual selection of astronauts and even to crew selection. It would be unfortunate if NASA decided to ignore the advances in psychiatric and psychological assessment technology, which could provide valuable data for the selection of future space travelers, and to revert back to a screening system that does not conform to an acceptable standard of psychiatric or psychological practice. Those individuals who have always believed in the traditional right stuff cannot be pleased to see it altered to conform with reality, and those who believe in the infallibility of astronauts may not be happy to see that their heroes are human, after all.

In the next several chapters the methods of psychiatric and psychological selection used in the space programs of Europe, Japan, and the former Soviet Union will be explored. The European and Japanese programs have both been somewhat influenced by their participation in the NASA Working Group. But each organization and country has its own unique method of astronaut psychological selection. Interestingly, the U.S.S.R. appropriated many of the techniques used in *Mercury* by George Ruff and his colleagues and still use those techniques today.

## NOTES

1. Tom Wolfe, *The Right Stuff* (New York: Farrar, Straus & Giroux, 1979), 23.
2. C. J. G. Perry, "Psychiatric Selection of Candidates for Space Missions," *Journal of the American Medical Association* 194 (1965): 841–844.
3. R. L. Helmreich, "Pilot Selection and Performance Evaluation: A New Look at an Old Problem," in *Proceedings of the Tenth Symposium: Psychology in the Department of Defense*, edited by G. E. Lee (Colorado Springs: U.S. Air Force Academy, USAFA-TR-1, 1986), 12–34.
4. R. L. Helmreich, L. L. Sawin, and A. L. Carsrud, "The Honeymoon Effect in Job Performance: Delayed Predictive Power of Achievement Motivation," *Journal of Applied Psychology* 71 (1986): 1085–1088.
5. R. L. Helmreich, J. R. Hackman, and H. C. Foushee, *Evaluating Flight Crew Performance: Policy, Pressures, Pitfalls and Promise*, NASA/University of Texas Technical Report 86-1 (Austin, Texas 1986). R. L. Helmreich and J. T. Spence, "The Work and Family Orientation Questionnaire: An Objective Instrument to Assess Components of Achievement Motivation and Attitudes toward Family and Career," *JSAS Catalog of Selected Documents in Psychology* 8 (1978): 35, ms. 1677. J. T. Spence and R. L. Helmreich, "Achievement-related Motives and Behavior," in *Achievement and Achievement Motives: Psychological and Sociological Approaches*, edited by J. T. Spence (San Francisco: W. H. Freeman, 1983), 10–74. J. T. Spence, R. L. Helmreich, and R. S. Pred, "Impatience versus Achievement Strivings in the Type A Pattern: Differential Effects on Students' Health and Academic Achievement," *Journal of Applied Psychology* 72 (1987): 522–528.
6. S. Gregorich, R. L. Helmreich, J. A. Wilhelm, and T. R. Chidester, "Personality Based Clusters as Predictors of Aviator Attitudes and Performance," in *Proceedings of the Fifth International Symposium on Aviation Psychology*, edited by R. S. Jenson (Columbus: Ohio State University, 1989), 686–691. R. L. Helmreich, W. E. Beane, G. W. Lucker, and J. T. Spence, "Achievement Motivation and Scientific Attainment," *Personality and Social Psychology Bulletin* 4 (1978): 222–226. R. L. Helmreich, J. T. Spence, W. E. Beane, G. W. Lucker, and K. A. Matthews, "Making It in Academic Psychology: Demographic and Personality Correlates of Attainment," *Journal of Personality and Social Psychology* 39 (1980): 896–908. R. L. Helmreich, J. A. Wilhelm, and T. R. Runge, "Study of Achievement Motivation and Job Satisfaction in Supertanker Crews," California Maritime Academy, Vallejo, California Seventh Annual Marine Industry Symposium: Motivation Organization and Satisfaction Aboard Ship, 1981.
7. P. T. Costa and R. R. McCrae, "Major Contributions to Personality Psychology," in *Hans Eysenck: Consensus and Controversy*, edited by S. Modgil and C. Modgil (Philadelphia: Falmer Press, 1986), 63–72.
8. P. T. Costa and R. R. McCrae, "Objective Personality Assessment," in *The Clinical Psychology of Aging*, edited by M. Storandt, I. C. Siegler, and M. F. Elias (New York: Plenum Press, 1978), 119–143.
9. L. R. Goldberg, "Language and Individual Differences: The Search for Uni-

versals in Personality Lexicons," in *Review of Personality and Social Psychology,* vol. 2, edited by L. Wheeler (Beverly Hills, Calif.: Sage, 1981), 141–165.

10. P. T. Costa and R. R. McCrae, *The NEO Personality Inventory Manual* (Odessa, Fla.: Psychological Assessment Resources, 1985), 1.

11. P. T. Costa and R. R. McCrae, "Concurrent Validation after 20 Years: Implications of Personality Stability for Its Assessment," in *Advances in Personality and Assessment,* vol. 4, edited by J. N. Butcher and C. D. Spielberger (Hillsdale, N.J.: Erlbaum, 1985), 31–54. L. R. Goldberg, "An Alternative Description of Personality: The Big-five Factor Structure," *Journal of Personality and Social Psychology* 59 (1990): 1216–1229. R. R. McCrae and P. T. Costa, "Validation of the Five Factor Model of Personality across Instruments and Observers," *Journal of Personality and Social Research* 52 (1987): 81–90.

12. J. M. Digman and J. Inouye, "Further Specification of the Five Robust Factors of Personality," *Journal of Personality and Social Psychology* 50 (1986): 116–123. O. P. John, "The 'Big Five' Factor Taxonomy: Dimensions of Personality in the Natural Language and in Questionnaires," in *Handbook of Personality Theory and Research,* edited by L. A. Perrin (New York: Guilford, 1990), 66–100. R. R. McCrae and P. T. Costa, "Clinical Assessment Can Benefit from Recent Advances in Personality Psychology," *American Psychologist* 41 (1986): 1001–1003. R. R. McCrae and P. T. Costa, "More Reasons to Adopt the Five-factor Model," *American Psychologist* 44 (1989): 451–452. J. S. Wiggins and P. D. Trapnell, "Personality Structure: The Return of the Big Five," in *Handbook of Personality Psychology,* edited by S. R. Briggs, R. Hogan, and W. H. Jones (Orlando, Fla.: Academic Press, in press, 1993).

13. Helmreich and Spence, "The Work and Family Orientation Questionnaire." Spence and Helmreich, "Achievement-related Motives and Behavior." Spence, Helmreich, and Pred, "Impatience Versus Achievement Strivings in the Type A Pattern."

14. Helmreich, Wilhelm, and Runge, "Study of Achievement Motivation and Job Satisfaction in Supertanker Crews."

15. Helmreich, Beane, Lucker, and Spence, "Achieving Motivation and Scientific Attainment." Helmreich, Spence, Beane, Lucker, and Matthews, "Making It in Academic Psychology."

16. Helmreich, Wilhelm, and Runge, "Study of Achievement Motivation and Job Satisfaction in Supertanker Crews."

17. R. S. Pred, J. T. Spence, and R. L. Helmreich, "The Development of New Scales for the Jenkins Activity Survey Measure of the Type A Construct," *Social and Behavioral Science Documents* 16 ms. 2769 (1986): 51–52.

18. R. Diaz-Loving, R. Diaz-Guerrero, R. L. Helmreich, and J. T. Spence, "Comparacion transcultural y analysis psicometrico de una medida de rasgos masculinos (instrumentales) y femeninos (expresivos)," *Revista de la Asociacion Latinoamericana de Psicologia Social* 1 (1981): 3–37.

19. R. L. Helmreich, *Pilot Selection and Training* (Washington, D.C.: American Psychological Association, 1982).

20. K. A. Matthews, R. L. Helmreich, W. E. Beane, and G. W. Lucker, "Pattern A, Achievement-striving, and Scientific Merit: Does Pattern A Help or Hinder?" *Journal of Personality and Social Psychology* 39 (1980): 962–967.

21. J. Barling, S. D. Bluen, and V. Moss, "Type A Behavior and Marital Dissatis-

faction: Dissecting the Effects of Achievement Striving and Impatience/irritability," *Journal of Psychology* 124 (1990): 311–319. S. D. Bluen, J. Barling, and W. Burns, "Predicting Sales Performance, Job Satisfaction and Depression by the Achievement Striving and Impatience/irritability Dimensions of Type A Behavior," *Journal of Applied Psychology* 75 (1990): 212–216.

22. T. R. Chidester, "Trends and Individual Differences in Response to Short Haul Flight Operations" (Ph.D. diss., University of Texas, Austin, 1988).

23. R. L. Helmreich, "What Changes and What Endures: The Capabilities and Limitations of Training and Selection." Paper presented at the *Aer Lingus/Irish Airline Pilots Association Flight Symposium,* edited by N. Johnston (Dublin: 1983).

24. S. Gregorich, R. L. Helmreich, and J. A. Wilhelm, *The Structure of Cockpit Management Attitudes,* NASA/University of Texas Technical Memo 89-1 (Austin, Texas: NASA/UT Project, 1989). J. R. Hackman and R. L. Helmreich, "Assessing the Behavior and Performance of Teams in Organizations: The Case of Air Transport Crews," in *Assessment for Decision,* edited by D. Peterson and D. B. Fishman (New Brunswick, N.J.: Rutgers University Press, 1987), L283–313. R. L. Helmreich, "Social Psychology on the Flight Deck," in *Proceedings of the NASA Workshop on Resource Management Training for Airline Flight Crews,* NASA CP-2120, edited by R. S. Jensen (NASA Ames Research Center: NASA-ARC, 1979). R. L. Helmreich, "Exploring Flight Crew Behavior," *Social Behavior* 21 (1987): 583–589. R. L. Helmreich, "Cockpit Management Attitudes," *Human Factors* 26 (1987): 583–589. R. L. Helmreich, "Theory Underlying CRM Training: Psychological Issues in Flightcrew Performance and Crew Coordination," in *Cockpit Resource Management Training: Proceedings of the NASA/MAC Workshop,* edited by H. W. Orlady and H. C. Foushee, CP2455 (San Francisco: NASA Ames Research Center, 1987), 14–21. R. L. Helmreich, T. R. Chidester, H. C. Foushee, S. E. Gregorich, and J. A. Wilhelm, *How Effective Is Cockpit Resource Management Training? Issues in Evaluating the Impact of Programs to Enhance Crew Coordination,* NASA/University of Texas Technical Report 89-2 (1989). R. L. Helmreich, H. C. Foushee, R. Benson, and R. Russini, "Cockpit Management Attitudes: Exploring the Attitude Performance Linkage," *Aviation, Space and Environmental Medicine* 57 (1986): 1198–1200. R. L. Helmreich and J. A. Wilhelm, "When Training Boomerangs: Negative Outcomes Associated with Cockpit Resource Management Training," in *Proceedings of the Fifth International Symposium on Aviation Psychology* (Columbus: Ohio State University, 1989).

25. T. R. Chidester and H. C. Foushee, "Leader Personality and Crew Effectiveness: Factors Influencing Performance in Full-mission Air Transport Simulation." Paper presented at the 66th Meeting of the Aerospace Medical Panel on Human Stress: Situations in Aerospace Operations (The Hague: Advisory Group for Aerospace Research and Development, 1988).

26. H. C. Foushee and R. L. Helmreich, "Group Interactions and Flightcrew Performance," in *Human Factors in Modern Aviation,* edited by E. Weiner and J. Nagel (New York: Academic Press, 1988), 189–277. R. L. Helmreich, J. A. Wilhelm, and S. E. Gregorich, *Notes on the Concept of LOFT: An Agenda for Research,* NASA/University of Texas Technical Manual 88-1 (Austin, Texas: NASA/UT Project, 1988).

27. J. T. Spence, R. L. Helmreich, and C. K. Holahan, "Negative and Positive

Components of Psychological Masculinity and Femininity and Their Relationships to Self Reports of Neurotic and Acting-out Behaviors," *Journal of Personality and Social Psychology* 37 (1979): 1673–1644.

28. J. T. Spence and R. L. Helmreich, *Masculinity and Femininity: Their Psychological Dimensions, Correlates and Antecedents* (Austin: University of Texas Press, 1978).

29. Costa and McCrae, *The NEO Personality Inventory Manual.*

30. J. E. Nardini, R. Herrmann, and J. E. Rasmussen, "Navy Psychiatric Assessment Program in the Antarctic," *American Journal of Psychiatry* 119 (1962): 97–105.

31. Y. Amir, Y. Kovarsky, and S. Sharron, "Peer Nominations as Predictors of Multi-stage Promotions in a Ramified Organization," *Journal of Applied Psychology* 54 (1970): 462–469. R. G. Downey and P. J. Duffy, *Review of Peer Evaluation Research,* Technical Paper No. 342 (Fort Knox: U.S. Army Research Institute for the Behavioral and Social Sciences, 1978). M. D. Mumford, "Social Comparison Theory and the Evaluation of Peer Evaluations: A Review and Some Applied Implications," *Personnel Psychology* 36 (1983): 867–881. R. M. Rose, "Predictors of Psychopathology in Air Traffic Controllers," *Psychiatric Annals* 12 (1982): 925–933.

32. R. L. Helmreich and P. A. Santy, "Preliminary Psychological Data Analysis of 1989 Astronaut Selection" (unpublished presentation to NASA Life Sciences Management, 1989).

33. T. J. McFadden and R. L. Helmreich, "Personality Predictors of Performance in Astronauts" (presentation to the Aerospace Medical Association, Miami, 1992). T. J. McFadden, R. L. Helmreich, R. M. Rose, and L. Fogg, "Predicting Astronaut Effectiveness: A Multivariate Approach" (in press, 1993). R. M. Rose, L. Fogg, R. L. Helmreich, and T. McFadden,"Psychological Predictors of Astronaut Effectiveness" (in press, 1993). R. M. Rose, R. L. Helmreich, L. Fogg, and T. J. McFadden, "Assessments of Astronaut Effectiveness" (in press, 1993).

34. Rose, Helmreich, Fogg, and McFadden, "Assessments of Astronaut Effectiveness."

35. McFadden, Helmreich, Rose, and Fogg, "Predicting Astronaut Effectiveness."

36. Rose, Fogg, Helmreich, and McFadden, "Psychological Predictors of Astronaut Effectiveness."

37. Ibid.

# 7

# An Invitation to Orbit: Part 1—The European Space Program

[With the development of its own space plans and an acceptance of the U.S. proposal] . . . Europe both committed itself to an ambitious future space program of its own and accepted, subject to the negotiation of acceptable terms and conditions, the U.S. invitation to participate in what ESA insisted on describing as an "international space station," rather than a U.S. station with foreign participation. (The differences in these two characterizations were more than semantic, since the degree of non-U.S. participation and the consequent share in the content and control of station development and operations were unsettled issues as far as Europe and other participants were concerned, while within the United States a decision that America must have the dominant station role had already been made.)

—John Logsdon[1]

## THE EUROPEAN SPACE AGENCY

After the *Apollo* program ground to a halt, the United States extended an invitation to Europe to participate in future U.S. space efforts. The Space Shuttle was on the drawing board, and the United States suggested several alternative cooperative endeavors related to Shuttle for the Europeans to consider, the first being joint ventures between U.S. and European companies. A second option was for the European countries to take on the development of an orbital transfer vehicle, while a third was that the Europeans develop a research module for the Shuttle payload bay.

Many European leaders were unhappy with what they visualized as a "secondary" role in such cooperation, and there was considerable dissatisfaction with the three limited options.[2] In spite of this, Europe spent a large amount of money studying which option might be most advantageous for it, but before the Europeans could even decide which approach to take, the U.S. position changed and, "on a 'take it or leave it basis,' [all that was offered to the Europeans] was the development of the Research and Applications Module."[3] This module was later named Spacelab in 1973.

Some European countries, particularly France, questioned the value of participating in expensive cooperative projects with the United States, preferring instead to focus on developing European capabilities for space exploration. The major advantage seen by those countries who desired to participate with the United States was the chance it offered to gain experience in aerospace system engineering and technology. A compromise position both to cooperate with the United States and to develop a strong European space effort was finally agreed on, and the European Space Agency (ESA) was created in 1975 to pool resources for both ventures.

ESA was made up of eleven western European countries: the United Kingdom, France, West Germany, Italy, the Netherlands, Belgium, Sweden, Denmark, Switzerland, Spain, and Ireland. Ireland, along with Austria, Norway, and Canada, initially had observer status in the fledgling organization. France and West Germany already had strong interests in space exploration and technology, and they each intended to further develop their own space technological capabilities. The French Space Agency (CNES) and the German Space Agency (DLR) both planned to have a separate cadre of national astronauts who would be available to fly with either the Americans or the Russians.

In the pact that led to the formation of ESA, each member country's autonomy was recognized, and it was understood that each could use its own methods to select astronaut candidates for ESA missions. While they were urged to have their selection criteria compatible with the criteria developed by ESA, only at the point when national candidates were put forward for consideration by ESA would the ESA selection criteria be used. It should be remembered that each of the European member countries had its own unique language, culture, and firm sense of national pride. Bringing them together to agree on something as complex as astronaut selection entailed a degree of diplomacy and patience unparalleled in modern times. From its very beginning, however, ESA recognized the cultural diversity of its members and, particularly with regard to astronaut selection, has tried to make sure that many viewpoints and philosophies are included in selection procedures.

In 1977, ESA requested that the German Space Agency (now called the DLR) Institute for Flight Medicine, Department of Aviation and Space Psychology, in Hamburg, West Germany, take the lead to establish psychological selection criteria for the first European payload specialists who would fly in Spacelab as part of the U.S. Shuttle program. The main field of work at the institute at that time was the psychological selection of operators for complex technical systems. This included the selection of pilots, flight engineers, and dispatchers for European airlines; air traffic controllers; and saturation divers.

## EUROPE BEGINS TO SELECT ASTRONAUTS

Spacelab was scheduled to fly in the Shuttle payload bay on STS-9, and, with it, would go one or more European payload specialists, who would perform scientific and technical experiments in the microgravity of space. The first European candidates for the position of payload specialist were due to be selected in two steps. The first step was a series of national selections in each of the member countries of ESA. The members could propose up to five candidates each by September 1977. The proposed candidates would then enter the ESA selection stage, which would be concluded by the end of 1977.

A total of fifty-two national candidates were put forward, and from that group, four scientists were eventually selected. Dr. Klaus-Martin Goeters, a DLR behavioral scientist involved in the ESA psychological selection process, later wrote: "The philosophy for the psychological selection of payload specialists is that this group consists of scientists who should qualify as operators of complex technical systems. So this group is comparable with other high level operators (e.g. pilots or air traffic controllers)."[4]

Since the DLR had been asked to develop a list of psychological criteria for the ESA selection, it is hardly surprising that the criteria adopted by ESA were almost identical to the specific German criteria. The Institute for Flight Medicine already had a set of psychological selection criteria for pilots and navigators. These criteria were measured by objective psychometric and cognitive test methods that had been developed by scientists at the institute. The ESA criteria for the 1977 selection cycle (and the German psychological criteria) are listed in Table 7.1.

Parallel to the German activity, the French, who were the most vociferous in their reluctance to participate with the United States, set up their own set of psychological criteria and selected their own astronauts, which would independently fly on U.S. and Soviet space missions. Table 7.2 lists the French psychological selection criteria. One major difference in the two countries' approaches that can be seen from examining the two lists is that the French scientists were highly concerned with *group* behavior and *group* adaptation in selecting individual astronauts. In this, they were very similar to their American counterparts in the NASA In-House Working Group.

Table 7.1
ESA and German Psychological Selection Criteria

| Performance Capabilities | Personality Traits |
|---|---|
| English proficiency | Motivation |
| Physico-technical knowledge and comprehension | Mobility |
| Mathematico-logical thinking | Rigidity |
| Memory | Vitality |
| Perception and attention | Extroversion |
| Spatial orientation | Personal warmth |
| Psychomotor coordination and dexterity | Aggressiveness |
| Multiple task capacity | Dominance |
| | Emotional stability |

## SELECTION OF GERMAN ASTRONAUTS – 1977

Of the 695 West Germans who submitted applications for the position of ESA payload specialist, 375 were immediately rejected because they did not meet the educational requirement of completion of a technical or scientific course of study at the university level. The remaining applicants were screened for scientific and technical suitability on the basis of information they provided in their applications. Of the remaining 320 applicants, only 164 were found to be basically qualified, all of whom were invited to participate in the initial psychological screening in Hamburg.[5]

The German selection highlights a fundamental difference in selection philosophies between the Europeans and the Americans. This difference arises from the high importance that is placed on psychological factors by the former, as compared with the latter. As we have seen, NASA had no formal psychological select-in evaluation and only recently permitted research, the results of which might convince NASA managers of the importance of implementing formal psychological testing to identify the "best" astronaut applicants (see Chapter 6). This is not the case in the German, French, or ESA selections. A formal psychological select-in evaluation is done on all applicants before they are even evaluated medically or psychiatrically. Thus, the sequence of events in the German selection was as follows: first, a psychological select-in evaluation; second, the medical evaluation, which includes a comprehensive psychiatric assessment; and third,

Table 7.2
**French Psychological Selection Criteria**

*Cognitive*

- general intelligence
- verbal ability (learning new languages)
- spatial ability
- memory
- attention
- decision-making

*Psychomotor*

- coordination
- tracking

*Personality and Motivation*

- equilibrated personality and flexible defense mechanisms
- strong motivation, need of achievement

*Group Adaptation*

- good communication abilities
- sense of leadership
- team orientation

*Public Relations*

- good contact, attitudes, and behaviors with the media

a job evaluation and interview. In the U.S. program, only the second and third steps are currently part of routine operations.

Of the 164 applicants invited to take the psychological tests, only 103 accepted the invitation and presented themselves to the institute in Hamburg for testing. Most of the candidates held advanced degrees in physics or engineering.

The tests administered to these 103 candidates, plus an additional 9 Austrian national candidates who were evaluated at the same time, are listed in Table 7.3.[6] There are no similarities between the German battery of psychological tests and any psychological tests used in the *Mercury, Gemini,* or *Apollo* programs. Since the NASA In-House Working Group on Psychiatric and Psychological Selection of Astronauts was not formed until 1988, there was no possibility of coordinating the testing procedures. This is unfortu-

Table 7.3
Psychological Tests Used in the German Selection

| Test Area | Test Method | Reference |
|---|---|---|
| Biographical information | Biographical Data | ESA |
| Knowledge of English language | Verbal meaning test | 6 |
| | English test | DFVLR/DLH |
| Technical knowledge | Technical comprehension | DFVLR/DLH |
| | technical quiz | DFVLR/DLH |
| Mathematico-logical thinking | Mathematics test | DFVLR/DLH |
| | IST Calculation tasks | 7 |
| | LPS 3 + 4 | 8 |
| Calculation capacity | Mental arithmetic | DFVLR/DLH |
| Memory | Memory test | 9 |
| | Clearance Test | DFVLR/DLH |
| | HAWIE-ZN | 10 |
| Concentration, perception and attention | KBT | 9 |
| | Letter cancellation test | DFVLR/DLH |

| | | |
|---|---|---|
| | APG | 11 |
| | LPS 10 | 8 |
| | LPS 11 + 12 | 8 |
| Spatial orientation | Way-figures test | 9 |
| | EGF | DFVLR/DLH |
| Psychomotorial traits | BET 9 + 10 | 12 |
| | BET 11 + 12 | 12 |
| | Two-handed coordination | 13 |
| | Tracking simulator | 14 |
| *Personality Traits* | | |
| Motivation | | |
| Flexibility | | |
| Mobility | Temperament Structure Scale (TSS) | 15 |
| Vitality | | |
| Spoilt ("being pampered") | | |
| Extroversion | | |
| Warmth/Coldness | | |
| Aggressiveness | | |
| Dominance | | |
| Emotional stability | | |

nate because comparison of the German and American data would have been useful. The first European astronaut selection occurred during a time when NASA had particularly downplayed the importance of psychological issues in selection.

### Results of the German Selection Process

The German candidates were evaluated in three basic areas by the institute psychologists: (1) biographical data, (2) performance tests, and (3) personality tests.

*Biographical Data.* All the applicants were male with an average age of 35.2 years. Twenty-five percent were single, 68 percent were married, and 7 percent were divorced. Thirty-one percent of the applicants held a private pilot's license. The professions of the applicants were primarily physicists, engineers, chemists, geologists, and medical doctors. Fourteen percent listed other occupations, including astronomer, biochemist, biologist, flight engineer, mathematician, meteorologist, mineralogist, and pilot.

*Performance Tests.* The German psychological selection tests emphasized cognitive and psychomotor functioning as a method of prescreening applicants. Redundant testing (meaning more than one test was administered for each function assessed) was performed in each of the performance fields of interest. The Germans were less interested in very high scores and, instead, gave preference to those candidates who demonstrated an ability to perform adequately on all tests. Test values were expected to be within the range of normal, and any weakness—as demonstrated by low score—in any performance field resulted in the rejection of the candidate. As it turned out, many candidates who were disqualified displayed weaknesses in more than one field. The primary causes of rejection were in the areas of concentration and perception (43% of candidates rejected), memory capacity (27% rejected), spatial orientation (23% rejected), and psychomotor functioning (10% rejected).

*Personality Tests.* The personality test used by the Germans is called the Temperament Structure Scale (TSS) and was developed at the institute by H. Kirsch, Klaus-Martin Goeters, and R. Ewe.[7] Personality traits assessed by this test are achievement motivation, emotional instability, rigidity, extroversion-introversion, aggressiveness, vitality, dominance, empathy, spoiltness, mobility, and openness. As can be seen these scales correspond exactly to the personality criteria identified by the DLR psychologists for select-in screening of payload specialist applicants. Appendix 7 contains further information on the TSS.

Scores on each personality scale were compared to a predetermined tolerance range in STANINE units (standard nine, i.e., a range of 0–9) defined for each characteristic. Applicants were rejected if they scored outside the range on any characteristic—either too high or too low. Table 7.4 lists the percentages of candidates who were disqualified because of personality

Table 7.4
Personality Traits as a Cause for Rejection

| Personality Trait | % Rejected Due to Extreme Ratings |
|---|---|
| Achievement motivation | 22 |
| Flexibility | 12 |
| Mobility | 12 |
| Spoilt | 10 |
| Extroversion | 15 |
| Warmth/Coldness | 6 |
| Hostility | 16 |
| Dominance | 16 |
| Emotional Stability | 22 |
| Vitality | 29 |

scores. As can be seen, many applicants failed to score in the predetermined ranges on the psychological tests.

After the completion of the performance and personality testing, only 30 candidates from the original 112 remained. These candidates were described by Goeters and his colleagues as "introverted, undemanding, enterprising, and highly motivated."[8]

Goeters and his colleagues then compared the cognitive and personality test scores of the payload specialist applicants with a group of pilot and flight engineer applicants. On most tests, the payload specialist applicants scored significantly higher than did the aviation applicants. In the area of personality, the payload specialist applicants were significantly more highly motivated, less spoiled (pampered), and less extroverted.

## SELECTION OF FRENCH ASTRONAUTS–1977 AND 1985

French astronaut (or "spationaut") candidates were of two types. The first type was called space craft operator, and those selected for this position would fly on the Soviet space station. The second type was a payload specialist who would fly on the American Shuttle. The same psychological criteria were used in evaluating each type and are listed in Table 7.2.

French applicants were prescreened for educational and professional background, age, sex, and initial medical information. Bernard Comet and Guy Veron, two of the French scientists involved in the selection, summarized the process in a series of reports.[9] The specific psychometric tests used included the following:

- *Cattell's 16-Personality Factor (16-PF).* This American test was used in the Soviet space program and is discussed more fully in Chapter 9. It assesses aspects of normal adult personality.
- *1-P9 (CERPAIR).* This is a computerized French personality test designed to assess normal personality (rather than psychopathology). It is used extensively in recruitment, selection, and retention in many personnel settings (including aviation). The test has three parts: a general personality inventory, a questionnaire on "decisional" personality, and a leadership or administrative "style" questionnaire. The specific personality traits that are evaluated by the tests are activation, anxiety, depressive tendencies, fatigue, histrionics, neuroticism, obsessiveness, sensitivity, somatiation, and tendency toward isolation.
- *Rorschach.* This projective test was described in detail in Appendix 1.
- *Group Selection.* This test was developed by Jean Rivolier and his associates at the University of Reims.[10] It involves the observation of small groups of applicants to determine their interactions for decision making, social abilities, leadership, and so on. It will be described in more detail later in this chapter when the European Space and Technology Center (ESTEC) study is discussed.
- *Motivation.* Several questionnaires were used to assess general knowledge about space, aviation, and technical matters.
- *Cognitive Functioning.* Cognitive functioning was assessed using a variety of tests of visual and auditory memory, reading comprehension, attention, spatial ability, verbal ability, logic, and reasoning ability.
- *Psychomotor Tests.* These tests included dual and multiple tracking tasks, as well as traditional pilot psychomotor tests (e.g., broomstick and rudder tests).

## SELECTION OF GERMAN ASTRONAUTS—1986–1987

The 1977 selection resulted in the identification of German candidates who flew on the Deutsche-1 (D1) Spacelab mission in late 1985. A second German selection was undertaken in 1986 for candidates for the Deutsche-2 (D2) Shuttle mission scheduled to fly sometime in the nineties.

The 1986 selection proceeded along identical lines with the 1977 German selection. The only tests used in the D1 evaluations and eliminated from the D2 evaluations were several commercially available tests, which the German behavioral scientists decided did not discriminate among applicants. Thus, only the internally developed DLR tests were used for the personality and performance evaluations. The D2 selection resulted in the selection of six German astronauts who were qualified to fly on both the U.S. Shuttle and the Russian *Mir* Space Station.

## THE EUROPEANS JOIN THE NASA WORKING GROUP

The DLR classifies its psychological tests as confidential and proprietary. As a result, the tests not generally available for use by other scientists. Since the tests are proprietary and have never been freely discussed in the scientific psychological literature (except by the test developers), a serious question regarding their reliability and validity arises. Goeters and his colleagues have repeatedly demonstrated test reliability in the populations for which the institute uses the tests (e.g., pilots and astronauts), but this has not been confirmed independently. The validity of the tests also remains unknown, since Goeters and his co-investigators have not published validity studies, despite encouragement from American and European colleagues that such studies need to be done.

Goeters was an active participant in the NASA In-House Working Group. In November 1988, he presented a comprehensive summary of the criteria and methodology used at the institute in Hamburg.[11] His presentation emphasized that the tests included in the DLR battery have been used successfully at the institute and have been shown to provide reliable differentiation among subjects. Although he maintained that they show "general validity" for operational work environments, no specific data were presented to support this assertion. Goeters discussed the institute's policy of redundant measurement for all traits and the conversion of test scores to the STANINE format. STANINE scores of 5 and above are considered acceptable, while those below 5 are considered critical and may result in rejection of a candidate.

Several phases of psychological testing occur that reflect the difference in philosophy about psychological issues between the Europeans and the Americans. The first phase includes tests of English, aptitude, and personality, and those who make it through this phase are invited to return for the second phase, which includes tests of vigilance and psychomotor performance. All of these tests are done prior to the medical selection, since the Germans have found it cost-effective to put only candidates who have already passed several levels of evaluation through the costly medical examinations. (NASA could learn a lesson from this procedure.) Candidates in the European selection are free to disqualify themselves for any reason and may reschedule the testing; but once the tests are taken, individuals cannot retake them or even re-apply for consideration. (This would be a problem in the United States because of legal considerations.)

Dr. Goeters' presentation elicited some controversy among other members of the Working Group. The two U.S. astronauts (Carter and Baker) believed that too much emphasis was being placed on psychomotor and cognitive tests to select-in applicants. Were there data to justify such an approach? In other words, did psychomotor skill predict success as either a pilot or an astronaut? In the view of the astronauts, the psychomotor skills necessary to function effectively as an astronaut can be acquired during astronaut training. All that is really needed is an aptitude for acquiring the skills. Their primary concern was that potentially excellent candidates from a personality and motivational perspective would be screened out by such tests, which, in their opinion, had little to do with the actual skills necessary to succeed as an astronaut.

Another issue involved the legal implications if the German methodology were implemented in the United States. For example, German candidates were prohibited from re-applying. In the United States, such a prohibition might result in extensive litigation over an individual's "right" to be an astronaut. (As a case in point, an applicant who had a history of alcoholism, treated by hospitalization, believed he had the right to become an astronaut now that he had been "dry" for several years and threatened to sue when the prescreening procedures at NASA eliminated him from even obtaining an interview.) Also, in the U.S. program, many candidates are chosen to come to the medical and psychiatric screening as a way to "look them over." It is expected that those who make it through the medical screening and are interviewed at NASA, but are not selected, will re-apply. Some of those candidates are even offered jobs at the Johnson Space Center in areas related to their scientific or engineering expertise. Some individuals in the U.S. program are not selected until their third or fourth interview, and some are given three or more interviews and then are still not selected.

Goeters also discussed the difficulties of dealing with the culturally diverse countries of ESA. The challenge to his group when they first developed their psychological selection procedures was to design so-called culture-fair tests—tests whose results do not depend on language or cultural experience. He pointed out that the Europeans have always taken a more serious view of the psychological aspects of space flight than have their American counterparts:

> Since the psychological evaluation of the earlier U.S. astronauts made no substantial contribution to the selection process, much ground had been gained in the USA by the view that the psychological assessment of astronaut applicants was basically only of subordinate importance or that one could even forego it altogether. This attitude was particularly prevalent as regards the payload specialists, who were often regarded as "second-class astronauts". We have never been able to share this point of view, and have always been of the opinion that in the selection of payload specialists one would have to expect a large number of unsuitable applicants, because in this case one would not be concerned with such a strongly preselected group of persons as was the case of the earlier US astronauts.[12]

It seemed clear from Dr. Goeters' presentation to the Working Group in 1988 that the DLR tests of performance and personality at the time of selection constitute a select-in battery. Again, the issue of validating such tests against real or simulated operational performance was brought up. It was universally felt by the Working Group that the German approach was thoughtful, comprehensive, and laudable. But since it had never been operationally validated, there was no possibility that the methods could be adopted in the U.S. Space Program. Robert Helmreich urged the German behavioral scientists to permit the use of some parts of their battery in the validation study that was then being planned in the United States, and Goeters graciously agreed.

## THE EUROPEAN ASTRONAUT PSYCHOLOGICAL CRITERIA WORKING GROUP

Partly due to the NASA Working Group's attempt to more clearly define psychological selection criteria for space missions, ESA in 1990 appointed a European Astronauts Psychological Criteria Working Group (EAPCWG), analogous to NASA's group. The members of the EAPCWG were Dr. Jan Smit (chairman) from the Netherlands, Dr. Dietrich Manzey from the DLR in Germany (and a colleague of Goeters), Dr. Antonio Peri from Italy, and Dr. Guy Veron from France. Dr. Franco Rossitto was the ESA representative who supervised the work of the group. During the EAPCWG's first meeting, Goeters presented a summary of the U.S. Working Group's efforts, as well as a review of his own department's work in the previous European astronaut selections.

The EAPCWG re-evaluated in detail the 1977 criteria used by ESA and all the national psychological criteria used by ESA member countries. Additionally, the psychological selection criteria developed by the NASA Working Group were discussed along with Soviet psychological requirements for the *Mir* Space Station. A complete review of any relevant psychological selection criteria used for analogous professions or environments (such as Antarctica) was also done. The goal of the group was to derive a set of standardized psychological criteria based on all the scientific literature. The criteria would then be used to identify suitable psychological tests to be used in forthcoming European astronaut selections. In essence, the work of the EAPCWG was identical to that of the U.S. group.

But although the process was the same, the results differed. This is actually easy to understand. From a scientific perspective, the European astronauts would be very few in number and would fly in space only infrequently at best. The dilemma faced was this: Do we do a prospective study that does not have a chance of providing enough data for validation in the foreseeable future, or do we do the best possible job in selecting our relatively few astronauts with the knowledge available right now and hope for the best? The EAPCWG was fully aware of the select-in research being conducted by Robert Rose and Robert Helmreich in the U.S. program. Their research project was possible only because NASA did not intend to formally assess the psychological characteristics of astronaut applicants or use psychological information to determine which candidates to select. But the European group was under some administrative pressure to develop psychological procedures that could be used to identify the best-qualified applicants now.

The EAPCWG summary[13] identified two major psychological factors that would be evaluated in all future European astronaut candidates: operational aptitudes and personality factors. Methodological recommendations in the summary about how the psychological assessment process was to be conducted noted that any psychological tests used had to be culture-fair, measured in a redundant manner, and reliable; that the constructs tested had to have the ability to differentiate among individual applicants; and that the tests used had to have adequate norms available for comparison.

### Operational Aptitudes

As previously mentioned, the European philosophy of selection for pilots and astronauts has a noticeable bias for cognitive and psychomotor capabilities, and the complexity of the tasks required of astronauts in the space environment perhaps justifies the formal testing of such capabilities. The hypothesis underlying this philosophy is that those individuals who obtain high scores on the cognitive and the psychomotor tests will have improved performance in the space environment. It is a reasonable hypothesis, but only a hypothesis, since no one has yet demonstrated that this is so.

The specific operational aptitudes that are tested for include logical reasoning, mental arithmetic, visual and auditory memory function, attention, perception, spatial comprehension, and psychomotor coordination. A complex multiple-task test was included to demonstrate multiple-task abilities.

### Personality Factors

The personality factors suggested by the EAPCWG as relevant for the space environment include the following:

- *Motivation* is the disposition to develop, direct, regulate, and maintain energy in order to reach an objective despite obstacles or difficulties. Relevant aspects of motivation are achievement needs (comparable to achievement striving) and occupational motivation (comparable to goal-oriented behavior).

- *Social capability* is the capability to develop, maintain, and enjoy contacts and relations with other persons. It is manifested by a team orientation, sensitivity, and tolerance of individual and cultural differences. The extremes of introversion (a need to be alone) and extroversion (a need to be with other people) should be balanced. Dominance (i.e., leadership and ability to take on responsibility) is important, as is empathy (the ability to relate to the experiences and emotions of others). Another psychological characteristic included in this cluster is aggression, which must be controlled in group situations.
- *Stress coping* is the capability of coping with external and/or internal stressors in such a way that efforts can be effectively directed in order to maintain control and reach the objective. Stress tolerance requires emotional stability, a readiness to bear privations, flexibility, and good coping and stress management skills.
- *Work orientation* reflects a positive attitude toward work and occupational demands. The specific indicators of this psychological characteristic are vitality (a positive attitude toward physical activities), mobility (a readiness to accept and practice new activities and to take risks), readiness to acquire new knowledge and skills, and acceptance of responsibility.

The EAPCWG also included the psychological characteristics of loyalty, ethical integrity, and sense of humor, but did not suggest how these characteristics might be measured, except by biographical history. Table 7.5 compares the final EAPCWG psychological select-in criteria with those of the NASA Working Group.

## PSYCHOLOGICAL TESTING OF ASTRONAUT CANDIDATES FOR COLUMBUS MISSIONS

The DLR Institute of Aerospace Medicine was again given the task in 1990 of testing the EAPCWG psychological criteria and procedures. They, in turn, requested experts from the University of Reims in France—Drs. Jean Rivolier and Claude Bachelard—and from the University of Bergen—Dr. Holger Ursin—to assist them in the project. These researchers are well known for their work in the psychological evaluation of personnel for Antarctic and undersea habitats. The inclusion of Drs. Ursin and Rivolier was an attempt by ESA to include other perspectives on the problem, rather than utilizing only the German approach. Both Dr. Ursin and Dr. Rivolier believed strongly that the German approach left much to be desired and, in fact, agreed with their American colleagues about the lack of scientific validation data to support the German tests. While these tests are culture-fair, and apparently very reliable, their construct validity is questionable. Additionally, the German battery does not include tests of group compatibility or functioning; nor does it include non-objective test measures (i.e., projective measures).

Thus, it was decided that the three groups of behavioral scientists (German, French, and Norwegian) would each develop psychological testing

Table 7.5
Comparison of American and European Personality Criteria

| American Criteria | European Criteria |
| --- | --- |
| *Motivation* | *Motivation* |
| Achievement/Goal-oriented | Need of Achievement |
| Hard-working/Self-starting | Occupational Motivation |
| Mastery | |
| Persistence | |
| Optimism | |
| Healthy Sense of Competition | |
| No Excessive Extrinsic Motivation | |
| Capacity to Tolerate Boredom | |
| Mission Orientation | |
| Healthy Risk-taking Behavior | |
| *Sensitivity to Self and Others* | *Social Capability* |
| Emotional Maturity | Extroversion/Introversion |
| Self-esteem | Dominance |
| Expressivity | Empathy |

| | |
|---|---|
| Appropriate Assertiveness | Aggression |
| Ability to form Stable and Quality Interpersonal Relationships | |
| Insight and Self-awareness | |
| Cultural Sensitivity | |
| *Aptitude for Job* | *Work Orientation* |
| Intelligence and Technical Ability | Vitality |
| History of Professional Success | Mobility |
| Flexibility | Readiness to Acquire New Skills |
| Adaptability | Acceptance of Responsibility |
| Team Player | |
| Ability to Represent NASA | |
| Stress/Discomfort Tolerance | *Stress Coping* |
| Ability to Function in Danger | Emotional Stability |
| Ability to Compartmentalize | Flexibility |
| Ability to Tolerate Separation from Loved Ones | Readiness to Bear Privations |
| Ability to Tolerate Isolation | Stress Management |
| *Sense of Humor* | *Sense of Humor* |
| | *Loyalty* |
| | *Ethical Integrity* |

methods based on their own research. The German approach has already been discussed in some detail in previous sections, and the following sections will outline the French and Norwegian contributions.

### Tests for Dynamic Group Assessment

The French, as we have noted previously, were particularly concerned with the group functioning of potential astronauts. Rivolier's group at the University of Reims proposed a set of tests to evaluate candidates in group situations. These tests included the following:

- *The Small Group Method for the Selection of Individuals.* Subjects are placed in small groups and given a specific set of tasks to accomplish. Members of the group are rated by trained observers on empathy, extroversion, aggressiveness, dominance, self-control, honesty, rigor, leadership, team behavior, and communication ability, and are also given an overall global rating.
- *The Matrix of Intra- and Interpersonal Processes in Groups (MIPG).* This test lists a number of interpersonal configurations, or how the subject sees himself and how he believes others see him.
- *The Videotaped Interpersonal Distance Measure (VIDM).* Candidates are videotaped in a group setting, interacting with others. Measurements are later taken of the actual physical distances between the individual and others he or she is interacting with.
- *The Test of Decisions (TD 9).* This test has nine scales: sincerity, impulsivity, taste for games of chance, interpretive projection, speed of decision, level of aspiration, dynamism, tolerance of the non-control of the environment, and risk-taking.
- *The Defense Mechanism Inventory (DMI).* This paper-and-pencil test assesses the major defense mechanisms used by the subject.

### Identification and Documentation of Test Methods for Psychodynamics under Stress

This portion of the test battery was supervised by Dr. Holger Ursin of the University of Bergen. The first test included was Helmreich's Personal Characteristics Inventory (PCI) (discussed in Chapter 6). The second test included was a projective test known as the Kragh Defense Mechanism Test (DMT). The DMT was under investigation by Ursin and his associates at the university because of its reported predictive ability in pilot selection. The test measured the ten major Freudian defense mechanisms: repression, isolation, denial, reaction formation, identification with the aggressor, introaggression (turning against the self), introjection of the opposite sex, introjection of another object, projection, and regression. The types of defenses used by an individual while taking the DMT are claimed to be based on stable characteristics and the individual's life experiences and to generalize to all situations where the individual requires a defense against anxiety.[14] This test has been extensively evaluated in Scandinavian pilot populations, and it has been shown that perceptual defense mechanisms—as tested by the DMT—predict flight performance. In particular, the test seems to predict fatal accidents and loss of aircraft. Pilot subjects who scored high on the DMT had a greater likelihood of being killed in plane crashes than did subjects with low scores.

Similar results have been reported with Danish attack divers, Norwegian parachute jumpers, and Norwegian divers.[15] Other studies done, however, have shown no correlations between the DMT and performance, and this was why Ursin and his group were re-evaluating the test and trying to standardize the collection and analysis of data from the DMT. The lack of standardized and reliable administration of the test may have contributed to the results reported outside of the Scandinavian countries. Or it may be that the defense mechanisms interfere with behavior only during extreme threats or anxiety. However, the existence of data relating defense mechanisms to pilot performance strongly supported the inclusion of the DMT in the European test battery and made it the only test used with data available on its validity as a predictor of performance. Miranda Olff and Ursin had also recently developed standardized procedures for administering and scoring the DMT.[16]

## THE ESTEC STUDY

In order to test all the psychological procedures proposed by the three groups of investigators, ESA sponsored a study using all the recommended tests. This study was conducted at the European Space and Technology Center (ESTEC) in Noordwijk, The Netherlands, in the summer of 1990. The primary goals of the study were to "shake down" all the procedures, as well as to determine the culture fairness and reliability of the proposed tests. Ninety-seven employees of ESA (fifteen women and eighty-two men) from the ESTEC volunteered to participate in the testing of the European Astronaut Selection Psychological Battery. The study took place from May 29 to July 4, 1990, with groups of about twenty-four volunteers taking the tests three days during the week.

The ESTEC volunteers were recruited because it was believed that, as a group, they would be very similar to individuals who might apply for the ESA astronaut positions, particularly since the volunteers were professionals in science or technology already working in the space area. More than 50 percent of the subjects were French or Italian nationals. The other 50 percent came from twelve different countries. Only two non-European subjects, both Canadian, took part in the study. The youngest participant in the study was twenty-four years old, and the oldest was thirty-eight.

About 25 percent of the volunteers intended to apply for the real astronaut selection the next year.

The ESTEC study was a monumental organization problem because data were collected on over fifty-four independent variables. Only the results of the personality assessment methods will be discussed.

### Personality Findings in the ESTEC Sample

The results of all the personality assessment methods (excluding the DMT) were factor analyzed, and seven final factors were extracted. Based on loadings greater than .40, the seven final factors and their interpretation by the researchers are listed below:

- Factor 1 is labeled aggressive behavior, since scales that assess aggression and conflict treatment load on this factor.
- Factor 2 is termed neuroticism; scales of anxiety, emotional instability, and irritability load on this factor.
- Factor 3 includes scales of achievement motivation and endurance.
- Factor 4 includes scales of risk taking and sensation seeking.
- Factor 5 includes scales related to empathy.
- Factor 6 represents the aspect of extroversion with several subscales. It can be interpreted as the level of social interaction. This factor included a questionnaire as well as real-life data (the assessment of the group-discussion behavior or the global rating score in the small-group method).
- Factor 7 seemed to represent a negative self-concept, which is indicated by the difference either between the self and the ideal concept or between the self-concept and the concept of how others see the subject. Factor 7 correlates positively with the DMI scale on turning against self and negatively with the TSS scale on dominance.

The extracted factors represent general personality components and include aspects of work orientation (factor 3, achievement motivation and endurance), stress resistance (factor 2, neuroticism; and factor 4, risk taking and sensation seeking), and social interaction (factor 6, level of social interaction; factor 1, aggressive behavior; and factor 5, empathy). Factor 7 was considered to be in the social interaction category, since self-concept applies directly to behavior in group or social settings.

This analysis merely demonstrates the significant personality factors found in this study population. It will be interesting to see if the factors later correlate with performance.

### The DMT Results

The DMT was not included in the above analysis, since it did not correlate with any of the items in the personality inventory. This lack of correlation with personality questionnaire data is not surprising, since the DMT measures of defense have meaning only in an anxiety-provoking context. In effect, the DMT (like all other projective tests) measures unconscious factors (i.e., those factors not within conscious control). As such, they represent an independent criterion, distinct from self-report questionnaires about personality, which may be separately evaluated for its relationship to later performance.

## Culture Fairness of the ESTEC Battery of Tests

A major goal of this study was to determine the culture fairness of the test methods, and, consequently, the ESTEC investigators analyzed the data for differences in test scores due simply to national origin and to difficulties with the English language.

In the Helmreich battery, three scales—negative communion, impatience/irritability, and work orientation—revealed significant effects for nationality. Verbal aggressivity, negative communion, and work orientation were also found to show a significant effect for English competency, while the other eight scales showed no effect. There were other minor problems with a number of the tests, which also used words with some cultural bias. However, in the final summary of the study, it was recommended that almost all the tests be included in the forthcoming European astronaut selection.

The study did not find significant national differences in the areas of operational aptitude and personality. However, English language skill varied considerably and played a role in the fifteen operational aptitude tests. Specifically, those subjects with better English skills performed better on these tests than did those with poor language skills. It was decided that these tests would be removed from the final ESA selection battery in order to ensure the use of culture-fair testing procedures. The reliability of the tests was found to be excellent (in most cases above 0.8).

The ESTEC study data collection took place over several weeks. What was most impressive in observing this study being done was the enthusiasm and commitment of the behavioral scientists from all three investigative groups (German, French, and Norwegian). While there were and still are significant differences among them in their philosophies of selection, a real camaraderie was obvious, as well as a real excitement for the task. Even more surprising was the degree of support these scientists received from ESA to pursue the development of the most comprehensive and reliable *psychological* test battery ever used for astronaut selection. (The numerous tests given during the Project *Mercury* selection were primarily for psychiatric or select-out screening.)

As Dr. Goeters earlier pointed out, the Europeans have, from the beginning of their own space exploration efforts, believed in the importance of the psychological evaluation to identify the best-qualified applicants. Since this represents a considerable philosophical difference compared to NASA, it was stimulating to see the data collection and later to participate in analyzing the data from this hallmark study. Only one part of the results has been presented, and the interested reader is referred to the original document written by Goeters and Christoph Fassbender, which contains all the data collected.[17]

It is important to remember that this study did not *validate* the psychological constructs used. Validation would require the determination of the relationship of the seven final factors analyzed from all the data collected with an objective method of assessing performance. This has not yet been done, but hopefully will be in future studies.

## RESULTS OF THE ESA ASTRONAUT SELECTION – 1991

In the summer of 1991, ESA began its search for ten astronaut candidates who would be specialists in science and space plane operations. The psychological evaluation of all candidates was again performed at the DLR Institute of Aerospace Medicine in Hamburg, Germany, under the direction of Fassbender and Goeters.[18] Also participating in the psychological evaluations were representatives of the Department of Biological and Medical Psychology of the University of Bergen (Holger Ursin's group).

The ESTEC study had only been a preliminary shakedown for the actual astronaut selection. The study established normative data and demonstrated to ESA the culture fairness of the test methods. It also enabled an estimation of the tests' reliability.

As in the 1977 selections, the first phase of the 1991 process was completed in the individual ESA member countries. After the national preselection, fifty-nine applicants were referred for the psychological evaluation in Hamburg. The procedures followed were almost identical to those for the ESTEC study done the prior year. The final results of this selection were initially presented at the Aerospace Medical Association Meeting in Miami in May 1992 by Dr. Dietrich Manzey.

Fifty-nine applicants from the member nations took part in the ESA selection. Each applicant underwent a medical assessment, which included a psychiatric evaluation and a technical interview. Most of the applicants for space plane operator (N = 12) were military pilots, and those for the position of laboratory specialist were scientists who had considerable experience in research or space-related activities.

Applicants were classified into three categories: Psychologically not recommended, recommended with reservation, or recommended. Of the 59 subjects, 13 were classified as not recommended, 20 as recommended with reservations, and 26 as recommended. Applicants not recommended were found not to meet the minimum psychological requirements. Applicants recommended with reservation were found to meet the minimum psychological requirements but had performance, personality, or biographical problems that needed to be considered individually. For example, a person's leadership style and team capability in combination with this individual's general personality profile was foreseen to pose a definite risk in some group constellations but to have a positive effect within a group of persons that were chosen to balance this. Or, for example, the current family situation may momentarily pose a definite risk for being an astronaut candidate, however, it is foreseen that the situation will be solved within a reasonable time span.[19]

A comparison of the ESA applicants and the ESTEC study subjects was done, and, not surprisingly, the astronaut applicants (which included professional pilots) performed significantly better than did the study subjects on all tests of operational aptitude.

In the area of personality there were also significant differences ($p < 0.001$), with astronaut applicants having significantly more vitality and extroversion, while the ESTEC subjects showed significantly higher scores on aggressiveness and emotional stability. The pilots in the astronaut pool were separately compared to the study subjects and were found to have higher levels of aggressiveness ($p < .05$) and dominance ($p < .05$).

## THE ISSUE OF FEMALE CANDIDATES IN EUROPE

Females were always free to apply in the European astronaut selection campaigns, but until the 1986 German selection, no female applicant succeeded in being chosen. It is generally supposed—particularly for operational environments like aviation and space—that the probability of being accepted in a selection process is lower for women than for men. The main reason for this supposition is the belief that females, on average, score lower on tests of spatial orientation and mechanical comprehension, as well as on complex psychomotor tests.[20] However, researchers who have reviewed the literature thoroughly have concluded that "differences in spatial skills [between men and women] are quite small—accounting for no more than five percent of the variance."[21] And, in fact, if one looks at the full range of performance scores on spatial ability tests for a mixed population of males and females, at least 95 percent of the variation is due to individual differences that have nothing to do with gender. There is considerable evidence that supports the idea that visual-spatial and mechanical skills are, in part, *learned* skills. It is highly likely—even probable—that, given the different learning, play, and career opportunities, by the time they have reached adulthood, men have had the opportunity to become more practiced at such skills than have their female counterparts. In fact, studies that allow girls to practice on particular psychomotor tests almost always show that they easily catch up to the boys in skill level, while the skill level of the boys remains the same.

Thus, there would appear to be a very simple explanation as to why more women do not get selected for aviation careers: The selection testing emphasizes skills that female candidates are less likely to have learned. If these skills can be learned, then their inclusion in a selection battery effectively discriminates against female candidates.

In order to determine if complex psychomotor skills are related to astronaut performance, one of the European battery's complex psychomotor tasks (the Test of Multiple Task Performance, or TOM) was included in the Rose and Helmreich study on U.S. astronauts in 1990. The results are presented in the next section.

## VALIDITY OF THE TOM IN PREDICTING ASTRONAUT PERFORMANCE

The TOM assesses the capacity to perform simultaneously several cognitive and psychomotor tasks under time pressure. Coordination of both hands while each is performing a separate task is essential, as is an integration of quick reactions and simple arithmetic problems. Attention must simultaneously focus on several control dials, each requiring constant adjustment to maintain performance. The test is language-free and culture-fair and lasts about thirty minutes.

Goeters and his colleagues at the DLR Institute of Aerospace Medicine graciously permitted the use of the TOM apparatus in the U.S. study. The purpose of including the TOM in the study was to assess its variables as possible predictors of differences in astronaut peer and supervisory ratings. Members of the U.S. astronaut corps were willing to be tested on the TOM to determine if such skills were really a factor in astronaut performance.

The results demonstrated that the TOM did indeed discriminate on the basis of sex. Female astronauts scored significantly lower on the tasks than did male astronauts. But the most important finding was that the TOM did not show significant differences among individuals—male or female—who were rated high by their peers in the various rating categories and those individuals who were rated low by their peers in the same categories. Thus, the TOM was found not to predict success as an astronaut, as measured by the peer and supervisory ratings of astronauts.[22]

## PSYCHOLOGICAL TRAINING

The German space program has instituted a vigorous psychological training program for their national astronauts as part of the overall biomedical training that the group receives. The goals of this psychological training are to "improve individual social competence of astronauts; to evoke and support a team-building process within the astronauts' team; and to increase the general efficiency of astronauts by developing general performance skills."[23]

Manzey and Albrecht Schiewe noted that most psychological training that exists (e.g., in the U.S.S.R. program) targets only long-duration space missions, but "even a short-term shuttle mission must be regarded as a long-term task for the astronauts assigned to it. . . ."[24] Additionally, the German perspective is that the everyday demands of the astronaut's professional life require the continuous provision of psychological support and training. To that end, a detailed number of group sessions (with all astronauts present) are undertaken to facilitate communicating, managing stress, and coping with operational demands. The astronauts work on group problem-solving techniques and on social and performance skills.

What is amazing is the fact that the German astronauts have a uniformly positive feeling about the training, and several who have discussed it with me endorse it wholeheartedly. This is in definite contrast to the many attempts by behavioral scientists in recent years to get NASA to implement formal psychological training in the one-year study curriculum of newly selected astronauts. A focus on group psychological factors in training may effectively counteract the lack of emphasis placed on group skills in the German selection process.

Manzey and Schiewe conclude: "Data which objectively prove the success of the training-program cannot be provided. Because of the small group size (n = 6) and a lacking control group, an empirical evaluation of the training program has not been performed. Without such data, an evaluation of the training success, of course, remains a matter of faith."[25] To lessen the "faith" required to accept psychological training, it is increasingly important that objective data—even if only before and after attitudes toward such training—be obtained to support the relevance and usefulness of psychological training programs for astronauts.

In the best of all possible worlds, when the psychological selection is rigorous and the psychological strengths and weaknesses of selectees are formally identified, psychological training programs can focus specifically on shoring up the weaker aspects of the personality and may also assist program managers in determining individuals to select for particular crews. A considerable amount of research has to be done before this can realistically occur, but, nevertheless, the potential benefits of such research in crew health, performance, and productivity would make it a worthwhile endeavor.

## FINAL COMMENTS

The various space programs of the western European nations have established psychological selection methodologies quite separate and distinct from that of NASA. For the most part, the European acceptance of the importance of psychological and behavioral factors in determining the success of space missions is much more advanced than in the United States.

The European procedures have placed almost equal emphasis on psychomotor/cognitive skills and on personality traits for selection of their astronauts. It remains to be seen whether an emphasis on psychomotor and cognitive skills is discriminating when determining astronaut performance in space, since in the U.S. study by Helmreich and Rose, one of the Europeans' major psychomotor tests was not at all predictive of high performance ratings by peers. Since compelling data to dispute Rose and Helmreich's findings do not exist, the conclusion that psychomotor tests unfairly discriminate against female candidates is unavoidable.

A major concern with the ESA test battery is the secrecy surrounding the use of the German tests. Since they are not available to the rest of the world, it is unlikely that such tests can ever be validated independently. This matters primarily from a scientific perspective. No one doubts the professionalism of the German scientists who developed and currently use the tests, but the proprietary nature of those tests will continue to detract from their credibility in the eyes of the rest of the world.

While the European methodologies have been shown to be practical, culture-fair, and reliable, they have yet to be demonstrated to be valid, that is, actually predictive of performance in the space environment. In fact, few data exist to justify their validity, even in the aviation environment from which they were derived. It may be that the major contribution of the Rose and Helmreich study lies in determining the predictive validity of a specific set of astronaut psychological selection criteria.

The importance of obtaining real measures of astronaut performance *in the space environment* cannot be overemphasized. To date, no data have been published on astronaut performance. Such data are essential if behavioral scientists are to understand individual and group psychological factors as they relate to individual and crew performance in space. Without objective data to clarify these relationships, even the best guesses about what psychological criteria are critical in selecting astronauts remain only guesses.

## NOTES

1. John M. Logsdon, *Together in Orbit: The Origins of International Participation in Space Station Freedom* (Washington, D.C.: Space Policy Institute, George Washington University, December 1991), 92–93.

2. John M. Logsdon, "U.S.-European Cooperation in Space: A 25-year Perspective," *Science* 223 (January 6, 1984): 11–16.

3. Logsdon, *Together in Orbit,* 10–11.

4. Klaus-M. Goeters, "The Recruitment and Organizational Integration of Space Personnel," *Acta Astronautica* 17 (1988): 227–229; quotation at 227.

5. Klaus-M. Goeters, E. Schwartz, C. Budczinski, M. Nordhausen, and B. Repp, *Psychological Selection of Spacelab Payload Specialists: The Evaluation of German Applicants,* ESA Technical Translation ESA-TT-586 (Bonn-Bod Gottesberg, West Germany: DFVLR, 1979).

6. R. Armthauer, *Intelligenz-Struktur-Test* (Gottingen, Germany: Hogrefe, 1953). S. Fichtbauer, *Ein mobiler Kleinsimulator fur Tracking und Mehrfacharbeitsversuche,* DFVLR Internal Report IB-355-75-06 (Cologne: DFVLR, 1975). W. Horn, *Das Leistungsprufysytem* (Gottingen, Germany: Hogrefe, 1969). H. Kirsch, *Selektions-strategie und psychologische tests bei der eignungsuntersuchung von Bewerbern fur die fliegerishche Ausbildung bei der DLH,* DFVLR Internal Report IB-355-76-04 (Cologne: DFVLR, 1976). H. Kirsch, K. M. Goeters, and R. Ewe, *Faktorenanalyse eines neuen mehrdimensionalen Personlichkeits-Fragebogens,* DLR FB 75-20 (Hamburg: DLR, 1975). A. Muller, *Das Aufmerksamkeitsprufgerat.* Hamburg, 1956. H. Schmale and H. Schmidtke, *Der Berufseignungstest* (Stuttgart: Huber, 1966). R. Seifert, "Neue gerate zur untersuchung der psychomotorik," *Diagnostica* 12 (1966): 6–16. L. L. Thurstone and T. M. Thurstone, *SRA Primary Mental Abilities* (Chicago: Science Research Associates, 1949). D. Wechsler, *Die Messung der Intelligenz Erwachsener* (Stuttgart: Huber, 1956).

7. Kirsch, Goeters, and Ewe, *Faktorenanalyse eines neuen mehrdimensionalen Personlichkeits-Fragebogens.*

8. Goeters, Schwartz, Budczinski, Nordhausen, and Repp, *Psychological Selection of Spacelab Payload Specialists.*

9. B. Comet, "Medical Standards for Selection and Annual Certification of French Astronauts" (unpublished manuscript, 1986). G. Veron, "French Selection Criteria for Astronaut Candidates" (paper presented to the European Astronaut Psychological Criteria Working Group, 1989).

10. J. Rivolier, C. Bachelard, G. Cazes, H. Mathian, C. Shaw, and G. Veron, "WP630 and WP730: University of Reims," in *Definition of Psychological Testing of Astronaut Candidates for Columbus Missions* (Hamburg: German Aerospace Research Establishment, Department of Aviation and Space Psychology, 1991), 117–150.

11. K. M. Goeters, Presentation to the NASA In-House Working Group on Psychiatric and Psychological Selection of Astronauts (December 12, 1988).

12. Goeters, Schwartz, Budczinski, Nordhausen, and Repp, *Psychological Selection of Spacelab Payload Specialists,* 4.

13. EAPCWG, "Proposed Psychological Criteria for the Selection of European Astronaut Candidates" (final draft, 1990).

14. C. L. Cooper, *The Stress Check* (Englewood Cliffs, N.J.: Prentice-Hall, 1981).

15. U. Kragh, "The Defense Mechanism Test: A New Method for Diagnosis and Personnel Selection," *Journal of Applied Psychology* 44 (1960): 303–309. U. Kragh, "Predictions of Success of Danish Attack Divers by the Defense Mechanism Test (DMT)," *Perceptual and Motor Skills* 15 (1962): 103–106. R. Vaernes, "The Defense Mechanism Test Predicts Inadequate Performance under Stress," *Scandinavian Journal of Psychology* 23 (1982): 37–43.

16. M. Olff, G. Godaert, and H. Ursin, eds., *Quantification of Human Psychological Defence* (Berlin: Springer-Verlag, in press, 1993).

17. K. M. Goeters and C. Fassbender, *Definition of Psychological Testing of Astronaut Candidates for Columbus Missions,* ESA 8730/90/NL/IW (Hamburg: ESA, 1991).

18. C. Fassbender and K. M. Goeters, "Psychological Evaluation of European Astronaut Applicants: Results of the 1991 Selection Campaign" (submitted for publication to *Aviation, Space and Environmental Medicine,* 1993).

19. Ibid.

20. C. Kruger, K. M. Goeters, and H. Eissfeldt, *Geschlechtsspezifische Unterschiede im Leistungs—und Personlichkeitsbereich bei Bewerbern fur den Gehobenen Flug—Verketirskontrolldienst,* DFVLR FB-316-88-01 (Hamburg: DLR, 1988).

21. A. Fausto-Sterline, *Myths of Gender: Biological Theories about Men and Women,* New York: Basic Books, 1985, 13–60; quotation at 21.

22. R. M. Rose, Memo to DLR: Preliminary data analysis on TOM variables, October 26, 1990.

23. D. Manzey and A. Schiewe, "Psychological Training of German Science Astronauts," *Acta Astronautica* 27 (1992): 131–138.

24. Ibid., 149.

25. Ibid., 153.

# 8

# An Invitation to Orbit: Part 2—Japan Enters the Space Age

> Most Japanese, especially the young, are familiar with and embrace American culture; but how many Americans know anything about Japanese culture?
>
> —Chi Sekiguchi, NASA Flight Surgeon

> [A]ccording to one informed observer, Japan—politicians included—does not want to miss the boat. The Japanese space community wants to participate.
>
> —John Logsdon[1]

Japan's National Space Development Agency Law was enacted in October 1969, thus establishing the National Space Development Agency of Japan, called NASDA. This was Japan's first official entrance into the space business, and as usual, the Japanese were very thorough. NASDA's responsibilities were to design and construct both satellites and launch vehicles and then to launch and track them. In order to do this, methods, facilities, and hardware had to be developed. After only six years, NASDA launched its first satellite, Engineering Test Satellite-1, also called KIKU. Since then, the agency has successfully launched meteorological satellites, communications satellites, and broadcasting satellites (using U.S. Delta rockets) and has invested a considerable amount of research and development into new technologies required for future satellite and launch vehicles. NASDA even has its own shuttle program (called *Hope*) on the drawing board.

Sixty kilometers northeast of Tokyo is a large facility that is the center of Japan's space efforts. This facility is actually a city in itself, half the size of Tokyo, and it includes institutes and universities entirely devoted to science education and training. Tsukuba Science City was completely planned under government supervision, and, in addition to the large national research institutes found there, there are also private research and educational facilities. Prominent among all the institutes is the Tsukuba Space Center.

One of the major responsibilities undertaken by NASDA has been to coordinate the Japanese participation in the U.S. Space Station with NASA. As one of the major international partners in the *Freedom* project, NASDA has developed a pressurized module called the Japanese Experiment Module (JEM). Since 1988, when an intergovernmental agreement was signed by Japan, the United States, European countries, and Canada, Japan has been an enthusiastic and committed member of the *Freedom* team. The JEM flew in September 1992 as a payload in a NASA Shuttle/Spacelab-J mission (STS-48), along with a Japanese payload specialist, Dr. Mamoru Mohri. Mohri, however, was not the first Japanese citizen to fly in space. He was preceded by a Japanese news reporter from the Tokyo Broadcasting System (TBS).

TBS reportedly paid the Soviet Union $11.3 million to fly the reporter on *Mir.* Until the last possible moment, the contract with the Soviet Union negotiated by TBS was kept secret from the Japanese government. TBS thought up the idea to celebrate the network's fortieth anniversary, and over 100 of the network's employees applied for the two positions. However, TBS had no intention of limiting applicants to its own employees and opened up the "competition" to "any healthy adult, man or woman, up to the age of 45. . . ."[2] According to the terms of the contract with the Soviets, TBS would select six semi-finalists and send them to Moscow where two finalists would be chosen to train with a Soviet space crew. The two candidates selected would be paid employees of the network for two years.

A detailed health examination was done on the applicants in Japan, and it included a one-week hospitalization for numerous physical tests. There is no documentation of any specific psychological evaluation, but candidates were asked to demonstrate their ability to describe the experience of going into space, since their activities on *Mir* would include broadcasting live from space via satellite. The winner would receive training on handling TV cameras and equipment and was expected to write a book and lecture on his or her experiences after the flight. TBS billed the competition as a search for the first "ordinary person" to fly in space.

A spokesman for TBS, when asked why the Soviets were approached rather than the United States, stated, "First, because the Americans do not sign agreements with private firms and, second, because they have had several major accidents recently and Soviet space technology appears more reliable."[3]

The individual finally selected in the competition was Toyohiro Akiyama, the editor of TBS's Foreign News Division. He rocketed into space with the Soviets in December 1990 for eight days on *Mir*. His flight with the Russians caused considerable controversy in the Soviet Union and abroad because of the buying and selling of trips into space. It underlined the desperate efforts of the Russian agency Glavkosmos to obtain hard currency to pay for its space efforts.

Mohri was one of three government-sponsored Japanese payload specialists selected by NASDA after an extensive, nationwide search for qualified applicants in 1985 for the Spacelab-J mission. Also selected were Dr. Chiaki Mukai (a cardiovascular surgeon, the only female and the only physician payload specialist from Japan) and Dr. Takeo Doi.

In December 1988, Dr. Mukai and Dr. Chi Sekiguchi, her flight surgeon, presented the procedures that were used for the psychiatric and psychological evaluations of Japanese candidates to the NASA In-House Working Group on Psychiatric and Psychological Selection of Astronauts.[4]

## MEDICAL SELECTION OF JAPANESE ASTRONAUTS FOR SPACELAB J

The Japanese medical selection procedures for the Spacelab J selection were similar to NASA's in the early 1980s, but also included medical challenges that NASA had not required since the early *Mercury* or *Gemini* selections. Japanese applicants went through a three-stage process. First, each of sixty-four applicants received an in-depth medical examination in Japan, based on the NASA medical standards, and this included a psychiatric interview by a Japanese psychiatrist. Twelve applicants made the cut to the next set of medical exams, which included trials on the rotary chair, the rocket sled, and the lower body negative pressure apparatus. These tests were designed to assess the ability of each applicant to deal with the physical challenges he or she would experience in the space environment. As we have seen, such tests were a big part of NASA's early selection procedures, but were later dropped in the interest of time.

Seven applicants proceeded to the third phase of the medical examination, which was conducted at the Johnson Space Center in Houston, Texas. These seven candidates underwent the entire NASA medical and psychiatric evaluation (the latter in English) that was described in Chapter 3. All but one of the candidates were found to be qualified by the NASA flight surgeons and psychiatrist. NASDA eventually selected three individuals from this group, including Dr. Mukai.

The specific procedures used in the psychiatric and psychological evaluation for this first Japanese selection are not well documented, but were reportedly very thorough and comprehensive.

## RESULTS FROM THE JAPANESE SPACELAB J SELECTION

The Japanese translation of the Minnesota Multiphasic Personality Inventory (MMPI) has been used since 1963, when M. Abe first translated the MMPI items and conducted studies on the item-content applicability.[5] The response patterns of Japanese normals on some MMPI scales differed greatly from the response patterns of American normals, and Abe restandardized the test for Japanese populations. Scores for two scales, depression (D) and schizophrenia (Sc), were over two standard deviations above the U.S. mean, and nearly all other scales were one standard deviation above it.

Figure 8.1 compares the MMPI scores of normal Americans and normal Japanese. As can be seen from this profile, Japanese are, as a group, higher on most MMPI scales on the average than are Americans.

The MMPI was obtained on all sixty-four Spacelab J applicants during the first Japanese selection. Figure 8.2 compares the MMPI scores of Japanese Spacelab J applicants and U.S. astronaut applicants. What is interest-

**Figure 8.1**
**U.S. Male Norm MMPI versus Japanese Norm MMPI**

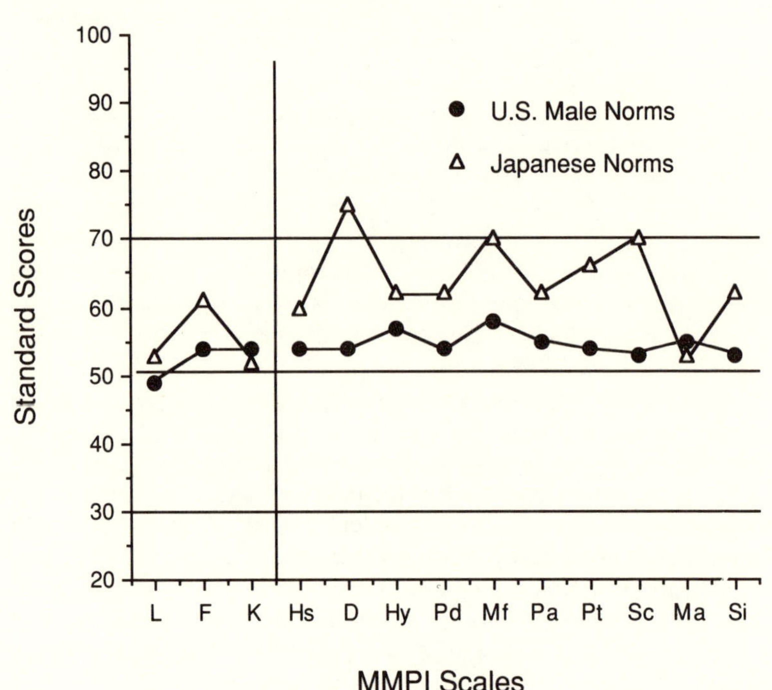

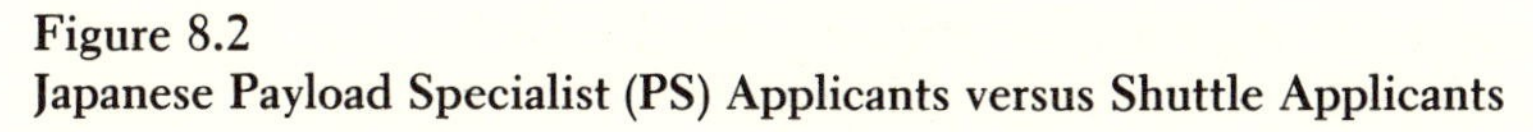

**Figure 8.2**
**Japanese Payload Specialist (PS) Applicants versus Shuttle Applicants**

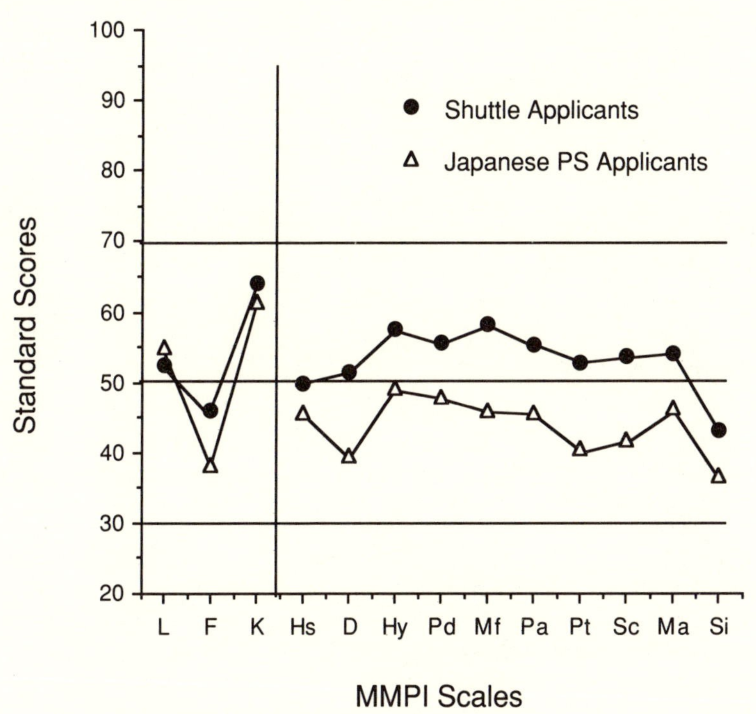


ing is the fact that although Japanese as a group tend to have higher MMPI scores than do Americans, the Japanese payload specialist applicants had lower scores on all scales, compared to Shuttle applicants, although the profile is very similar. This suggests that the Japanese applicants were unusual individuals, compared with the rest of the Japanese population and more closely resembled U.S. astronaut applicants.

## HEALTH CARE FOR THE JAPANESE ASTRONAUTS

The timing of the Spacelab J mission was greatly affected by the *Challenger* disaster, which pushed the Japanese mission (along with a great many others) back several years. In order to ensure that the payload specialists stayed in good health during that time, each one of the three came to the Johnson Space Center for an annual NASA medical examination. However, they also received a semi-annual medical examination by Japanese flight surgeons. Interestingly, the Japanese also required a psychological interview and psychological tests every six months, to follow the mental health of their astronauts. This more than any other factor demonstrates the commitment that NASDA had toward monitoring the psychological issues of space flight.

Japanese behavioral scientists were justifiably concerned by their observation that "[astronauts and cosmonauts] occasionally behave dangerously." They cited the anecdotal reports of a cosmonaut crew that refused to communicate with ground control for several days and an astronaut who left the spacecraft without a safety rope.[6] These kinds of reports had been circulating in the international space establishment for some years, and it is nearly impossible to know if they have any underlying truth. The reports are possibly given more credence than they deserve due to the official secrecy imposed by both the U.S. and the U.S.S.R. space programs. Their refusal to discuss such problems openly suggests to many observers that something is being covered up.

## THE PSYCHOLOGICAL STANDARDS WORKING GROUP

In planning for their participation in the U.S. Space Station, NASDA organized through one of its contractors the Psychological Standards Working Group, which was parallel to the U.S. Working Group. The Psychological Standards Working Group's goals were to develop psychiatric select-out criteria and procedures, after reviewing all potential psychological problems that might occur on the Space Station; to define the medical and psychological peculiarities of Japanese people; to develop psychological select-in criteria and procedures; to develop psychological training programs for the selected astronauts (including training in understanding group dynamics); and to develop a comprehensive medical and psychological health care program for their astronauts.

Members of the Japanese Working Group became very involved with their counterparts on the NASA Working Group, which was proceeding along parallel lines. Dr. Chi Sekiguchi was appointed by NASDA to oversee the Japanese work and was very helpful in arranging the interactions between the Japanese and American behavioral scientists. Dr. Minoru Kume, a distinguished psychologist from Waseda University in Tokyo, was one of the major consultants employed by NASDA. Other consultants were Dr. Toshi Kitamura from the Japan National Institute of Mental Health and Dr. Shigenobu Kanba from Keio University School of Medicine. Dr. Kitamura was particularly well known in the United States and had worked extensively in the past with Dr. Jean Endicott.

Cooperation between the two groups flourished, and in February 1989, Dr. George Ruff led a delegation from the U.S. Working Group (which included Robert Rose, Al Holland, and John Patterson) to attend one of the Japanese Working Group's meetings in Tokyo.

## PSYCHOLOGICAL AND PSYCHIATRIC "PECULIARITIES" OF JAPANESE PEOPLE

Drs. Sekiguchi and Kume reviewed a number of important cultural characteristics of the Japanese people for the NASA Working Group.[7] The purpose of the presentation was to familiarize the members of the Working Group with some of the fundamental differences between traditional Japanese culture and American culture. Sekiguchi wryly commented at one point that "Most Japanese, especially the young, are familiar with and embrace American culture; but how many Americans know anything about Japanese culture?"[8]

### Food

Japanese history and culture demonstrate significant differences in diet compared to Americans. The staple food in Japan is rice, noodles, or bread. Younger Japanese like and have access to American food such as hamburgers and fried chicken. Japanese also traditionally use chopsticks instead of forks or spoons.

### Housing

The Japanese are much more used to small places for living. The typical Japanese house has a low ceiling. Shoes are not generally worn in the house. Many older individuals still prefer sitting on their legs on a mat on the floor rather than on chairs.

### Education

Education is mandatory for the first nine years. However, 90 percent of all students go on to college or a special school. Thus, Japan maintains a very highly educated population. Dr. Sekiguchi described the excessive valuing of an academic background and the terrible stresses young people were under to get into a "famous" school, college, or university. In general, science and technology are the fields favored by the students. Japanese students undergo a unique experience called "Examination Hell," which occurs when they must take the all-important exams that will determine their future. Suicide is not an uncommon occurrence if a student does not do well on these exams.

In general, the Japanese education system is believed to limit the development of individual creativity. A Japanese saying is relevant: "The nail that stands up will be pounded down." Contrast that saying to an American proverb: "The squeaky wheel gets the grease."

### Employment and Management Practices

In recent years, American business has come to know some of the key aspects of Japanese employment and management techniques, since some have been adopted by American companies seeking to improve worker productivity. In Japan, as a rule, employment is considered a lifetime arrangement. Changes from one company to another are rare, as are transfers of personnel from one section of a company to another. The Japanese management style is often familial and emphasizes seniority rule, with wages determined by seniority. The hallmark of Japanese business is the Ringi system, a group decision-making process. Typically this means that a group consensus must be reached before a particular policy is adopted.

### Religion

Another major cultural difference between Japanese and Americans is religion. The three major religions in Japan are Buddhism, Shintoism, and Christianity. Japan has many special days during the year that entail religious obligations (e.g., Buddhist all-souls' day). On the whole, Japanese are probably more superstitious about certain days, designating them as "lucky" or "very unlucky" (this may correspond to Americans' dislike of the number 13 and avoiding its use in buildings, etc.).

### Personality Characteristics and the Development of Psychiatric Illnesses

Some important personality differences exist between typical Japanese and typical Americans. For example, there is a propensity in the Japanese culture to rely on others as opposed to encouraging self-reliance. The Japanese culture prioritizes group logic and process and tends to discourage too much individuality. This is demonstrated by the "principle of equality by the ignorance of individuality." Only mild self-assertiveness is tolerated in the Japanese culture. Value is placed on the virtues of effort, diligence, moderation, obedience, and gentleness.

There are no specific differences in the incidence and prevalence rates for psychiatric illnesses between Japanese and Western people. From a psychological perspective, however, a number of typical personality traits are commonly found in Japanese people. These include:

- a discrepancy between their principles and their own real intentions
- a defensive, autistic, false attitude or set

- superficial conforming behavior
- an exclusive attitude, critical of those outside the culture
- nervousness and emotional instability

Japanese often tend to respond to questionnaire items on the basis of the social desirability rather than their own real behavioral tendencies.

## JAPANESE PSYCHOLOGICAL SELECTION CRITERIA AND SELECT-IN EVALUATION

The Japanese Psychological Standards Working Group completed its deliberations in early 1990. According to Kume,[9] the following are the desirable psychological attributes in the Japanese mission specialist selection.

### Personality Traits

The Japanese selection includes the evaluation of both the personal and the social aspects of personality.

The personal aspects to be evaluated are the ability to maintain self-control and self-restraint and mental and emotional stability. Undesirable aspects of personality that will be excluded are

- egocentrism
- autism
- emotional instability
- depression
- impulsiveness
- past history of behavioral disorders
- any abnormal habits or tendencies
- weakness of will
- lack of stress tolerance

From a social perspective, the Japanese personality criteria stress social maturity, as evidenced by sensitivity in social situations, empathy, and obedience. Mild self-assertiveness (i.e., low aggression) was preferred. In the area of interpersonal relations, affiliation, sociability, dominance, cooperativeness, and conformity are emphasized. Candidates were also expected to demonstrate evidence of leadership. Characteristics that are undesirable in the social domain are

- insensitivity
- autism
- egocentrism

- interpersonal avoidance
- homophobia
- fear of foreigners
- aggression
- impulsiveness
- tendencies toward excessive self-display

### Group Activity

These psychological traits emphasize an ability to communicate and participate in group decision making and problem solving. Candidates should be able to participate in cooperative activity and be aware of their role in the group. While a candidate must have the ability to express him/herself, this should not be done in an aggressive or excessive manner. Undesirable traits include a need for dominating others and anxiety or fear of rejection.

### Motivation

High levels of energy, vitality, positiveness, and optimism are desirable. Candidates must have a high achievement need and a sense of purpose, mission, and duty. They must exhibit endurance, patience, tenacity, and diligence and must be stress tolerant. Undesirable qualities include apathy, asthenia, helplessness, indifference, and avoidance of problems.

### Personal Background

Candidates should display an awareness of and have been exposed to other cultures. They should have a positive view of life, with concomitant values and beliefs, and their life history and family environment should support the presence of these traits. Applicants should have a sense of humor. The family of the candidate, as well as fellow workers, should approve of the candidate's application for astronaut. (Family members were interviewed to assess this aspect of the applicant.)

### Psychomotor Activity

This area concerns good hand, finger, and foot dexterity and general psychomotor coordination.

### Performance Capability

Candidates must have demonstrated good technical performance in their occupation and various performance aptitudes (e.g., cognitive and multiple task aptitudes). Applicants must be trainable and have a history of accurate and stable performance and attentiveness.

### Mental Activity

This general criterion includes intelligence, basic knowledge, technical knowledge, and the ability to discriminate.

### Physiology and Psychophysiology

An appropriate arousal level must be present, and candidates must also have normal circadian rhythms. Insomnia, hypersomnia, and other sleep problems are undesirable.

### Physical Fitness

Applicants must be physically strong and have good balance and regulation. A medical history that demonstrates a weak constitution is undesirable.

As can be seen from the above list of criteria, the Japanese behavioral scientists were interested in a far broader range of psychological characteristics than were their U.S. or European counterparts. One of the more interesting aspects of the Japanese psychological criteria is the emphasis on obedience, mild self-assertiveness, and family support. It would be hard to imagine such characteristics as psychological criteria for any professional selection in the West. It will be interesting to evaluate the effect these factors might have on astronaut performance.

The Japanese Psychological Standards Working Group identified at least one method for assessment of each of the criteria listed above. Table 8.1 lists the psychological characteristics and the tests that were used to evaluate candidates for that characteristic.

## THE JAPANESE PLAN TO PSYCHIATRICALLY AND PSYCHOLOGICALLY EVALUATE APPLICANTS

Kitamura and Kanba summarized the Japanese psychiatric select-out procedures in several papers.[10] They viewed the psychiatric procedures as having three purposes: to select out, to select in, and to select up.

> The select-out procedure is to screen out those applicants who have had, do have or will have any of the disqualifying psychiatric disorders. The select-in procedure is to identify those applicants with special psychological ability and attributes that

Table 8.1
NASDA Psychological Selection Criteria and Assessment Tests

| Psychological Criteria | Psychological Tests |
|---|---|
| Personality (Personal Aspects) | *Projective tests:* |
| | Rorschach Test |
| | Sentence Completion Test (SCT) |
| | *Personality questionnaires:* |
| | Environmental Adjustment Test |
| | Personal Characteristics Inventory (PCI) |
| | Cattell Anxiety Scale |
| | 16-Personality Factor (16-PF) |
| | MMPI |
| Personality (Social Aspects) | Encounter Group Method (T-Group) |
| | Stress Tolerance Scale |
| | Measurement of Stress with Video |
| | Live-in Observational Method |
| | Application Questionnaire |
| | Interview |
| Group Activity | Human Assessment Methods |
| | Encounter Group Method (T-Group) |

| | |
|---|---|
| | Group Discussion |
| Motivation | PCI<br>SCT<br>Composition (from Application)<br>Interview |
| Personal Background | Social Adjustment Scale<br>SCT |
| Structured Psychiatric Interviews | Family History Interviews |
| Psychomotor Activity | General Aptitude Test Battery (GATB) |
| Performance Capability | GATB<br>Kraepelin Performance Test |
| Mental Activity | IQ Tests (R-100 and WAIS)<br>English Ability Tests |
| Physiology and Psychophysiology | Various Sleep Tests |
| Physical Fitness | Various Sports Tests |

make them suitable as U.S. Space Station astronauts. The select-up procedure is to identify those applicants with special resilience to an isolated environment.[11]

Specific disqualifying psychiatric conditions include most Axis I and Axis II disorders listed in the revised third edition of the American Psychiatric Association's *Diagnostic and Statistical Manual of Mental Disorders.*

In the first stage of the selection, several psychometric instruments were recommended to be given to applicants: the General Health Questionnaire (GHQ)[12] to cover Axis I disorders, the Personality Diagnostic Questionnaire (PDQ)[13] to cover Axis II disorders, the Social Desirability Scale[14] to examine social desirability, and an unpublished questionnaire to evaluate sociodemographic variables. The tests could be sent to all applicants to fill out and return.

Kitamura proposed that between the first and second stages of the selection 200 applicants be selected randomly and invited for an interview in order to conduct a validity study on the questionnaires. Participation by applicants would be voluntary, and their informed consent would be obtained before they were interviewed. Those conducting the validation study would then carry out all the interviews scheduled to be performed in the second stage of the selection. Such a study would permit the calculation of cutoff points for the questionnaires, and discriminant function analysis could be performed to identify those cases showing significant levels of pathology and those that do not.

The second stage of the proposed psychological selection process would utilize the Schedule for Affective Disorders and Schizophrenia–Lifetime version (SADS-L, which is similar to the NASA interview; see Chapter 5) to evaluate for Axis I psychiatric disorders. It was recommended that, instead of the Personality Assessment Schedule (PAS), the Japanese—who were not constrained by time factors—use the longer and more detailed Structured Interview for DSM-III-R Personality Disorders (SIDP-R).[15]

During the third stage, the applicants not eliminated would be re-examined by different interviewers using the SADS-L and SIDP-R. In addition to those interviews, the candidates would also have to provide information on family history during two interviews—the first using the Family History–Research Diagnostic Criteria (FH-RDC)[16] and the second the Family History of DSM-III Personality and Anxiety Disorders (FHPD).[17] The applicant would then identify at least one "informant" who, with written consent from the applicant, would be invited for an interview about the applicant, using the SIDP-R (third-person version), FH-RDC, and FHPD. Using third persons to assess an applicant has a significant advantage over just interviewing the candidate. It offers a unique opportunity to get a rather more objective interview about a candidate for astronaut than that candidate is likely to give. In the best of all possible worlds, this would eliminate some of the concerns when interviewing a highly motivated and defensive applicant.

## RESULTS OF THE MOST RECENT JAPANESE SELECTION FOR A SPACE STATION MISSION SPECIALIST

A second Japanese astronaut selection for a mission specialist trainee to fly on *Freedom* took place in late 1991 and early 1992. The individual chosen by the new psychological selection procedures would become part of the U.S. astronaut corps and train in the United States. In May 1992, Chi Sekiguchi presented a paper at the Aerospace Medical Association meeting in Miami, summarizing the psychological results of that Japanese Astronaut Selection.[18]

An announcement was issued to the Japanese public in July 1991, soliciting applicants for the mission specialist position. The basic qualifications required were Japanese citizenship and a minimum of a bachelor's degree in natural science, with at least three years of professional experience. The applicant had to be less than thirty-five years of age and able to perform the training activity and space flight activities of an astronaut. All applicants were expected to be able to pass a NASA class II medical examination (which was required for all U.S. mission specialists).

A total of 372 individuals applied, and 233 of these applicants began the three-phase examination process.

### Phase I Selection

The 212 male and 21 female applicants took the tests for English ability, general knowledge, scientific and technical knowledge, and psychological traits. The actual tests used during this phase included the Cattell Anxiety Scale, the Kraepelin Performance Test, the Environmental Adjustment Scale, and the GHQ. Psychological traits were evaluated using Helmreich's Personal Characteristics Inventory (PCI).

### Phase II Selection

By phase II, the field of candidates had narrowed to forty-three males and three females. These individuals received a complete physical examination, equivalent to the NASA class II exam. Their English ability was further probed using conversational tests. Additional psychological tests were given these candidates, including the General Activity Test Battery (GATB), Human Assessment test, the psychiatric interviews (SADS-L and SIDP-R), and additional intelligence tests. During this phase, the candidates also were interviewed by officials at NASDA.

### Phase III Selection

This phase included only special medical evaluations. Only 6 candidates (all male) remained from the original group of 233. All six were sent to the

Johnson Space Center in Houston, Texas, to undergo the NASA medical and psychiatric evaluations. After passing tests at NASA, the candidates then had an executive interview at NASA. The final decision in May 1992 was to select only one applicant.

While the exact procedures outlined by Kitamura and Kanba were not followed, the basic proposals made by the Japanese Psychological Standards Working Group were followed. Japanese behavioral scientists were able to collect a considerable amount of data in their evaluations, which, if looked at carefully will be seen to be a careful and comprehensive integration of U.S., European, and uniquely Japanese components. Of all the astronaut selections taking place in the world today, the Japanese probably have developed the most valid, reliable, and comprehensive system for the psychiatric and psychological evaluation of astronaut applicants.

The first annual meeting of the Japanese Association of Space, Aviation, and Environmental Psychiatry was held in Tokyo on October 13, 1991. The enthusiasm of the Japanese psychiatrists was truly astonishing, and their interest in developing the most scientific methods for space flight selection and training was inspiring. During the meeting, the psychiatrists who were to evaluate the forty-six Japanese candidates in phase II later that month met at the NASDA center in order to observe a training session on how to give the SADS-L interview to subjects.

## UNRESOLVED ISSUES

There remain a number of unresolved issues related to international cooperation in the selection of astronauts. In Chapter 4, the philosophy of selecting out versus selecting in was discussed in detail.

When it comes to select-out procedures, those psychiatric procedures used to identify psychopathology, the Americans, Europeans and Japanese are able to agree on methodology. Selecting out is not particularly controversial, since whatever procedures are developed will only exclude 1 to 3 percent of astronaut applicants at best. While this is an important and necessary step in the screening process, a psychiatric examination per se is not particularly helpful in determining which applicants to actually select. However, for the purposes of determining the likelihood that psychiatric and emotional problems will develop in the space environment among a group of highly selected individuals, it will be important in the future for all countries to pool their data on the incidence of psychiatric disorders among astronaut applicants—and, in fact, it would be useful to follow those highly motivated and professionally successful cohorts over time for just such a purpose.

The scientific evidence available does not support any particular psychological select-in criteria, and, as can be seen from the discussions of the U.S., European, and Japanese selection procedures, the utilization of psychological data to select candidates varies considerably from country to country. The U.S. program has undertaken a strictly research approach on this issue, primarily because NASA management did not believe that psychological data would be useful or better than routine interview procedures. This negative attitude toward psychology on the part of NASA has actually resulted in something positive. The U.S. Space Program has the largest number of astronauts, so it offered the best opportunity to validate psychological selection criteria, primarily because such criteria were not a part of the operational selection procedures. The study that ensued (described in Chapter 6) demonstrated for the first time in the history of space exploration that psychological factors may be used to predict which candidates will perform better as astronauts. While there is much work to be done, the American researchers' results are very promising.

The Europeans and Japanese have repeatedly pointed out that they cannot wait for validation of psychological criteria to select their astronauts. They argue—quite reasonably—that they must rely on a "best guess" of psychological select-in criteria, rather than a "just pick anyone" philosophy. But the Japanese, in particular, are interested in the results of the American research, since they are not wedded to their procedures if more predictive methods can be developed.

On the other hand, the European selection of astronauts has long been dominated by the German behavioral scientists, and it is only recently, with the implementation of the ESTEC study (described in Chapter 7), that other psychological selection philosophies have been included in European astronaut selection. While the German psychological test battery has many fine points, it has no possibility of being validated, since the tests used are proprietary and cannot be used by other scientific researchers to resolve the issue. Although Dr. Klaus-Martin Goeters and his colleagues have always been helpful and involved in the international efforts (including permitting the use of the vitality scale and the TOM multiple-task simulator in the American study), they are unlikely to change procedures that are financially advantageous.

It is vitally important for future research on the psychological factors of space exploration that there be an integration of all the international psychological criteria and evaluation methods. This is not a one-way street. NASA has a rather hard-line attitude toward procedures other than its own, but, as we have seen, it can learn much—particularly in the area of psychological evaluation—from the practices of its international partners.

Other issues that must be addressed in future collaborative efforts are the multicultural factors of space flight. This includes the language problems that international crews will have to address, as well as the diversity of ethnic backgrounds and cultures of future crew members. It is not a trivial problem, since both these factors will play a role in the success or failure of international space missions. It seems unlikely, with the national and ethnic conflicts that exist in the world today—some of them unresolved for centuries—that cooperative efforts will be successful unless the issues are addressed forthrightly in advance. National pride and ethnic fervor are still overriding concerns for politicians in many countries. A few of the operational problems that might develop in space are considered in recent articles in *Aviation, Space and Environmental Medicine.*[19] These articles present information obtained from astronauts and cosmonauts who have flown on international space missions and who have reported a wide variety of misunderstandings, miscommunications, and significant inflight conflicts.

Now that the formerly powerful Soviet Union has broken up into a collection of squabbling countries, the future of its once proud and dominant space program is literally and figuratively "up in the air." The Russians appear to be willing to facilitate cooperative efforts in space with the Americans, Europeans, and Japanese, and several joint projects have been officially sanctioned. Will the Russians and their space expertise join the international community in the coming years, and if they do, how much will we be able to learn, particularly in the psychological realm, from their experiences in space?

The psychological selection of Soviet cosmonauts has a unique and fascinating history that will be explored in the next chapter.

## NOTES

1. John M. Logsdon, *Together in Orbit: The Origins of International Participation in Space Station Freedom* (Washington, D.C.: Space Policy Institute, George Washington University, 1991), 96.

2. A. R. Curtis, *Space Almanac* (Woodsboro, Md.: Arcsoft Publishers, 1990), 175.

3. Ibid., 176.

4. C. Mukai and C. Sekiguchi, "Japanese Medical Selection" (presentation to the NASA In-House Working Group on Psychiatric and Psychological Selection of Astronauts, 1988).

5. M. Abe, K. Sumita, and M. Kuroda, *A Manual of the MMPI,* Japanese standard ed. (in Japanese) (Kyoto: Sankyobo, 1963).

6. George E. Ruff, "Report on the NASA In-House Working Group on Psychiatric and Psychological Selection of Astronauts Meeting with the NASDA Psychological Standards Working Group, Tokyo, February 9–10, 1989" (unpublished, 1989).

7. C. Sekiguchi and M. Kume, "Cultural Characteristics of Japanese People" (presentation to the NASA In-House Working Group on Psychiatric and Psychological Selection of Astronauts, 1988).

8. C. Sekiguchi, personal communication, 1988.

9. M. Kume, "Present Status of Japanese Psychological Astronaut Selection" (report presented at the Joint NASA/NASDA Working Group Meetings, Tokyo, February 9–10, 1989).

10. T. Kitamura, *Psychiatric Selection Procedure of Japanese Astronauts for the NASA Space Station: I. The Selection Framework* (Tokyo: National Institute of Mental Health, 1989). T. Kitamura and S. Kanba, *Psychiatric Selection Procedure of Japanese Astronauts for the NASA Space Station: II. Suggested Select-out Procedures* Japan National Institute of Mental Health, 1989.

11. Kitamura, *Psychiatric Selection Procedure of Japanese Astronauts for the NASA Space Station*, 1.

12. D. P. Goldberg and V. F. Hiller, "A Scaled Version of the General Health Questionnaire," *Psychological Medicine* 9 (1979): 139–145.

13. S. E. Hyler, R. O. Rieder, J. B. W. Williams, R. L. Spitzer, M. Lyons, and J. Hendler, "A Comparison of Clinical and Self-report Diagnoses of DSM-III Personality Disorders in 552 Patients," *Comprehensive Psychiatry* 30 (1989): 170–178.

14. D. P. Crowne and D. Marlow, "A New Scale of Social Desirability Independent of Psychopathology," *Journal of Consulting Psychology* 24 (1960): 349–354.

15. B. Pfohl, "Structured Interview for DSM-III-R Personality Disorders (SIDP-R)," draft ed. (Iowa City: Department of Psychiatry, University of Iowa, 1989).

16. J. Endicott, N. C. Andreasen, and R. L. Spitzer, *Family History–Research Diagnostic Criteria (FH-RDC)*, 3d ed. (New York: New York State Psychiatric Institute, 1978).

17. J. Reich, R. Crowe, R. Noyes, and E. Troughton, *Family History of DSM-III Personality and Anxiety Disorders (FHPD)* (Iowa City: Department of Psychiatry, University of Iowa, 1982).

18. C. Sekiguchi, S. Yumikura, M. Kume, and N. Okada, "Japanese Astronaut Selection Psychological Results" (paper presented at the Aerospace Medical Association Meeting, Miami, May 10–14, 1992).

19. A. D. Kelly and N. Kanas, "Crew Member Communication in Space: A Survey of Astronauts and Cosmonauts," *Aviation, Space and Environmental Medicine* 63 (1992): 721–726. P. A. Santy, A. W. Holland, L. Looper, and R. Marcondes-North, "Multicultural Factors in the Space Environment: Results of an International Shuttle Crew Debrief," *Aviation, Space and Environmental Medicine* 64 (1993): 196–200.

# 9

# The Soviet Right Stuff

> Poyekhali! (Off we go!)
>
> —Yuri Gagarin (1961)

> Somewhat incompatible people can go into space on a flight of up to several weeks duration. On flights of a month, the factor [of compatibility] already starts to show up, and on longer flights it becomes essential. One must give the psychologists their due: Recently, in the course of preparation, they have learned to explain to us just who your partner is and you yourself, what our character traits are and how we must deal with them. In the case of Yuri Romenanko and me, we had been acquainted only a short time before the flight, but over a period of three-months in orbit we never once quarreled.
>
> —Georgi Grechko[1]

## A GREAT SHAME

In the summer of 1966, George Ruff got a call from Washington asking him to go to an international aerospace meeting in Prague, Czechoslovakia. NASA gave him permission to talk about whatever he wanted regarding astronaut selection while he was there. In the Great Hall of Prague's Charles University after a formal academic procession, punctuated by trumpets and fanfare, the minister of health, wearing a colorful cap and gown, opened the academic conference—the first such conference to host both American and Soviet space scientists.

Ruff remembers a strange man approaching him and asking if he would like to meet Colonel Kuznetsov. Ruff wondered who that was, but he followed the stranger to an impressive-looking individual wearing a Soviet military uniform. Colonel Kuznetsov was, it turned out, Ruff's counterpart in the Soviet space program.

During the peaceful days of the conference, the two psychiatrists met numerous times, talking and socializing, and eventually they came to trust one another. One day, Kuznetsov confided in Ruff that the Russians were extremely suspicious of all the important research NASA was doing on the psychological aspects of space flight. Where, Ruff asked in puzzlement, did he get that impression? It was simple, Kuznetsov responded. The United States was "pretending" it was not doing anything in the psychological area. Therefore, they must be doing some very important things that they did not want the Russians to find out about.

It took a while for Ruff to convince Kuznetsov that there really was no research being done in the area by NASA scientists. The Soviet psychiatrist was skeptical. He pointed out that George Ruff and Ed Levy had published several papers on their work, but then suddenly, there was nothing. What did that mean if not that they now wanted to keep things secret? With some difficulty, Ruff finally convinced him of the truth. The colonel was astonished.

"What a great shame," Kuznetsov exclaimed. "These missions are so expensive! Just think what could be done together!" Enviously, Ruff then listened to him talk about the intense interest in psychological factors and the input that psychology and psychiatry had in the selection and training programs for cosmonauts. None of it, of course, was published in the scientific literature because the Russians did not want the Americans to get any ideas from their work.

It is ironic that the space program of a closed society like the U.S.S.R. was more open to studying the potential mission impact of psychological factors than the U.S. Space Program was. In spite of a pervasive and paranoid need for secrecy, Soviet managers were unable to believe that psychological factors—particularly in selection and training—were unimportant. And, interestingly, despite this acknowledgment and the subsequent integration of behavioral scientists into Soviet space operations, the U.S.S.R. program has still been plagued by a significant number of emotional and interpersonal problems in space.

Georgi Grechko, during a 1989 visit to the Johnson Space Center, related that it was a mistake to think that Soviet psychological experts were better than Americans when it came to selecting and preparing cosmonauts for flight. After all, he pointed out, the Soviets use only American psychological tests. And when a crew is formed, he added, while each crew member is briefed on the psychological strengths and weaknesses of his colleagues, the psychologists also give advice on crew composition, but are rarely listened to.[2]

## THE STATE OF SOVIET PSYCHIATRY

In 1961, the year Yuri Gagarin became the first human being to rocket into space, Soviet psychiatry and psychology had already developed an international and infamous reputation because of the politicalization of psychological and psychiatric concepts and the elimination of psychodynamics—particularly Freudian concepts—from the field. Psychodynamic formulations of behavior focused on the individual mind as the main site for understanding psychological processes, and this could not be reconciled with Communist political doctrine, which asserts that "being determines consciousness."

In Sidney Bloch and Peter Reddaway's 1977 book on Soviet psychiatric abuse, Vladimir Bukovsky states:

> As Socialism has been built in the USSR, and Communism is being built, the consciousness of people must be exclusively communist. Where, then, can belief in God appear from, if for 60 years atheism has been propagated and the preaching of religion outlawed? And from where does an opponent of Communism come—in a Communist society? Within the confines of Communist doctrine there are only two possible explanations: the cause must lie either in subversive activity directed from abroad—i.e., every dissenter has been bought or recruited by the imperialists; or in mental illness: dissent is just a manifestation of pathological processes in the psyche.[3]

The Soviet abuse of psychiatry has been well documented by the World Psychiatric Association (WPA), which for years had heard testimony from patients and physicians locked into the Procrustean bed of Communist psychiatric theory. In the early sixties, numerous scientists from around the world began to collect evidence that psychiatry was being used by the Soviet leadership as an instrument to control political dissent. Individuals with points of view that differed from the Soviet leadership were labeled as "mentally ill." Psychiatrists who might have protested received the same label and also tended to disappear into the mental hospital back wards.

It is precisely because psychiatry and psychology are inexact sciences that they can be so abused. Many countries around the world, even the United States, have documented cases of the *political* misuse of psychiatry (such misuse is not to be confused with incompetence or even a lack of concern for patients—a crime that psychiatrists and psychologists of any country might be guilty of). But it was only in the Soviet Union that political misuse had been systematic and widespread.

In spite of compelling documentation and many individual witnesses, many world psychiatric leaders were unwilling to believe that such abuses could possibly occur in a country as "advanced" as the Soviet Union. Those of us who have witnessed the disintegration of Soviet society over the last few years have trouble understanding why it took so long for the WPA to do anything. The World Psychiatric Congress chose to ignore the entire problem at their 1971 and 1977 meetings. It was not until 1983 that enough momentum was generated in the international psychiatric community to condemn Soviet psychiatric practices, and to forestall condemnation from the world, the All-Union Society of Psychiatrists and Narcologists of the Soviet Union resigned from membership in the WPA, rather than face imminent expulsion.[4]

Six years later, in 1989, the WPA voted in Athens to re-admit the Soviet psychiatrists—with some conditions attached. One stipulation was that the Soviet psychiatric society had to issue a statement acknowledging the charges of psychiatric abuse. Such charges had always been self-righteously and vehemently denied in the past. The statement issued by the Soviet society reads in part: "Previous political conditions in the U.S.S.R. created an environment in which psychiatric abuse occurred for non-medical, including political reasons."[5]

What does this history of psychiatric abuse have to do with the U.S.S.R. space program? It is important to put the work that Soviet psychiatrists and psychologists have done in the space program in the proper political perspective. One must congratulate them for their foresight in understanding that psychological factors were going to be an important part of space exploration. But the political context of their work and the complete absence in their work of any understanding of psychodynamic principles lend it a unique bias that makes it fundamentally suspect. For example, in the first chapter of a serious book on pilot and cosmonaut selection, the following discussion ensues:

> Marxism recognizes the existence of unequal individual abilities, viewing this difference as the result of unequal natural potential of personal characteristics and features. This position is accepted even in a communistic society. About this position V.I. Lenin wrote: "To expect equality in strength and abilities in people in a communistic society is absurd."
>
> A fruitful general theory of aptitude development depends on the application of dialectical materialism's position and the basic methodological principles that promote the development of a single theoretical concept that includes such scientific research directions as the examination of aptitudes within a general social framework. . . . The word "aptitude" is used by the founders of Marxism-Leninism when they state "from each according to ability".[6]

The author continues in this discussion for many pages, obviously trying hard to figure out how to reconcile dialectic materialism with a focus on an *individual's* aptitude and personality. Or, in another section of the same chapter:

There are a number of important shortcomings that must be noted in personality tests (inventories and projective tests) which has engendered justified criticism. Each of these tests borrowed from Western personality theory is associated with a particular system for interpreting the data, which to a greater or lesser extent is influenced by the personality theory espoused by its developer. At the same time it is clear that no personality theory can be divorced from a particular ideological system. *For this reason, in the interpretation of any personality data we must be strictly guided by Marxist methodology which demands that we study a specific individual as someone who lives in particular social and historical conditions, which consist of particular social relationships to other individuals.*[7]

This sort of meaningless (particularly in scientific terms) and mindless adherence to political doctrine is the hallmark of many Soviet scientific publications, particularly in the behavioral sciences. Because of the absence of fundamental psychodynamic theories of behavior, Soviet scientists seem to emphasize factors such as psychophysiological compatibility, a theory determining which cosmonauts are most compatible to fly together. The theory is based on matching the biorhythms of basic physiological functions (like heart rate, respirations, and blood pressure).

There is no evidence in the scientific literature outside the Soviet Union that such an approach to determine who is compatible with whom works in the real world. We will discuss this specific theory in some detail later, but it exemplifies the typical approach that the Soviets have taken toward psychological problems in space flight, including cosmonaut selection. Although such work is presented here, it does not imply that the author considers it reliable or useful for future operations or research. However, now that scientists in the former Soviet Union are no longer working for the Communist party, and now that their work can be evaluated by the rest of the world scientific community, it is presented in the hopes that there is something that might be learned from their approaches.

## SOVIET PILOT SELECTION

The academician S. P. Korolev, the father of Soviet space science, believed that "[a] pilot, especially a fighter pilot is most qualified for [manning a spaceship]. . . . He is a universal specialist. He is a pilot, a navigator, a communications specialist, and a flight engineer. As an experienced soldier, he can be characterized as moral, organized, disciplined, and goal-oriented."[8] And so it was that the Soviet Union developed a policy similar to that of the United States by which it selected only active military test pilots.

The personality assessment of Soviet pilots has been rather limited in scale, particularly for the purposes of predicting flight personnel performance. For the most part, U.S.S.R. behavioral scientists have had to use and adapt foreign (generally American) psychometric instruments for assessing personality for this purpose. Again, the primary reasons for this relate to the political pitfalls of running afoul of Communist doctrine by focusing too much on the individual.

Several well-known U.S. personality tests have been used for pilot (and later cosmonaut) assessment and selection. They are the Minnesota Multiphasic Personality Inventory (MMPI), the 16-Personality Factors (16-PF) of Cattell, and the Thematic Apperception Test (TAT), a projective test used in the early *Mercury* and *Gemini* selections.

Of particular interest for this discussion is the Soviet use of the MMPI, which was renormed by Soviet behavioral scientists. The traditional scales were made "non-clinical," and the entire inventory was finally renamed the Standardized Method of Personality Testing (Russian abbreviation: SMIL).[9]

By altering the loading of scales 1, 4, 6, and 8 (on the traditional MMPI these scales correspond to hypochondriasis, psychopathic deviation, paranoia, and mania), which the Soviets believed mainly reflected the differences between flight personnel and random subjects used for standardization, V. A. Bodrov and his colleagues obtained normative data on 1,000 subjects for the use of the MMPI with pilots. Data were then analyzed using a previously published Soviet method for interpreting the MMPI based on a psychological approach, rather than a psychiatric or pathological approach.[10] This unauthorized alteration in the MMPI may have had some impact on the instrument's usefulness in determining psychopathology (see Chapter 1) and is at least interesting, if not reliable or valid. A comparison of MMPI and SMIL scales can be found in Table 9.1.

In a study of pilots done in the 1980s using the SMIL, TAT, and 16-PF, Soviet behavioral scientists correlated fifty-two personality characteristics derived from those tests with what was called "the pilot's level of professional qualification," Table 9.2 lists the correlation coefficients and significance levels for those personality factors found to be important in the study. None of the correlation coefficients is greater than 0.32, but they may reflect some significant variance in outcome.

When the SMIL data collected during the study are analyzed, some very interesting results begin to emerge, particularly when compared to U.S. pilot data. Figure 9.1 shows the MMPI profile of U.S. pilots versus the SMIL profile of U.S.S.R. pilots. Figure 9.2 compares the Soviet pilots' profile with the Soviet-developed norms. The most obvious difference between the two countries' pilot profiles lies in the LFK scale configurations. The American pilot LFK configuration suggests that the individual is attempting to avoid or deny unacceptable feelings, impulses, and problems and is trying to present himself in the best possible light. This profile occurs most frequently in defensive, but usually normal, individuals, particularly job applicants who tend to deny psychopathology. However, the Soviet pilots have a completely different LFK configuration. When the L and K scales are less than a T-score of 50 and the F scale is greater than 60, it usually represents an individual who is admitting to personal and emotional difficulties or is exaggerating symptoms in order to get help. Such individuals are generally open to psychological interventions. In this pattern, the Soviet pilots are very different from most pilots around the world, who tend to follow the American pattern of defensiveness.

In another study done by the same authors, student pilots' personalities were assessed. The "best" students were then compared with those students who flunked out of pilot's school and those who were expelled for lack of discipline. These three groups have very different SMIL profiles, as can be seen from Figure 9.3. Again, all three groups of student pilots demonstrate an LFK profile that suggests the individuals are admitting to significant emotional difficulties, particularly the student pilots who "failed." The "failed" pilots also have significant elevations, suggesting a high degree of emotional lability and anxiety. The "dismissed" student pilots are also emotionally labile and have a marked elevation on scale 9, mania (or, as the SMIL calls it, activity/optimism). Even the "best" student pilots have scores on most scales at least one standard deviation above U.S. pilots. These findings suggest significant cultural differences between Soviet and American citizens, particularly in the propensity to tolerate and/or express emotions.

## COSMONAUT PSYCHOLOGICAL SELECTION

Since the early days of the Soviet space program, the Soviets have included a vigorous social and psychological testing and training program before flights and have monitored the psychological reactions of cosmonauts in flight.

### Psychological Criteria

Specific psychological criteria for the selection of cosmonauts were described by M. A. Novikov:[11]

- low anxiety level
- emotional stability and balance
- extroverted personality
- strong intellectual and perceptual abilities
- resistance to boredom and repetitive tasks
- vigilance and ability to focus attention
- flexibility
- memory
- ability to control one's own reactions (self-control)

Table 9.1
A Comparison of MMPI and SMIL Scales

| MMPI Scales | | SMIL Scales |
|---|---|---|
| L | A validity score that, if high, may indicate evasiveness | L |
| F | A validity scale measuring the tendency to present one's self in an overly favorable or highly virtuous light | F |
| K | A validity scale that measures defensiveness | K |
| *1–Hypochondriasis (Hs)*<br>High scorers are described as cynical, defeatist, preoccupied with self, complaining, hostile, and presenting numerous physical problems. | | 1–Hypochondria |
| *2–Depression (D)*<br>High scorers are moody, shy, despondent, pessimistic, and distressed. This scale is the most frequently elevated in clinical syndromes. | | 2–Depression |
| *3–Hysteria (Hy)*<br>High scorers tend to be repressed, dependent, naive, and outgoing and to have multiple physical complaints. Expression of psychological conflict occurs through vague and unbased physical complaints. | | 3–Emotional lability |
| *4–Psychopathic deviation (Pd)*<br>High scorers are often rebellious, impulsive, hedonistic, and | | 4–Impulsivity |

| | |
|---|---|
| antisocial. They often have difficulty in marital or family relationships and trouble with the law or authority in general. | |
| *5–Masculinity-Femininity (Mf)*<br>High-scoring males are described as sensitive, aesthetic, passive, or feminine. High-scoring females are described as aggressive, rebellious, and unrealistic. | 5–Masculinity/Courage |
| *6–Paranoia (Pa)*<br>Elevations are associated with suspiciousness and being aloof, shrewd, guarded, worrisome, and overly sensitive. High scorers may project or externalize blame. | 6–Rigidity |
| *7–Psychasthenia (Pt)*<br>High scorers are tense, anxious, ruminative, preoccupied, obsessional, phobic, and rigid. They frequently are self-condemning and feel inferior or inadequate. | 7–Anxiety |
| *8–Schizophrenia (Sc)*<br>High scorers are often withdrawn, shy, unusual, or strange and have peculiar thoughts or ideas. They may have poor reality contact and in severe cases bizarre sensory experiences—delusions or hallucinations. | 8–Individualism/Nonconformity |
| *9–Mania (Ma)*<br>High scorers are described as sociable, outgoing, impulsive, overly energetic, optimistic, and in some cases amoral, flighty, confused, and disoriented. | 9–Activity/Optimism |
| *0–Social Introversion-Extroversion (Si)*<br>High scorers tend to be modest, shy, withdrawn, self-effacing, and inhibited. Low scorers are outgoing, spontaneous, sociable, and confident. | 10–Introversion (Si) |

Table 9.2
Correlation Coefficients and Significance of Personality Traits

| Instrument | Trait | Correlation Coefficient | Significance Level (p) |
|---|---|---|---|
| SMIL | Emotional stability during stress | 0.32 | .001 |
| 16-PF | Energy, risk-taking | 0.34 | .001 |
| SMIL | Intuitiveness, diminished self-preservation instinct, lack of fear of death | 0.31 | .001 |

Figure 9.1
U.S. Pilot MMPI versus USSR Pilot SMIL

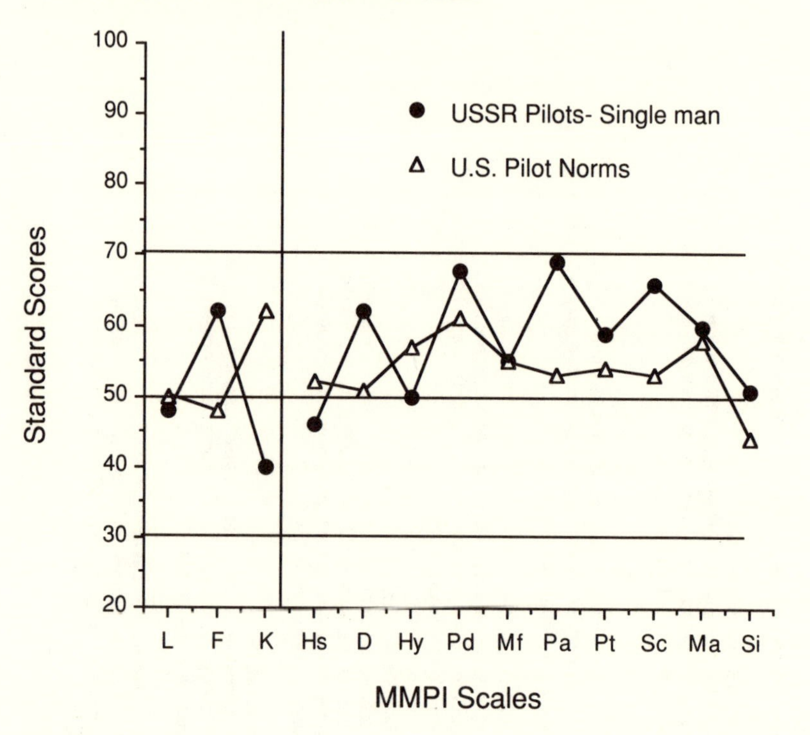


Another paper written by G. T. Beregovoy and his colleagues adds the following traits:[12]

- high moral and ideological level and overall maturity
- self-critical and tolerant of others
- able to learn quickly

**Figure 9.2**
**USSR Jet Pilots versus USSR Norms**

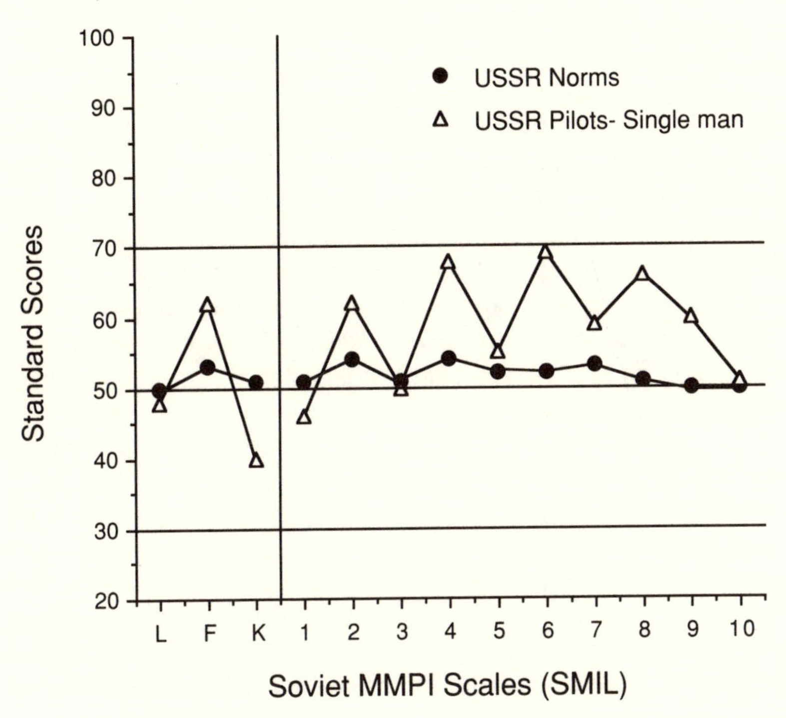


1- Hypochondria
2- Depression
3- Emotional lability
4- Impulsivity
5- Masculinity/courage
6- Rigidity
7- Anxiety
8- Nonconformity/individualism
9- Activity/optimism
10- Introversion

- a well-developed imagination
- a sense of humor
- a strong, balanced mobile type of higher nervous activity (Pavlov's Classification)
- reliability
- ability to work under noisy conditions

## Psychological Tests for Selection

Psychological tests are done during cosmonaut selection to evaluate the following characteristics: (1) perception, (2) intellectual abilities, (3) psychomotor efficiency, (4) personality traits, and (5) physiological indices of emotional reactivity.[13] The purpose of this psychological testing is to reveal

Figure 9.3
Soviet Student Pilots: "Best" versus "Failed" versus "Dismissed"

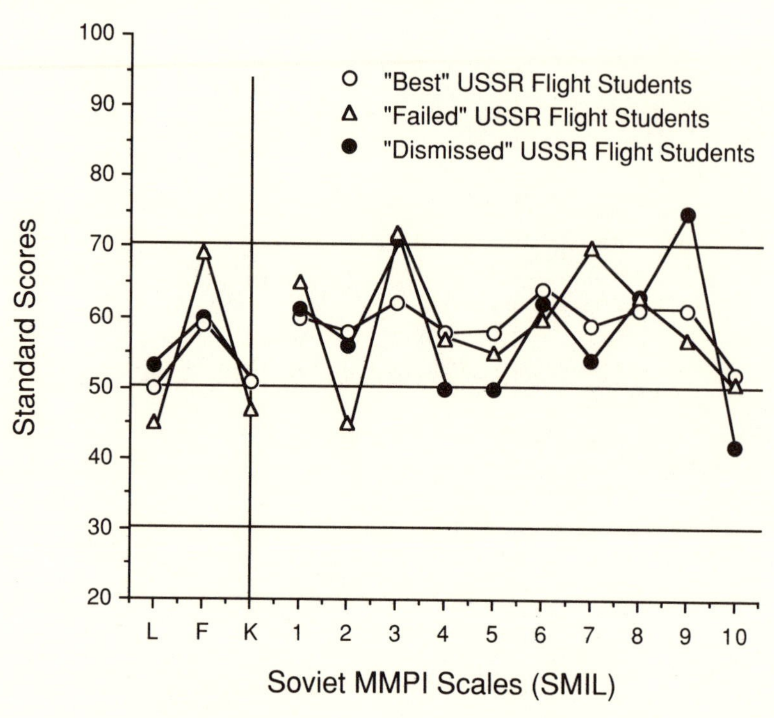


hidden psychopathology and abnormalities in the neuropsychic area, as well as to define personality structure, motivation, and values. Soviet testing of cosmonauts also includes the study of stress tolerance, endurance, behavioral peculiarities, and interpersonal interaction. The Soviets make no distinction between select-out and select-in methods.

The Soviets have traditionally used a combination of projective and nonprojective methods of personality and motivation testing. Soviet behavioral scientists have paid particular attention to acute stress responses and group interaction. It is not clear from published reports the extent to which psychological selection methodology has changed over the years. One has the impression that it has remained fairly constant, and this is supported by Y. I. Senkevich and M. M. Korolayev, whose article concludes: "Thirty years of successful spaceflight experience confirm the accuracy of the chosen selection system."[14]

### The Psychiatric Interview[15]

The psychiatric interview (both a long and a short version) is given in a semi-structured manner. The format of the long interview is as follows:

*Biographical data.* Name, year of birth, ethnic group, permanent and current address; occupation or profession before military service; membership in Komsomol youth organization; family status, number of children; education, where obtained, grade average, favorite subjects, participation in extracurricular activities; reasons for choosing to become a pilot, time and circumstances of initial interest in aviation, and how it manifested itself (e.g., clubs, societies, enrollment in young cosmonaut school, reading).

*Moral and political traits.* How does the candidate understand the concepts of Soviet patriotism, military duty, hatred of the Fatherland's enemies, vigilance, and socialist internationalism? Does he read the newspapers regularly?

*Background data.* Parents' occupations and their relationship; composition and financial status of the family; characteristics of school days (academic success, discipline, interests); work experience (discipline and focus on work, interest in it, tendency to fatigue, awards, reprimands); athletic participation; use of alcohol, tobacco, or other narcotics, physiological reactions to them, at what age and under what circumstances did the candidate start to smoke.

*Hobbies.* Interests, e.g., music, theater, sports, literature, theoretical disciplines, mathematics, physics; interest in volunteer work; participation in Komsomol projects, positions held and success in them; leisure and recreational activities, including mental (chess, word puzzles) and physical hobbies (sports, physical work); interest in do-it-yourself projects and crafts.

*Character and temperament.* Has he had to overcome significant adversities to achieve a personal goal? What specifically? Did he succeed? Behavior in situations of danger or hardship? How easily does he experience fatigue after working, studying; is he bothered by noise, distractions? Does he need quiet or other special conditions in order to work well? How does he typically work (fast or slow, at an even pace, systematically or in spurts)? In everyday life is he serene or irritable (dominant mood: elevated, composed, or depressed; even-tempered or prone to mood fluctuations)? How does the candidate tolerate the need to wait (calmly, with moderate impatience, or with extreme irritation)? Have there been instances of gross violation of discipline or serious conflicts in school or at work, and what form did they take? Can he readily shift attention from one type of work to another, or is this difficult? Does he show initiative and creativity in his work, or does he need guidance? Does he get nervous before he must make reports or presentations, or take exams, for a long or short period? Is his nervousness mild or severe? How rapidly does he calm down? Does he retain the appearance of equanimity and confidence? Does he fall asleep and wake up slowly and with difficulty or rapidly and easily? Does he feel energetic and rested when he awakens?

*Characteristics of memory, thinking, and attention.* How does the candidate best remember material: presented visually or auditorially? What does he find easier to learn and retain: numbers, names, meaningful information, or something else? What is the dominant mode of thinking, practical or abstract (theoretical)? What sort of judgments does he voice about the profession of cosmonaut? How well does he analyze, generalize, and compare? How rapidly and for how long can he concentrate on the performance of some task? How carefully does he listen to the questions? Does he get distracted during the interview? Is he able to attend to everything said to him? Does he reply to all questions and remarks? How well does he switch from one topic or question to another?

After the interview, the candidate is assessed according to a nine-point scale, and the interviewer writes an evaluation reflecting his subjective impression and opinion concerning how promising a candidate is for training.

### Individual and Group Psychological Tests

Cosmonaut candidates are observed directly or indirectly under a variety of conditions. They participate in a number of individual and group-administered psychological tests to assess psychophysiological, electrophysiological, clinical, and physical fitness.[16]

Group-administered tests include analytic paper-and-pencil, verbal, spatial, and performance tests, such as compass, clock, copyreading, rule induction, alphanumeric clusters, scales, battery of special physical exercises, and others. Individually administered tests include analytic and synthetic tests and those using charts and apparatus, such as number search, number search with shifts, psychological selection apparatus, motor coordination and stress, and others.

One important group test is the mutual talking test, which assesses compatibility. Two candidates are designated as "partners" and then asked to take part in various experiments developed to simulate mutually dependent activity. When the partners begin to help each other, their pulses begin to synchronize.

> The higher the pulse rate, the greater the compatibility of the crew members and the more successful their cooperation in orbit will be. For Lyakhov and Ryumin, this factor was 0.75. For Kovalenok and Ivanchenkov it was 0.78. It has also been noticed that opposites can be compatible: the reserved and silent introvert can be paired with an extrovert. The prerequisite for compatibility is based on mutual understanding on both the intellectual and emotional level.[17]

Personality inventories are used to generate information about an individual's personality based on his responses concerning his likes, interests, sociability, behavior, and responses to various situations. Soviet behavioral scientists are very ambivalent about using Western personality tests for pilot and cosmonaut selection.

> Without a doubt, a positive attitude toward personality tests because of their suitability for solving a number of applied problems does not imply agreement with their ideological interpretation, which is generally accepted in the West. However the need for rapid solution of such important problems as psychological selection and certification of pilots does not allow us to wait for the moment when a battery of validated Soviet tests will be ready for these purposes. B.M. Teplov expressed himself very cogently with regard to the harsh criticism of Western methods of psychological diagnosis: "Our criticism of their methods gives us the right to fail to collect any data, but instead make do with common sense and random observation. This will not do. American differential psychology is nevertheless a scientific discipline, of course, not without faults, but this is still better than nothing."[18]

Despite the officially restrained attitude toward Western personality tests, it is virtually unanimously believed that in principle such tests can be used to identify real and enduring personality characteristics. Thus, from the beginning of the Soviet space program, Western psychological tests have been used almost exclusively.

Faithfully translating a psychometric test from one language to another and providing norms for that test from the country where the test is to be given comprise a delicate and required process. It is frequently necessary to restandardize normative data for the scales of the inventory for the appropriate professional group. This is precisely what the Soviets did when they appropriated such Western tests as the MMPI, the 16-PF, and others for the purposes of pilot, and later cosmonaut, selection.

*Minnesota Multiphasic Personality Inventory (MMPI).* The MMPI (renamed the SMIL) has been used in the Soviet Union for a variety of purposes besides cosmonaut selection (e.g., clinical diagnosis, selection of pilots and flight certification, and psychodiagnosis in sports psychology).

Experience with using the MMPI for selection to flight school and in various types of pilots was described in a previous section. The difficulty in the interpretation of U.S.S.R. SMIL data centers around the redefinition of scales and the interpretation of those scales. In reading descriptions of SMIL profiles, it is difficult to determine how the interpretation was made from the specific data. In order to compare the Soviet results with the American and Japanese results, the interpretive scheme generally used in the United States will be followed, although this may lead to erroneous conclusions.

*Cattell's 16-Personality Factor Test (16-PF).*[19] The scales of this widely used American test were developed a number of years ago, and it is a test of *normal* adult personality (remember, the MMPI was primarily developed to assess psychopathology, not normal aspects of personality). Personality traits measured by the 16-PF are warmth, intelligence, ego strength, dominance, impulsivity, group conformity, boldness, tender-mindedness, suspiciousness, imagination, shrewdness, guilt proneness, rebelliousness, self-sufficiency, compulsivity, free-floating anxiety, extroversion, anxiety, tough poise, and independence.

In the United States this test is primarily used in business settings to predict job-related criteria for employee selection, placement, and promotion. Interestingly, despite the fact that Raymond Cattell was an advisor to Ruff and his co-investigators, the 16-PF has not been used in astronaut selection in the United States.

*Eysenck Personality Inventory (EPI).*[20] This personality inventory was intended to measure two sources of personality questionnaire variance reported by Eysenck in 1963 from a large number of factor analytic studies. Two important characteristics of personality are measured: extroversion-introversion (E) and neuroticism-stability (N). The EPI uses nine items from the Lie (L) scale of the MMPI to determine whether the subject is answering questions in a socially positive light. For healthy (i.e., non-psychotic) samples, E and N are relatively orthogonal and have a minimal correlation. The test is not used much in the United States anymore.

*Thematic Apperception Test (TAT) and Rorschach.* These projective tests are the most commonly used in the United States. They were both used in the *Mercury* selections, and the Rorschach was also used in the *Gemini/Apollo* selections. See Appendix 1:A for further description of the *Mercury* psychometric tests.

*Rosenzweig Frustration Test.* This is another type of projective test that has been available for over forty years. This test is also known as the P-F Study. It consists of twenty-four cartoon-type pictures, each depicting two people in a very common, mildly frustrating situation. The subject is asked to provide a reply for the frustrated person in the picture. The basic purpose of the test is to assess the type of aggression and the direction of the aggression in the subject.

*The Potter Test.* This appears to assess the effect of temporary situational factors on Rorschach test results. Again, it does not seem to be much in use in the United States.

*Speilberger Anxiety Test.* The scales of this test measure state and trait anxiety characteristics.

*Luscher Color Test.* The Luscher Color Test had great popularity in the United States in the late sixties. It assesses personality and emotional characteristics by preferences for different colors. It never achieved great significance as a reliable predictor of personality style, although it appealed to many non-psychologists.

*Personality Scale of Taylor.* J. A. Taylor developed this personality scale in 1953 to assess manifest anxiety in subjects.[21]

Numerous other tests were developed to assess cosmonaut "psychic efficiency and readiness for operating activity," tolerance of extreme conditions, concentration, attention, memory, reasoning, spatial orientation, sensorimotor function, and ability to function under time stress.

## THE FIRST COSMONAUT SELECTION

The initial selection of the first group of Soviet cosmonauts was conducted in two stages. Yuri A. Gagarin, who was later to become the first human in space, described the process:

> I was called before a special medical panel. There were many physicians, each as strict as a public procurator. Appeals to decisions were not taken; the cosmonaut candidates were taken off the list with lightning speed. Internists, neuropathologists, surgeons and laryngologists made the rejections. Our hearts were of primary focus; the doctor "read" our biography by looking at the heart. Nothing could be kept secret. The complex equipment uncovered all, even the smallest flaws in our health. The drop-out rate was high. Out of ten people, only one would be selected. The committee chairman joked as he escorted the rejected candidates out saying, "you can fly, but not higher than the stratosphere." The ability to work under difficult conditions was examined. We had to produce mathematical rules with numbers that first had to be found in a special table. The speed of work and correct answers were noted. . . . The clinical and psychological examinations, begun by the first commission were then continued. In addition to evaluations of health, the doctors looked for hidden defects or lowered resistance to conditions common in space flight and evaluated the reactions to such conditions. The examinations were conducted with the use of all kinds of biochemical, physiological, electrophysiological, and psychological methods and special function tests. . . . We were also asked about our . . . family, friends, and community service. Along with health, our intellectual and social interests and emotional stability were evaluated. . . . All of this took several weeks. Again several people were excluded. I remained among the selected pilots of cosmonaut candidates. Among the other candidates were German Titov, Andrian Nikolayev, Pavel Popovitch, Valeri Bykovsky, Vladimir Dosarov, Pavel Belyaev, Alexei Leonov and other friends.[22]

It is noteworthy that Gagarin failed to mention Valentin Bondarenko, a fellow cosmonaut candidate who died in a flash fire in a hyperbaric chamber accident before he ever flew in space. It was many years later before the details of Soviet space disasters were enumerated.[23]

The clinical psychological selection of cosmonauts was conducted in two stages lasting nearly a month each. The Soviet psychological selection philosophy is that "it is an ongoing process that spans the entire cosmonaut training period starting at the first clinical psychological selection through preparation for an actual flight and the period between flights."[24]

Soviet psychiatrists believe that psychopathological conditions such as psychotic depression, hallucinations, and other major mental illnesses will not occur during long-term space flight if selection and training methods are accurate. It is probably too optimistic and too soon in the history of space flight to reach this conclusion, since such disorders occur with reasonable frequency in the general populations of both countries, and no selection process—however rigorous—can reliably rule out their development 100 percent in even the healthy population of aviators. It has been pointed out that many of these emotional disorders, including psychosis, are likely to be induced by environmental or toxic factors in the spacecraft.[25]

Specific psychometric results from cosmonaut selections have not been published, and the data are generally classified. Recently SMIL/MMPI data were obtained from a past cosmonaut selection. Since the data were passed on anonymously, they may or may not be what they purport to be, but since their publication violates no individual confidentiality of subjects, these data are reproduced in Figure 9.4. As in the case with the Soviet pilots, the data demonstrate a significant difference between cosmonauts and U.S. astronauts, particularly in the LFK configuration.

## CREW SELECTION

When the crew size is greater than one, it is clear that interpersonal interactions play a large role in ability to adapt to space flight. This was not a big concern for *Mercury*, but it is interesting that during *Gemini* the U.S. program did not invest any time, effort, or money in studying this issue. One explanation is that for missions of short duration, like *Mercury, Gemini,* and *Apollo,* the number of individuals interacting and interpersonal interactions are not of great operational significance.

But the Soviet space program expected that their cosmonauts would remain in the space environment for long periods of time. Hence, their program has emphasized research in group compatibility, psychophysiological compatibility, psychological training, and the provision of psychological support to space missions. Each of these aspects of the Soviet space program will be discussed separately.

## GROUP COMPATIBILITY

In the book *Space Academy,* the authors write: "In the assessment of group compatibility, such parameters as interaction, mutual understanding, communication, intragroup management, and cohesiveness are important. The study of the characteristics of group functioning plays a leading role in improving the compatibility of manned spacecraft crews."[26]

Individual personality factors also play a key role in determining group performance. When forming a crew, the Soviet behavioral scientists look to exclude such traits as "hypochondriacal sensitivity, depressive tenden-

**Figure 9.4**
**Soviet Cosmonaut SMIL Profile**

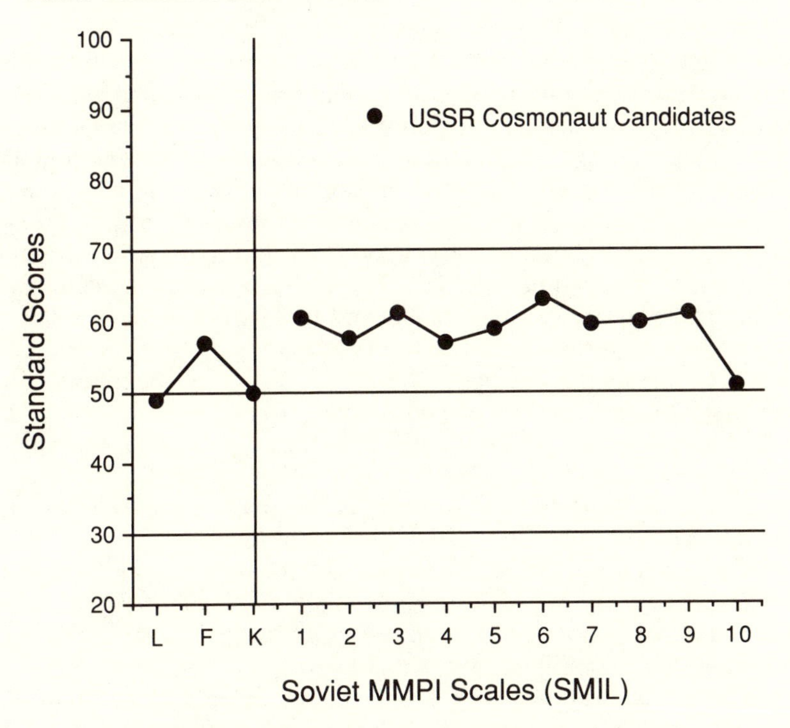


1- Hypochondria
2- Depression
3- Emotional lability
4- Impulsivity
5- Masculinity/courage
6- Rigidity
7- Anxiety
8- Nonconformity/individualism
9- Activity/optimism
10- Introversion

cies and emotionalism [which] lead to free-floating anxiety, affecting mental and physical health."[27]

The major psychosocial parameters considered in selecting U.S.S.R. space crews include

- similarity of crew members' values
- social and motivational attitudes toward performance of work
- a combination of complementary personality and character traits, with the commander having predominantly positive personality traits and qualities
- complementary objective/productive cognitive styles
- complementary job-related skills

- positive emotional attitudes of crew members toward each other
- crew members who learn rapidly and efficiently

To achieve group cohesion and facilitate intragroup understanding, crews are trained during five- to seven-day pressurized-chamber experiments, which simulate a set of psychoecological space flight factors (e.g., a three-day period of sleep deprivation with continuous work). The crew also undergoes survival training together in various climates and geographical zones, and periodic psychological examinations are done to assess effective group interaction. The Soviet experience in group training suggests that the length of time training together is one of the most important factors in increasing a space crew's compatibility and ability to work together. Thus, for short-term flights of one to two weeks, the group training period is usually at least six months; for flights of one to two months, the training time is increased to one year. For long-term flights from two to twelve months, the training period may take one and a half to two and a half years, depending on the specifics of the mission. Beregovoy states, "Only after comprehensive psychological study of the group development of a crew is the final decision made about its participation in a space flight."[28]

Psychological incompatibility among U.S.S.R. crews is a serious matter. The Soviets view psychological incompatibility within a crew as a reason to change the cosmonauts assigned to the crew. In the U.S. program, such psychological incompatibility, while it does exist, is generally ignored (even by the crew members themselves), since flights are of short duration. But Soviet crews have been grounded before flight because of psychological incompatibility. This does not necessarily mean that the cosmonauts in the grounded crew are grounded permanently. The Soviets take a rather practical view of such matters and believe that the cosmonaut may perform acceptably as part of another crew. Since selection is an ongoing operational process, the Russians consider the cosmonauts only trainees, and as such, they are only prospective cosmonauts.

Final selection is not made until just prior to the actual space flight. Various cosmonauts have mentioned that the prime crew for *Soyuz-13* in December 1973, composed of Leo Vorobyov and Valery Yazdovsky, was removed only one week before launch on advice from the psychologists. Both men were brilliant individuals, but they made a bad team. Georgi Grechko—another cosmonaut—explained that if nothing would have gone wrong on the flight, it would have resulted in tremendous success, but if any problems arose, it could have been a disaster because the two men did not work well together in emergencies. None of these reports, however, are official or published in the Soviet scientific literature.

When Soviet scientists select individuals to fly on a particular space crew, they have specific indications, and both relative and absolute contraindications, on which their decision is based. Potential crewmates must have similar values and goals, and none must have personal ambition or an overly assertive personality. All crew members must have positive emotional attitudes toward each other and demonstrate no aberrant traits or hostility toward other crewmates. Finally, potential crewmates must share a common desire to collaborate and learn quickly and efficiently together. The presence of rivalry among crew members or their inability to learn to work together is an absolute contraindication for forming a space crew.

## PSYCHOPHYSIOLOGICAL COMPATIBILITY

The concept of psychophysiological compatibility is an interesting one, and uniquely Soviet. "A psychophysiological approach allows for the unique formation of a small group based on the principle of psychological compatibility. In this, one assumes an accurate and objective recording of behavioral and psychophysiological reactions of both the group as an entity, as well as separate individuals in experimental situations modeling interdependent group activity."[29]

The basic premises for such an approach were first developed by V. M. Bekhterev, who wrote what are known as the Organic World Laws for groups and collectives. Bekhterev identified certain collective associative reflexes, which later Soviet researchers built upon, citing the "methodological paucity" of his work and the "limited psychological technology of his time"[30] as the reason why the theory was not fully developed. In the hands of later Soviet researchers, this technology has presumably been fully developed, in spite of the lack of reproducibility of the work in any other part of the world. There is a suspicion that much of the work in this area is "Soviet humbug" at worst (based more on the communist philosophy than on scientific principles); at best, some useful ideas might derive from the concept. Similar work in the United States and other countries, however, has never replicated the Soviet work. The concept is presented here because it has been used extensively in the psychological selection of cosmonauts and cosmonaut crews.

Novikov writes:

> The dissatisfaction with sociometric and other widespread socio-psychological methods in resolving problems in group selection of space crews, led to the development of integrative methods which model instrumental behavioral and verbal interactions, such as the "Homeostat" and "Pair-Matching Test," respectively. A valuable part of these methods is the possibility of recording physiological indices which reflect the level and peculiarities of the emotional-autonomic stress during experimental activity and the determination of the "physiological cost" in participating in an interpersonal process.[31]

### The Homeostat Test

This is a group test whose underlying principles were first described by F. D. Gorbov. Gorbov observed that when individuals using showers supplied from a common limited source of hot water were satisfied with only moderately warm water, the system rapidly reached a state of equilibrium. But if one individual decided to have hotter water, then the whole system was thrown off balance, and the other individuals would have to adjust the temperature of their own water, which was now cold.[32]

Cosmonauts undergo a simulation—the Homeostat Test—based on Gorbov's observation. In order to achieve an equilibrium, or homeostasis, the members of the group must coordinate and cooperate in their actions. Psychologists then interpret the results of such tests to identify individuals who may not be able to participate productively in a group situation.

### The Pair-Matching Test

This is a test to determine the compatibility of individuals based on similarities of word association among cosmonauts.

### Complex Living Conditions

Another group test situation is one termed complex living conditions. Complex living conditions include a variety of environmental simulations—isolated, confined quarters; open space; living in various climatic conditions; altered diurnal schedules; sleep deprivation; and continuous work performance—all with a rigid control of interpersonal relationships, to study cosmonaut candidate neuropsychological endurance. These types of experiments often result in dramatic decreases in performance and physiological tolerance of the human body during and after task performance. "Poor endurance of exposure to complex living conditions may take the form of clinically distinct psychoneurotic disturbances; elevated emotional tension and fatigue; worsened operator performance; accentuation of negative personality traits; or poor performance on tests requiring interdependent cooperative types of activities and joint actions by crewmembers."[33]

Based on the above tests, Soviet behavioral scientists have classified group structure into the following types:

*Nominal groups.* These types of groups are not considered real groups at all. They represent the association of several people gathered for an activity. These types of groups have no strong ties and no requirements to perform interactively. Prolonged association of members in nominal groups may lead to their developing into a connected group.

*Connected groups.* Further development of intragroup processes directed toward achieving homeostatic balance causes the connected groups to have increased intragroup communications and improvement in informational channels.

*Consolidated groups.* The consolidated group's capability or incapability to further develop is in direct relation to the group composition and the relationship of the psychophysiological peculiarities of its members. This ability is connected with the presence or absence of psychophysiological compatibility or conformity among the group members. This phase assumes, besides the emotional and informational communications, their optimum utilization.

*Integral groups.* The integral group is considered to be the most developed group type, and it assumes psychophysiological compatibility among its members. It is characterized by high group learning ability, a high level of group performance, mutual empathy, uniformity in emotional expression, flexibility, and the coincidence of similar goals, tastes, and habits among its members.

## ISOLATION CHAMBER TEST OF COSMONAUT STABILITY

The Soviets have frequently used isolation tests to measure the tolerance and stability of cosmonauts for long-duration space missions characterized by isolation and confinement. These studies also reveal how an individual will perform doing routine, daily, repetitive, and boring tasks. Depending on the scenario, isolation studies may also be done to demonstrate how well the individual can perform under stress. The most common stress imposed by the Soviet researchers is sleep deprivation. One expectation of cosmonauts is that they be able to maintain a high level of performance over a long period of time. These isolation tests have been used to select cosmonauts for long- versus short-duration flights, and also for training purposes. "Our long-term isolation chamber experiments have shown such personality training significantly increases neuropsychological tolerance in mastery of altered living conditions through anticipation of the conditions to be mastered and thus fosters the development of an individualized, efficient, flexible system of attitudes toward the environment."[34]

The long-term isolation experiments were designed by Gorbov and then modified by Kuznetsov, who was interested in studying personality under isolated, confined conditions.[35]

A U.S. congressional delegation visited the Institute for Biomedical Problems in 1985. Dr. Oleg Gazenko was, at the time, director of the institute. Gazenko was asked about the isolation tests done in Star City with cosmonauts. He stated that scientists there had discovered that "there are two types of people: those who can tolerate the isolation for long periods of time, and those who cannot."[36] Gazenko said that the isolation chamber was a good way to find out which type of person a cosmonaut was, and he implied that crew selection for long missions and short missions was made on the basis of the results, although this has not been officially documented by the Soviets.

## PSYCHOLOGICAL TRAINING

Over the years, the Soviets have developed an exceptionally thorough psychological training program. The purpose of the program is not so much to train crew members in psychological techniques as it is to facilitate their adaptation to real-life stressful situations so that they will have the confidence to deal with any type of situation when in space. The usual training of underwater tank dives in preparation for extravehicular activity, centrifuge, pressure chamber, and isolation testing is augmented by real-life activity—for example, remote survival missions and parachute training.

Psychological training is believed by Soviet space management to be one the major factors in resolving problems of space flight safety. "The basic component of psychological training is to mobilize the necessary psychic functions of a cosmonaut by means of pedagogical and psychological methods to help him actively form a conceptual model of the coming mission, to teach him to bring into effect his psychological reserves in the most efficient way."[37]

## PSYCHOLOGICAL SUPPORT GROUP

The psychological support team was first put together by the Soviets in response to a real operational problem during the 1977 flight of Yuri Romanenko and Georgi Grechko on *Salyut-6*. Victor Blagov, the deputy director of the 366-day mission of Musa Manarov and Vladimir Titov in 1989, told reporters in an interview about the origins of the psychological support concept.[38] During the three-month mission of Georgi Grechko and Yuri Romanenko, Georgi's father died. Grechko and Romanenko had been very close, and the mission managers were unsure how to handle the situation.

> What were we supposed to do? Tell the cosmonaut the bad news and give him the right to interrupt the mission? Or conceal the truth? We chose the latter. A group of technical experts, doctors, psychologists, and Georgi's relatives invented all kinds of stories to justify his father's absence on the line, though Georgi always wanted to talk to him. We managed to survive for the whole 96 days of the flight. When Vladimir Shatalov, who was responsible for the mission, told Georgi about the tragic event after the landing, the cosmonaut said we had been right. That was

the baptism, so to speak, of our psychological support team, though psychologists have been involved in the preparation of flights since the first one by Yuri Gagarin.[39]

The goal of the psychological support team has been to prevent the occurrence of emotional disorders during long-term space flights. As can be seen from the above example, their tactics may even include deception. The success of the mission is of primary importance—even more important than the individual cosmonauts in space.

The activities of the team range from preflight psychological training to the control of autonomic function using biofeedback, muscular control, and self-hypnosis.[40] Dr. V. I. Myasnikov, one of the key psychologists on the psychological support team, told a visiting U.S. psychiatrist that the concept of the "collective" must always be kept in mind. The psychiatrist later wrote: "Under this concept, an individual's personal goals and values are subordinated to the goals and values of society at large. Consequently, cosmonauts are aware that their activities not only affect each other but also reflect social principles that take precedence over their own individual desires and interests."[41]

Myasnikov and his colleagues have suggested that this awareness on the part of the cosmonauts has led to the voluntary suppression of interpersonal tension on Soviet space missions. Soviet culture has traditionally embraced the concept of the "collective" as superior to the individual, and Soviet citizens have had this drilled into them from early youth. Indeed, one obvious selection criterion for cosmonauts has been that they are "good Soviet citizens" (which usually means that they must be members since youth of the Communist party). Visitors to Star City, the center for cosmonaut training, are frequently taken on a tour of the cosmonaut museum, where cosmonauts have donated personal mementos for display in individual niches. Almost every niche included the cosmonaut's personal copy of the Communist Youth League Handbook.

Cosmonauts are reluctant to talk publicly about hostility among crew members. This is not unusual and is also characteristic among U.S. astronauts. But, typically, by thirty days into the mission, crews begin to show signs of hostility, in spite of the diligent testing to arrange personality compatibility. The interpersonal hostility is usually controlled, but the antagonism between crew members can spill over to space-to-ground communications. Cosmonauts sometimes hold back "confidential communications," deliberately hiding information and reactions, and show agitation at "unnecessary questions" put to them by ground personnel. Rechanneling anger and aggression toward persons on the ground (which is safer than focusing on a crew mate upon whom your life frequently depends) may be a typical emotional defense used by astronauts and cosmonauts on future long-duration space missions.

It is a myth that the forced closeness of an isolated, confined environment is always conducive to the development of deep friendships—although that happens. Usually the latter is more the exception. When asked how cosmonauts generally feel about their crew mate after an extended mission, one cosmonaut told me in 1989 that "most of us don't even want to go through postflight rehabilitation in the same city"—preferring, in many cases, not to see their comrade again. When the personalities are matched correctly, however, deep and long-lasting friendships have occurred.

During preflight training, the team is involved in helping cosmonaut crews learn how to live and work together in the most harmonious manner. Group dynamics has always been of concern to flight planners. In planning for a mission in 1971, the Soviets sent the *Salyut I* crew on a full-month auto tour of the Soviet Union to see if they could get along together.[42]

For practical reasons, the psychological support team has focused on a number of accepted human factor issues, including the planning of work/rest schedules, habitability, and an emphasis on ergonomic factors. The last two factors, however (from glimpses of the interior of the *Salyut*), would not impress even the most poorly trained human-factors psychologist in the rest of the world. By all accounts, the habitability of the *Salyut* Space Station was abysmal. Particulate matter floated freely throughout the cabin, which was cluttered on every surface with wires and work material. Privacy was limited, and as far as aesthetics were concerned, *Salyut* was no more than a messy, dirty, and depressing place to live. Matters apparently have improved somewhat on *Mir.*

The support team also helps cosmonauts spend their "free" time while living in space, providing movies, videos, shows from Soviet TV, and audiotapes. Probably the most important innovation was their introduction of two-way TV, enabling cosmonauts to spend time talking with members of their family, famous personalities, and friends. A great deal of thought is taken to make sure that each cosmonaut receives favorite items, letters, and other "surprises" when a visiting resupply capsule comes to the station. Cosmonauts are usually able to talk with family members once a week, although this pattern may change as the mission progresses. Of course, the support team is also available for other inflight problems and actively monitors the crew throughout a mission. Several methods are used for this purpose, including following vital signs such as heart and respiratory rate and galvanic skin response. Specialists also observe the crew interact and watch them individually for signs of imminent problems.

One of the more dubious methods used to evaluate psychological state is voice stress analysis. A large research effort in this area was undertaken by Soviet scientists who believe that the method is accurate and of great operational significance.[43] Unfortunately, individual variations in speech are generally so great that most U.S. researchers have long abandoned such an approach for assessing psychological stress. In truth, much of the psychophysiological monitoring done in the Soviet Union may give the appearance of a situation well in control, but in fact there has been little in this research area to warrant the Soviets' excessive claims of success for their methodology. In fact, it is well established that psychological and behavioral states are not in the least equivalent to physiological state, nor can they be reliably predicted by such data.

However, the notion of monitoring and identifying potential individual and group psychological problems before they become mission or operational problems is generally a useful one. The new Russian and the U.S. space programs must undertake a systematic research effort to determine the best methods to prevent problems. This has not been done, and, quite frankly, the Soviets' lavish praise of their efforts is somewhat contradicted by the fact that there have been a number of well-publicized (and some deliberately hidden) interpersonal and individual psychological problems on their long space missions.

For example, at a recent meeting at the Johnson Space Center, a Soviet flight surgeon confided to NASA flight surgeons that the real reason a cosmonaut was brought back early from a mission was not prostatitis (as was announced at the time)—a condition that could have been adequately treated in space—but was, in fact, what the flight surgeon termed psychoneurotic prostatitis. Apparently, the cosmonaut was obsessed with the idea that he was becoming impotent. Whether this story is true or not, it is typical of the subjective reports that frequently circulate in both the U.S. and the U.S.S.R. space programs, as opposed to documented facts. Other reports circulated about this same incident indicated that the cosmonaut in question had a high fever and was very ill physically (all of which would contradict the presence of a psychological problem as the basis of his behavior). The absence of formal reports about such incidents is extremely interesting, and, as mentioned before, this absence leads to rumors and innuendo. As Sherlock Holmes once commented in *Silver Blaze,* the fact that the dog did nothing in the nighttime was the most important clue in the case.

## SOME FINAL THOUGHTS ON U.S.S.R. PSYCHOLOGICAL SELECTION

The Soviet space program's willingness to look at individual and group psychological factors is certainly laudable. But, in spite of the special attention given to psychological compatibility when putting together crews, it is interesting that the number of anecdotally reported interpersonal or emotional problems on Soviet missions has been rather high (e.g., the stories of the space crew that refused to talk to ground control personnel and of the supposed "panic" symptoms developed by a Soviet spacewalker). One may conclude from this that either (1) the techniques used by the Soviets are not particularly good (in fact, there is no evidence in the scientific literature that demonstrates their effectiveness, reliability, or validity) or (2) in spite of good techniques, the psychological problems encountered by individual crew members and crews as a whole are so great that even more attention must be paid to the issue. Of course, even if the first assertion is true, it seems likely that the second is also true.

As the Soviets have gained more confidence in their psychological methods, they have repeatedly stated that their psychological selection procedures have been based on more and more restrictive psychological standards. At the same time, the general medical standards relating to cosmonaut selection were being made *less* restrictive. It has been exactly the opposite in the United States—where psychological and medical standards not only are less restrictive than initially, but also are frequently waived. The Soviets' tightening of the psychological standards—more than any other action—supports the assertion that psychological factors may indeed become the most important consideration in future space exploration, even more important than other medical disorders.

## NOTES

1. N. C. Chriss, "Psychological Training for the Cosmonauts," *Houston Chronicle,* January 16, 1988, p. 3.
2. G. Grechko, Comments made during a visit to the NASA Johnson Space Center (May 1989).
3. S. Bloch and P. Reddaway, *Psychiatric Terror: How Soviet Psychiatry Is Used to Suppress Dissent* (New York: Basic Books, 1977), 14.
4. "Soviets Reenter World Psychiatric Society," *Science News* 136 (1989): 278.
5. "Soviets Gain WPA Readmission with Conditions," *Psychiatric News,* November 3, 1989, p. 1.
6. N. F. Lukyanova, "Personality Tests," in *Psychological Selection of Pilots and Cosmonautics,* edited by V. A. Bodrov, V. B. Malkin, V. B. Pokrovskiy, and D. I. Shpachenko, vol. 48 of *Problems in Space Biology,* translated by Lydia Stone, edited by B. F. Lomov (Moscow: Nauka Press, 1984), 108–125. Emphasis added.
7. Ibid., 122.
8. V. A. Bodrov, V. B. Malkin, B. L. Pokrovskiy, and D. I. Shpachenko, "Development and Status of Psychological Selection of Pilots and Cosmonauts: A Short Historical Review," in *Psychological Selection of Pilots and Cosmonauts,* vol. 48 of *Problems in Space Biology,* translated by Lydia Stone, edited by B. F. Lomov (Moscow: Nauka Press, 1984), 9–37.
9. V. A. Bodrov, V. B. Malkin, B. L. Pokrovskiy, and D. I. Shpachenko, "Psychological Selection and Professional Performance of Pilots and Cosmonauts," in *Psychological Selection of Pilots and Cosmonauts,* vol. 48 of *Problems in Space Biology,* translated by Lydia Stone, edited by B. F. Lomov (Moscow: Nauka Press, 1984), 192–217.
10. L. N. Sobchik, *Manual for Use of the MMPI Psychological System* (Moscow: RSFSR, 1971).

11. M. A. Novikov, "Psychophysiological Selection, Crew Manning, and Training for Space Flight," in *Psychological Problems of Space Flight,* translated by N. Timacheff (Moscow: Nauka Press 1979), 196–204.

12. G. T. Beregovoy, V. N. Grigorenko, R. B. Bogdasherskiy, and I. N. Pochkayev, "Characteristics of Psychological Selection," in *Kosmicheskaya Akademiya* (Space Academy), translated by Lydia Stone (Moscow: Mashinostroyeniye, 1987), 16–25.

13. B. J. Bluth and M. Helppie, *Soviet Space Station Analogs,* 2d ed., NASA Grant NAGW-659, Washington, D.C., 1987, IV-48 to IV-57.

14. Y. A. Senkevich and M. M. Korotayev, "Medical Selection Methods and Criteria for Initial Qualification of Cosmonauts and Placement in Group Training for the Purpose of Standardizing Psychological and Physiological Examinations," translated by N. Timacheff (unpublished paper, 1991).

15. V. A. Bodrov, V. B. Malkin, B. L. Pokroskiy, and D. I. Shpachenko, "Research Methods of Experimental Psychology," in *Psychological Selection of Pilots and Cosmonauts,* vol. 48 of *Problems in Space Biology,* edited by B. F. Lomov (Moscow: Nauka Press, 1984), 90–108.

16. Bluth and Helppie, *Soviet Space Station Analogs,* IV-42–IV-47.

17. Ibid., IV-53.

18. Lukyanova, "Personality Tests," 306.

19. R. B. Cattell, *The Description and Measurement of Personality* (Yonkers-on-Hudson, N.Y.: World Book Co., 1946). R. B. Cattell, *Personality and Motivation Structure and Measurement* (Yonkers-on-Hudson, N.Y.: World Book Co., 1957).

20. S. Eysenck and H. J. Eysenck, "An Experimental Investigation of 'Desirability' Response Set in a Personality Questionnaire," *Life Sciences* 5 (1953): 343–345.

21. J. A. Taylor, "A Personality Scale of Manifest Anxiety," *Journal of Abnormal and Social Psychology* 48 (1953): 285–290.

22. Y. A. Gagarin, *The Road to Space* (Moscow: Nauka Press, 1978), 96–97.

23. J. E. Oberg, *Uncovering Soviet Disasters* (New York: Random House, 1988).

24. Bodrov, Malkin, Podrovskiy, and Shpachenko, "Development and Status of Psychological Selection of Pilots and Cosmonauts," 15.

25. P. A. Santy, "Psychiatric Support for a Health Maintenance Facility (HMF) on Space Stations," *Aviation, Space and Environmental Medicine* 58 (1987): 1219–1224.

26. Beregovoy, Grigorenko, Bogdasherskiy, and Pochkayev, "Characteristics of Psychological Selection," 4–5.

27. Ibid., 7.

28. G. T. Beregovoy, V. N. Grigorenko, R. B. Bogdashevskiy, and I. N. Pochkayev, *Kosmicheskaya Akademiya* (Space Academy), translated by Lydia Stone (Moscow: Mashinostroyeniye, 1987), 18.

29. M. A. Novikov, "Principles and Methods of Studying Psychophysiological Compatibility" (paper presented at the Twelfth U.S./U.S.S.R. Joint Working Group Meeting on Space Biology and Medicine, Washington, D.C., November 9–22, 1981), 1.

30. Ibid., 1.

31. Ibid.

32. Beregovoy, Grigorenko, Bogdasherskiy, and Pochkayev, "Characteristics of Psychological Selection," 5.

33. Ibid., 6.

34. Ibid., 7.

35. Personal communication to author, October 1985.

36. O. N. Kuznetsov and V. I. Lebedev, *Psychology and Psychopathology of Solitude* (Moscow: Nauka Press, 1972).

37. G. T. Beregovoy, I. V. Davydov, N. V. Krylova, and I. B. Solovyeva, "Psychological Training—One of the Most Important Factors of Enhancing the Safety of Space Flights." Paper presented at the Thirtieth Congress of the International Astronautical Federation, IAF79, Munich, September 17–22, 1979, 1.

38. "Space Psychological Support: An Interview with Victor Blagov," *Soviet Life*, March 1989.

39. Ibid.,

40. V. I. Myasnikov and O. P. Kozerenko, "Prevention of Psychoemotional Disorders in Prolonged Space Flight by Method of Psychological Support," *Space Biology and Aerospace Medicine* 2 (1981): 25–29. V. I. Myasnikov, O. P. Kozerenko, A. A. Gerasimovich, and E. V. Ryabov, *Psychological Support and Psychological Adaptation of Crewmembers in Flights of Long Duration* (Moscow: Kaluga, 1979), pt. II, 5–6. V. I. Myasnikov, O. P. Kozerenko, and F. N. Ukov, "Actual Problems in Providing Medico-psychological Support in Flights of Long Duration," in *Aerospace Medicine* (Dresden: 1980), 41–42.

41. N. Kanas, "Psychosocial Support for Cosmonauts," *Aviation, Space and Environmental Medicine* 62 (1991): 353–355.

42. B. J. Bluth, "Soviet Space Stress," *Science 81* 2 (1981): 30–35.

43. V. I. Myasnikov, E. F. Panchenkova, and F. N. Uskov, "Prospects of Using Radio and TV Communication Data in the Medical Supervision of Cosmonauts In-flight." (Paper presented at the Twenty-fifth International Congress of Aviation and Space Medicine, Helsinki, September 4–9, 1977). A. V. Nikonov, "Reflections of the Dynamics of the Human Operators' Psychophysiological State in Radio Exchange," in *Psychological Problems of Space Flight* (Moscow: Nauka Press, 1979), 335–343. V. Remek, "Communication Problems of International Crews." Paper presented at the Thirtieth Congress of the International Astronautical Federation, Munich, September 17–22, 1979. I. S. Zamaledtdinov, N. V. Drylova, S. A. Kiselev, and Y. V. Trufanova, "The Possibility of Identifying Levels of Emotional Stress According to Speech Indices in Space Flight Activity," in *Psychological Problems of Space Flight* (Moscow: Nauka Press, 1979), 131–148.

# 10

# Space Station, Lunar Base, Mars, and Beyond

> Humming to myself, I float through the station. Is it possible that some day I'll be back on earth among my loved ones, and everything will be all right?
>
> —Valentin Lebedev (after 116 days in orbit)[1]

## THE ELUSIVE MISSION OF SPACE STATION *FREEDOM*

NASA's embattled Space Station *Freedom* project has gone through many iterations in its relatively short history. It was only in 1991 that plans began to move off the drawing board and into the actual testing of components in the space environment during Shuttle missions. In 1993 the Space Station barely survived by a one-vote margin in Congress. *Freedom* poses two major difficulties. First is the cost. NASA has estimated that the construction over five years could exceed $30 billion, an impossibly large amount in these times of fiscal restraint and economic upheaval. Second, and possibly even more important, is the fact that *Freedom*'s mission is vague and ill-defined. What exactly will astronauts be doing during their 30- to 120-day tour of duty? What tasks could be so important that they would justify the mission's cost to Congress and the American public? Unlike a planetary mission, where the obvious goal is exploration of the unknown, an orbiting space station does not go anywhere, but passes repeatedly over the Earth.

NASA officials have been relatively bland and certainly shortsighted in their discussions about what activities would justify the cost. Initially, the agency discussed the "great benefits" of drug manufacturing and materials science research in microgravity. However, experts disputed NASA's contention that these activities would be profitable in the foreseeable future. As evidence of this, they justifiably pointed out that drug companies and those industries that might benefit most from research in material science processing in space were reluctant to spend any money on it, at least until the technology was more developed.

Hoping to capitalize on the new environmental consciousness of Earth's citizens, NASA has proposed the "Mission to Planet Earth" as another potential goal of an orbiting space station. From its perspective a few hundred miles above the Earth's surface, Space Station astronauts could provide invaluable information on the environment. This is undoubtedly true, but why we could not continue to do this on routine Shuttle missions is not explained.

Taking another tack, NASA pointed out the valuable medical research that could be done on *Freedom*. The acquisition of knowledge about how the human body adapts to the environment of space became the next justification for the Space Station's existence. Again, the critics could look to NASA's record. In thirty years of space exploration, didn't we already have a considerable amount of data on this? Actually, the answer is no, but the public does not realize how few life sciences data have been collected over the last thirty years, since NASA has always boasted about its medical research on the one hand, while on the other, it has consistently underfunded and cut medical research programs from its priorities.

Critics also ask about data from the Soviet space program on human adaptation to long-duration exposure to microgravity. Are those data not available? Why not use and build on the Soviets' data? Why not? Because the data that are available are not particularly helpful and also because NASA has not been overly interested in obtaining Soviet data, which might only lead to reductions in the budgets of its own researchers. Although the United States and the Soviet Union have had a Joint Working Group in Life Sciences for many years, the actual exchange of medical data has been miniscule, and much of that has never been published or reviewed in the scientific literature. The Soviet and U.S. space bureaucracies share in promulgating this dismal state of affairs because, as with most bureaucracies everywhere, data frequently remain buried internally, instead of being disseminated to the scientific establishment.

The papers that have come out of Soviet space medical research are for the most part the result of poorly designed and analyzed scientific experiments, many with the number of subjects (the N of the experiment) too small to make such incredible generalizations. Also, many of the papers only report the scientists' conclusions and never mention method or even present the data that were collected so they might be objectively reviewed by the world scientific community. Now, after the dissolution of the Soviet empire, it is unclear what will happen to any potentially usable space medical or psychological data. A number of entrepreneurial undertakings by some Russians suggest that such data may be sold to the highest bidder in the future, and, assuredly, that bidder will not be NASA.

For all of these reasons, the ill-defined nature of *Freedom*'s mission will not be helped by including the data from the former Soviet Union in the justification. So the question remains: What exactly will astronauts be doing up there, and will it be worth the cost?

Perhaps the best way to answer this question once and for all is to admit that the benefits of space exploration are long-ranging and far-reaching and not immediate. It is an investment in the future of our country, our world, and our species as a whole. Perhaps some of the seemingly insoluble problems of today have answers in tomorrow's exploration of space. Space exploration in and of itself is not a business—though business will ultimately benefit from it. It is a quest for a better life. Space explorers are like the pioneers of the American West, who built log cabins in the wilderness and—against all odds—not only survived, but also thrived. Space exploration is like the westward movement of a wagon train filled with people who believe that the unexplored holds a brighter tomorrow for them and their children. It means risking lives and losing lives—all in the name of exploring the unknown, and sometimes for no better reason than because it *is* unknown. And, ultimately, that is a far more powerful reason to do it than the glory of materials science processing or even the fleeting euphoria of national prestige. Ultimately, if the American public is to pay the cost of exploring space, then it will be for the American public to decide if the investment is worth it.

With the question of economics conveniently put aside, we can now focus on speculation about future space missions and implications for psychological selection.

## A TYPICAL SPACE STATION *FREEDOM* MISSION

A great deal of thought has been given to what typical tasks would be expected of *Freedom* crew members and what the living and working conditions on *Freedom* would be like. Let us first consider the living and working environment.

### *Freedom*'s Habitat Module

Since *Freedom* is designed and redesigned on an annual basis, many aspects of the crew habitat are as yet undecided. Several mock-ups of NASA's evolving concept exist and demonstrate some of the more important facets of the design. A cylindrical module, approximately 13.6 meters long and 4.5 meters in diameter, will provide living quarters for six to eight people, with private rooms for each crew member. Although the rooms provide some personal space for each astronaut, that space is actually quite small. Each room tentatively has a window and contains sleeping accommodations, personal entertainment, and a computer for personal use by the *Freedom* astronaut. The window is the focus of a heated debate between the engineers, who have the difficult task of designing a stable structure for the habitat and for whom windows represent a weak structural point, and the astronauts, who insist that they need to be able to see "outside" the station. How this issue will finally be resolved is unclear.

The habitat module has three separate living areas, referred to as the active, buffer, and quiet zones.

The active zone is for recreational activities, exercise, and eating and also provides an area where all crew can meet together. The galley and wardroom are located in this zone. The galley is fairly advanced and is equipped with an oven, refrigerator/freezer, trash compactor, handwasher, dishwasher, and water supply. The wardroom is a multipurpose area and can be used for video teleconferencing. It is also equipped with two twenty-inch-diameter windows for Earth viewing. Designed for eight crew members, the wardroom has tabletop panels that adjust to provide various seating arrangements for the entire crew.[2]

The quiet zone is at the other end of the module from the active zone and contains private crew quarters for sleeping and relaxing. By separating it from the active area of the module, noise can be minimized.

The buffer zone, located between the other two zones, is devoted to personal hygiene, stowage, and health maintenance. This zone has controls for monitoring the environmental conditions within the entire module.

The habitat module is parallel and adjacent to the laboratory module, where crew members will do most of the mission work. Both of these modules and the node connecting them will be fully pressurized and will comprise the portion of *Freedom* that is the manned base.

In spite of the great attention to detail in designing *Freedom*'s habitat, living conditions may still be somewhat primitive. Shuttle astronauts have frequently described their experiences as similar to "camping out." The Shuttle's zero-gravity toilet has consistently presented problems, particularly for the fastidious astronaut, because of the unpleasant characteristics it takes on in the microgravity environment. While conditions have improved dramatically over earlier *Mercury*, *Gemini*, and *Apollo* missions (where a simple plastic bag had to be used without any privacy at all), many astronauts try to avoid using the toilet as much as possible during the short Shuttle missions. This behavior has resulted in abdominal cramps or reduced food intake in some astronauts during some missions. On *Freedom*, it is hoped that a better functioning and more acceptable toilet will be designed, taking into account all that has been learned from the Shuttle experience.

Food will be fairly acceptable. Freeze-dried dishes, in combination with some fresh food supplements, appear to enjoy good acceptance among current astronauts. Meal preparation and consumption for crews will be considerably different on *Freedom* than of the Shuttle. For one thing, a more careful inventory of food items will be required on *Freedom*, since consumables are limited. *Freedom* crew members will also have a microwave oven and other benefits not currently available to Shuttle crew members. Shuttle missions have been short enough not to require regular bathing or extensive cleaning facilities, but because of the length of *Freedom*'s crew assignment, several designs for a zero-gravity shower are currently being considered for the use of *Freedom* astronauts.

Privacy on the Shuttle has not been a significant problem—even with mixed crews—because of the short missions. Any possible privacy concerns should be alleviated by the attention to personal space on *Freedom*, which provides for more privacy than has been available on any previous space flight.

Crew members' free time will be spent exercising to remain somewhat adapted to one-gravity or simply relaxing. Exercise facilities are primarily located in the node between modules to minimize noise and vibration. *Freedom* crew members will have access to a VCR, laser discs, sound systems, computer games, and even "books" on the computer.

### *Freedom*'s Laboratory Module

Work activities will take place in the laboratory module. The United States is expected to develop one module, and the Japanese and Europeans will also develop one module each for different kinds of scientific work. Depending on the design, the work modules will be connected via nodes that attach the basic modules together. At the present time, Shuttle crew work activities are fairly rigidly structured. On *Freedom*, it is likely that the crew will be permitted more flexibility in time management. Most of the documentation and checklists for all crew activities will be stored on computer. Terminals or work stations, named multipurpose application consoles (MPACs) will be located at various sites in the separate work module. The daily work plan will also be accessible by computer, significantly cutting down on the documentation that has had to be carried on every Shuttle mission.

## *FREEDOM*'S CREW

Crew members will fall into one of three categories: station operator, station scientist, and payload scientist. Members of the current astronaut office strongly subscribe to having a designated commander for each *Freedom* crew. Their rationale is that safe operations require a degree of discipline and coordination among the crew that is not achieved in the absence of a formal command structure. While this typical "military" model of leadership is not surprising, coming as it does from the predominantly military Shuttle astronauts, it is not intuitively obvious to many. Leadership style on long-duration space missions is an important question—one that should be the subject of scientific research to determine which model most enhances crew performance. It is interesting to note in this connection that *Salyut* and *Mir* crew members report that democratic leaders are much easier to live with than are autocratic ones.

It has been tentatively determined that the commander would come from the station operator category of astronaut, since that job is responsible for operating and maintaining *Freedom*'s systems. The station scientists' job responsibilities include operating and maintaining the payload and scientific systems and performing scientific experiments. Station operators and station scientists will be trained for extravehicular activities (EVA) since both their jobs may require it. Station operators will remotely pilot free-flyers such as the orbital maneuvering vehicle (OMV), while station scientists will normally operate the robotic arm to capture or manipulate experiments.

Since *Freedom* is an international effort, it is expected that crew members will come from a variety of countries in Europe, as well as from Canada and Japan. A work module called Columbus from the European Space Agency (ESA) and a Japanese Experiment Module (JEM) will eventually be added to the habitat and U.S. laboratory modules. These later modules are for doing scientific research and materials science experiments. Although it is unclear how many individuals will comprise a typical *Freedom* crew, it is likely that a minimum of four individuals will be assigned initially, with up to eight crew members working on two twelve-hour shifts when *Freedom* is permanently manned. One or more crew members will be payload scientists. Payload scientists will be only generally trained to live and work in *Freedom*'s environment and will not be trained to operate the major systems of the station. Unlike station operators and station scientists, payload scientists will not be professional astronauts, but rather scientists flying in support of a major payload system. They may remain on *Freedom* for a shorter period of time than the regular astronaut crew and may not be as well integrated into the group.

All station operators will probably come from a pool of career astronauts with prior experience on the Shuttle. One significant change from the Shuttle program is that on Shuttle missions, the commander has always been a pilot astronaut, but this will not necessarily be the case for *Freedom*, since piloting skills will not be required for its operation. All station operators will be required, however, to have a good technical background, permitting them to understand at least at a basic level all aspects of *Freedom* and its payloads.

Station operators will be in constant contact with the Space Station

Control Center in order to ensure the smooth operation of the station. The commander must have a general understanding of the nature of the ongoing payload activities, since these activities may have a profound impact on *Freedom*'s operations.

It is likely that there will be frequent communications with the control center on the ground. Crew members will continuously apprise ground controllers of the status of ongoing activities. One issue that has not yet been fully decided is the allocation of responsibilities between ground personnel and *Freedom* crew members. Ground personnel should probably have less control over the day-to-day activities and decisions of *Freedom*'s crew, particularly in emergency situations. The opposite situation has existed on all prior space flights.

The station operator's daily activities are expected to include routine maintenance of *Freedom*'s many operating systems. Appropriate tools and support for routine maintenance activities must be developed. Most of the station operator's duties will center around the nodes, where the systems' controls will be located. These nodes will have a cupola window area so that the operator is able to view EVA and to control items outside the base, such as free-flyers.

The station operator will also be responsible for decisions concerning any resource conflicts and will advise and direct shift operations, lead safety drills, and be the management interface in interchanges with the ground. Due to the presence of non-career crew members, crew members of both sexes, and crew members of various ethnic and national backgrounds, the commander must be sensitive and responsive to the psychological needs of the crew. He or she will have to have the ability to deal with the inevitable interpersonal tensions before those tensions become disruptive.

The station scientist's role will be similar to that of the Shuttle's mission specialist. Astronaut scientists will be generalists who also have a good technical background. The station scientist is responsible for maintaining payload systems and performing experiments. From time to time this might require a spacewalk. The educational background of the station scientist will most likely be in the natural sciences, life sciences, mathematics, or engineering.

Station scientists will interact with payload scientists to conduct scientific experiments. *Freedom* payload scientists are analogous to payload specialists in the Shuttle program. Because they are not members of the permanent astronaut corps, payload scientists will be less familiar with both the astronaut office environment and the astronauts themselves. But payload scientists will have payload duties comparable to those of station scientists and will also participate in routine "housekeeping" duties while assigned to a mission. Housekeeping chores will be rotated among all members of the crew.

This brief mission guide gives a limited picture of what will be expected of astronauts chosen for Space Station *Freedom* flights. In general, such missions will require a higher degree of personal responsibility (since the ground crew will have less to do with day-to-day and emergency situations), a greater degree of interpersonal sensitivity (to deal with the increased number of fellow crew members, as well as the diversity of crew members), and a broader technical and scientific background. It is precisely these three qualities that were identified in Chapter 6 as comprising the psychological right stuff.

## LUNAR BASE AND MARS EXPLORATION

President George Bush expressed his administration's commitment to the goal of human exploration of Mars in a 1989 speech on the twentieth anniversary of the *Apollo* moon landing: "First for the coming decade for the 1990s—Space Station *Freedom*—our critical next step in all our space endeavors. And for the next century, back to the Moon. Back to the future. And this time back to stay. And then, a journey into tomorrow—a journey to another planet; a manned mission to Mars."[3] Whether Bush's commitment will be carried on by future administrations is not clear at this point in time. As mentioned before, the *Freedom* project is currently faced with an ongoing battle for survival. Thus, it may be unrealistic to expect that a lunar base or a mission to Mars will be a real project in the next decade or so. Nevertheless, it is not too early to begin planning for these projects of the future.

The specific scientific goals of a manned Mars mission have all been based on the previous unmanned voyages to that planet: the *Mariner 4* and 9 flybys and the *Viking I* and *II* missions, as well as earthbound telescopic observations. To date, the United States is the only nation to fly a totally successful mission to Mars, although the Soviet Union sent an orbiter to investigate Mars and Phobos in 1988. All these unmanned missions addressed geological, geophysical, and climatological questions about the red planet. In addition to research in planetary science, some investigations must be planned that are directly related to sustaining and protecting people on Mars for long periods of time. These include exploration for readily recoverable supplies of water, development of the capability to monitor solar flare activity from Mars, completion of general engineering and geological studies to support surface operations, and study of the performance of closed ecological life-support systems. The flight to and from Mars will provide opportunities for astronomy and life sciences, among other studies.[4]

A manned flight to Mars requires the solution of a number of physiological problems associated with long stints of weightlessness. Over the years, space medicine has gathered no small amount of knowledge on the effects of weightlessness on the human body. More than 200 individuals have flown in space for varying lengths of time. Five Soviet cosmonauts have lived in space for more than 200 days each, and several cosmonauts have assembled rather impressive figures in terms of number of flights (Romanenko has spent a total of 430 days in zero-gravity).

A number of detailed plans and rationales for the exploration of Mars have been written in recent years.[5] The mission scenario for the trip to the red planet varies from report to report. It has been estimated that using a general chemical-powered drive, the trip to Mars could last up to nine or more months. The most recent scenario studied at the Johnson Space Center is that of a nuclear-powered drive, which would enable a Mars mission crew to reach the planet in a relatively short period of time. But because of the orbital constraints, crew members would be expected to remain on the planet for up to 500 days before the planetary alignments would be right to return to Earth. No matter what scenario is developed, the entire mission will probably last up to two or three years.

It is too soon to determine the job characteristics for a Mars astronaut. One would hope that the next twenty years will see the necessary research done to identify the really important psychological selection criteria for long-duration space flights. We must be humble, however, and remember, as Carlos Perry pointed out, that if it were left to psychologists or psychiatrists, the exploration missions of the past might never have been undertaken. We must also remember that these missions accepted and frequently had to deal with a considerable mortality rate among the men who, for purposes of money or glory or both, made the decision to explore the strange new worlds of Planet Earth. Do we dare accept the same challenge? Could we tolerate the loss of life that goes along with such endeavors?

Our methods have developed over the years to try to minimize the risk, to predict the problems and develop possible solutions in advance of risky undertakings. We have spent a lot of time on the engineering systems to make them as reliable, as fail-safe, as possible. Relatively less time has been spent on understanding the weaknesses and potential failures of the human system. Yet, as certainly one of the most critical of the mission systems, the human component must be selected to maximize success. The tremendous cost and complexity of space missions preclude the likelihood that any individual or small group will build his or her own spaceship and go off to colonize the universe—at least in the near future. Perhaps if it were individuals and not governments who managed the effort, it would be clearer that only the competent and the capable would survive.

Behavioral science was in its infancy when American pioneers headed westward. The fittest and most clever of those pioneers survived and successfully opened the western door of America. But thousands died in the wagon trains of those days, and if they made it, they still had to conquer the hazards of a hostile environment. What were the psychological profiles of those gallant men and women—both those who succeeded and those who did not? Would they have been at all similar to *Mercury* or Shuttle astronauts? Perhaps. One important difference between the pioneers of the past and present is that the pioneers of the past were a self-selected group of individuals. They were not chosen by bureaucrats—or by psychiatrists and psychologists—and the risks (and financial investment) they undertook were personal, not national. Those individuals who did not have what it takes to successfully complete the journey West did not survive.

It is really only in the context of a government-sponsored effort (i.e., one paid for by taxpayers) to explore space that questions about selecting "the right person for the job" come up because a rigorous selection process makes sense as a method of saving money in the long run. If astronauts were simply private individuals who financed their own way into space, then behavioral scientists could simply study which individuals were successful and which were not. Since it is not possible at this time for individuals to make a personal decision to explore space utilizing private resources, governments must do the best they can to minimize the risk and cost to the average taxpayer who bears the financial burden. The personalities of past and present pioneers may actually be quite different because of this fact, with past pioneers including a great many courageous individuals (in the sense that they were willing to make the dangerous journey and accept the risks), and yet very unsuitable individuals (in the sense that they were not able to survive the journey).

So discovering the best way to choose the right stuff for space missions makes sense if risk and cost must be minimized. What then have we learned about the best way to do this from the Soviet and U.S. experiences?

## LESSONS LEARNED FROM THE SOVIET SPACE STATION EXPERIENCE

The Soviet space program has already had two functional space stations: *Salyut* and *Mir.* With thousands of man-hours of experience living and working in space, the Russians are the most likely source of data on what kind of individuals do well in space over the long run. As mentioned previously, the distinguished researcher Dr. Oleg Gazenko from Moscow's Institute of Biomedical Problems was asked this very question in 1985. Gazenko said that there were two basic types of astronauts. The first type did well on short missions—that is, they performed well and tolerated the space environment well. The second type were the kind who were not easily bored and did well on long-duration missions. How do you tell them apart? Dr. Gazenko shrugged and responded that you simply put them in isolation and see how they do over time.

During a U.S. congressional visit to Star City in early winter of 1985, an isolation experiment with two cosmonaut trainees was just drawing to a close. The two trainees had almost completed one full year of isolation together. The delegation toured the outside of the isolation capsule and watched for a few minutes the continuous TV monitoring of the events inside. The two individuals who were part of the experiment were sitting in the common room reading. There was no interaction at all between them.

The Soviet scientists who were monitoring the pair were asked whether they had ever done such an experiment with two women or one female and one male together in isolation. The psychologists were quite startled by the question. They replied in a condescending tone that they had data demonstrating that women could not possibly tolerate the stresses of such an experiment. When asked if the data could be seen, "It's simply common knowledge" was the response. Have you ever done the isolation experiment with female subjects? Their answer was illuminating. No, they had not because it just would not be right to subject women to such conditions. So these "scientists" knew the answer to the question without having to bother to collect any data to justify the operational decision. How convenient.

The behavioral scientists were simply horrified at the suggestion that women might stay for an extended period of time in space. One scientist in Star City mentioned that occasionally they might possibly permit a woman to go up in one of the short *Soyuz* flights. But, alas, there were currently no women in cosmonaut training. When they had tried to include women, he continued, they all wanted to go home to their husbands and children because the work was too grueling. This irrational attitude—particularly on the subject of female astronauts—is actually very common, as we have seen from the practices of even the U.S. Space Program.

It is interesting that many sources in the medical literature suggest that the Soviets have solved the psychological problems that occur on long space missions. They point to the presence of the psychological support group as justification that the U.S.S.R. has devoted more resources to this subject than has the United States. This latter point is certainly true. But it does not prove that the problems are solved. In fact, as we have seen, rumors abound about the interpersonal conflicts, poor morale, and psychological issues that have arisen on Soviet missions. There have been instances when a crew member panicked during EVA and when a crew member who developed a prostate infection became convinced, despite reassurances, that he was going to become impotent if he continued in the space environment.

Although most of the cosmonauts deny that the psychological problems they have during flight are all that severe, one confided to me that the real psychological problems seem to occur after the return to Earth. Many cosmonauts have become convinced that their doctors are lying to them about their health. They have numerous psychosomatic complaints, and some have publicly said they have never felt they returned to normal after long periods of time in space. Another cosmonaut told me that he was convinced that his flight surgeon was lying about the extent of bone demineralization he had developed over the course of his stay in microgravity. The doctor had recently told him that his bones were now back to their preflight status. "But I didn't believe him," the cosmonaut said, "because I always still feel so weak. Not at all like my normal self. And it's been four years since my flight!" If these reports have any truth, the Soviets have never officially acknowledged it.

Again, anecdotal reports are fairly common in both the U.S.S.R. and the U.S. programs. The reason that such non-objective reports have any credibility at all is the reluctance of both programs to scientifically study what is really going on. The U.S.S.R. space program has clearly had the most opportunities to look at such incidents, to collect and analyze them, and even to try to correlate them with preflight psychological data. As we saw in Chapter 9, psychological data are routinely collected on cosmonaut candidates. And yet the correlations have not been done. Or, if done, nothing has been written and communicated in the scientific literature. Most available Soviet papers on psychological issues are reminiscent of the large flowing banners that used to be omnipresent in Moscow until the recent dissolution of the Communist government in that country. The banner over the Institute of Biomedical Problems proclaimed "Glory to Soviet Science." This is exactly what the papers that have been published emphasize. Entire books about the psychological selection of cosmonauts do not contain a single data point. Many of the papers simply discuss conclusions and operational recommendations without presenting any data at all.

But the optimistic discussions about psychological problems that have been "solved" are counterbalanced by the number of psychological and behavioral incidents that have occurred during space missions. When such problems are spoken of—if at all—it is always quickly pointed out that "it never had any effect on cosmonaut performance or the mission." How can this be known with such certitude if cosmonaut performance was never measured in the first place? Usually such generalizations are based on the cosmonaut's or astronaut's evaluation of his or her own performance. If one looks at studies on performance and its assessment that have been done in other environments, it can be seen that subjects are *never* accurate about measuring or assessing their own performance. They will maintain *subjectively* that they are performing at their usual level, while *objective* measures clearly demonstrate a decline in that performance. In other words, all we have is the astronaut's word for it. There is no reason to believe that cosmonauts, or astronauts, are any more accurate at subjective evaluation of their performance than any other individuals are. And, in fact, there is every reason to suppose that they might actually be worse at such evaluations, since they believe that their assignment to future space flights depends on how well they are perceived to do.

Numerous reviewers of space flight in the U.S. and U.S.S.R. programs have pointed out that astronauts from both nations have faced periods of depression, often alternating with periods of euphoria, anxiety, and other mood problems.[7] A crew mutiny on one of the Soviet *Salyut* missions, during which the crew cut off all communication with Earth for twenty-four hours, is a good example of a serious psychological and social disturbance.

In 1980, Valeri Ryumin wrote this in his diary: "O. Henry wrote in one of his stories that if you want to encourage the craft of murder, all you have to do is lock up two men for two months in an eighteen by twenty foot room. Naturally, this sounds humorous. Confidentially, a long stay even with a pleasant person is a test in itself."[8] Another cosmonaut, Valentin Lebedev, wrote: "Humming to myself, I float through the station. Is it possible that some day I'll be back on earth among my loved ones, and everything will be all right?"[9]

According to most sources, the Soviet Union was the world's leader in space medicine and space psychology. But this lead has little to do with actual research or data, particularly in space psychology. For example, Soviet psychologists recommend that a crew for a space mission sent to Mars be all male and limited to six or eight members. The best age is from thirty-five to forty-five years of age.[10] This conclusion is drawn from the experience of long-term orbital flights, including the recent year-long mission of Vladimir Titov and Musa Manarov aboard the *Mir* Space Station. But how can it be justified by any scientific evidence? Since the Soviets do not fly women on long-duration missions, they cannot logically conclude that a Mars crew should be all male. Perhaps it should be all female. Where are the data that support their assertions? Not surprisingly, none are offered. Their guess is as good as anyone else's.

In short, what we have learned from the Soviet experience is that there will be psychological problems in long-duration space flight. No psychological selection strategy by itself will exclude this possibility. It is to minimize the development of psychological problems as much as possible that the psychological right stuff must be determined.

## LESSONS LEARNED FROM THE HISTORICAL REVIEW OF U.S. ASTRONAUT CANDIDATE PSYCHOLOGICAL AND PSYCHIATRIC EVALUATIONS

The first few chapters of this book detailed as accurately as possible the history of the psychiatric and psychological evaluations of U.S. astronaut candidates. This history is important, not just from a behavioral scientist's point of view, but also from the point of view of anyone who wants to understand what happened to NASA, which went from an agency composed of highly talented workers and visionary leaders with a "can do" philosophy to an incompetent bureaucracy (which fortunately still had many talented workers) concerned primarily with its image and with vision only to the next funding cycle. The history of behavioral science in NASA is important because its decline within the agency directly parallels the development of what is now referred to as the "*Challenger* Syndrome." This is not a coincidence. The elimination of behavioral science research and the anti-psychological mentality at NASA, which began late in the *Mercury* era and reached its peak during the Shuttle period, were the products of psychological *denial* on an institutional level. *Denial* is defined as the refusal to acknowledge some aspect of external reality that would be readily apparent to others. It is the psychological process that resulted in the preventable *Challenger* tragedy—preventable if NASA managers had followed their own rules or listened to their competent employees and contractors.

When the traditional right stuff—exceptional technical competence, arrogance, and a sense of invulnerability—is not complemented by a modicum of self-awareness, insight, and an appreciation of one's limitations, there is always the danger that a tragedy will occur. Because of this, it has always been necessary for those who believe in the traditional concept to constantly redefine who has it. If someone you thought had "it" is killed in one of those common, but tragic, accidents that plague jet pilots, then, by definition, that pilot must not have had the right stuff. As Carlos Perry observed over thirty years ago, this type of denial only serves to underscore the importance and relevance of psychological factors in the aviation and space environments, particularly in the area of selection.

Recent research on the correlation between personality and performance has demonstrated that this can be a valuable (and cost-effective) approach. The preliminary results are promising, and one can only hope that this is a research area that NASA will continue to support.

## WHAT WE KNOW AND CAN LEARN FROM ANALOG ENVIRONMENTS ON EARTH

The principal characteristic of the space environment, or the environment of other planets in our solar system, is the fact that they possess extremely hostile physical conditions and require sophisticated engineering systems to support human life.

Not all such environments exist off of our own planet, however, and with the development of technical expertise, humans have been able to live and work in the extreme environments of Earth as well. These extreme environments are sometimes referred to as space analog environments, and they include such exotic locations as the Arctic and Antarctic, offshore oil platforms, and submarines or other underwater habitats. The societies that have been established in these hostile and isolated environments are sometimes called engineered societies.[11] They are often remote and extremely isolated from the rest of the world. The engineered nature of the society frequently means that individuals are restricted by the environment to a small and confined living and working space, within which the environment is controlled.

Much has been learned—and can be learned—from these environments about how to select the individuals and assemble the teams that can live and work most productively under the conditions imposed by the environment. Research has shown that isolation has very predictable emotional consequences for individuals and groups—including sleep disturbances, depression, anxiety, psychosomatic complaints, and an inevitable decline in group productivity.[12]

The selection techniques for these environments are generally much less stringent than are those for space, and this provides researchers with excellent opportunities to study a more heterogeneous group of explorers than astronauts or cosmonauts are. In the Antarctic, the psychological and psychiatric selection procedures used vary considerably from country to country, and even within countries. Most of the procedures are psychiatric, or select-out, in that a past history of psychiatric problems precludes assignment to Antarctic stations. However, occasionally psychological select-in screening is used in the selection of expedition leaders or of groups who will be living under special conditions. Table 10.1 summarizes the different procedures used by countries with active research programs in the Antarctic.

Many countries have also extensively studied psychological adjustment in Antarctic winter-over personnel, using questionnaires or debriefings conducted during the long winter period. There is rarely any systematic follow-up to these studies on adjustment, and there is no systematized collection of data, nor have any methods been standardized.

One researcher has summarized all the psychological results thus far obtained in the Antarctic as "unspectacular."[13] The unremarkable results may actually reflect the unimportance of psychological factors in these environments, but it may also be due to not consistently asking the right psychological questions. In essence, the question for the Antarctic and other hostile, isolated, and confined environs is this: What is the right stuff? What personality factors are predictive of the ability to perform well in the stressful conditions imposed by these environments? Does the right stuff vary from environment to environment (is it different for the Antarctic and space, for example), or do the individuals who do well share the same psychological profiles, no matter what situation they may be in? In other words, is the real psychological stuff found *only* in pilots or astronauts—or is it perhaps the quality that exists in successful men and women, no matter what their occupational pursuit may be?

**Table 10.1**
**Antarctic Psychological Selection Criteria**

| Psychological criteria | USA | Australia | France | Great Britain | Japan | South Africa | New Zealand | Poland |
|---|---|---|---|---|---|---|---|---|
| No history of psychopathology | x | x | x | x | x | | | x |
| No recent life crisis | x | | x | x | | | | |
| No history of alcohol abuse | x | x | x | x | x | | | x |
| Positive motivation | | | x | x | | | | |
| Intelligence | x | x | x | x | | | | |
| Emotional criteria | | | | | | | | |
| -Stable | x | | x | x | x | | | x |
| -No depression | | | x | | x | | | x |
| -No anxiety | | | x | | | | | |
| -Stress tolerant | | x | x | | | | | |
| -Independence | x | | | | | | | |
| -Low neuroticism | | | x | | | | | x |
| -Isolation tolerance | | | | | | x | | |
| Social criteria | | | | | | | | |
| -Stable | x | x | x | x | x | | | |
| -Compatible | x | | x | x | x | x | | x |
| -Acceptance of authority | | | x | | | | | |
| -Medium intro/extroversion | | | x | | | | | |
| Occupational criteria | | | | | | | | |
| -Task motivation | x | x | x | x | x | x | x | |

## SUMMING UP: WHAT YET NEEDS TO BE DONE TO IDENTIFY THE REAL PSYCHOLOGICAL STUFF?

What will the space missions of the future demand of the human participants? As we have seen, in order to develop an ideal astronaut selection program, the first step is to define as precisely as possible the job you expect the astronaut to do. Thus, in developing selection criteria for astronauts who will live and work on a space station or another planet, there must first be a clear understanding of how those space missions will be structured and what goals the humans selected to participate are expected to accomplish. Because the goals have not been clearly defined, and because the speculation about what astronauts will or will not be doing changes on a regular basis, almost all of the data reported here on the psychological characteristics believed to be critical for space exploration must be considered only preliminary. Additionally, since psychological and performance data have not been systematically collected on actual space missions, we cannot possibly know yet what characteristics will ultimately prove to define the real psychological stuff.

To determine the incontrovertible right stuff for space, a considerable amount of research needs to be done over the next twenty years, so that when the human species is ready to expand into the universe in the next century, we will take with us a new awareness of the human psyche and its relationship to the cosmos. Concerning our evolution from *homo sapiens* to *homo spacialis,* the universe will remain neutral, but absolutely unforgiving when we ignore truth.

A reasonable agenda for psychological research in the next several decades of space exploration would take into account the following points.

### Science versus Conjecture

Gross generalizations, capricious conjectures, and arbitrary conclusions are the hallmarks of "pop" psychology—not behavioral science. They should not be confused with the careful scientific research carried out by reputable behavioral scientists. The U.S. Space Program deserves more than a faddish approach to the very real human problems engendered by long-duration space missions. NASA officials should be extremely leery of anyone who claims to have "the answer" to all potential psychological questions. Psychiatry and psychology are both relatively new sciences. Because there is so much that is still unknown about the human mind and emotions, it is easy to believe that any plausibly presented psychological assessment method might be useful. But, unless the methods are backed up by data that demonstrate both the reliability and the validity of the methodology, they should be considered suspect.

### Collect Data on Astronaut and Cosmonaut Performance

To establish reliability and validity, real data on performance and behavior in the space environment must be collected and analyzed like every other kind of mission data. The reluctance of astronauts and cosmonauts to provide such data is unacceptable, and selection committees should make it a priority that those who apply for the position of astronaut clearly understand that the only reason to have a professional corps of astronauts in the first place is to expand our knowledge on human capabilities in the space environment. If they cannot deal with that simple fact, they should not be selected.

### Research Should Include Female Subjects

Medical and psychological research on women should become a priority. A recurring theme in this book has been the halfhearted—and sometimes absent—attempts to include women in the adventure of space exploration. It is not acceptable to plan to exclude them from Mars missions—or any mission—when there is plenty of time still remaining to do the research necessary to guarantee their physical and psychological health as members of such crews. The integration of women into space crews as fully qualified members of the team requires studies that document the strengths and weaknesses that women and men will bring individually and together to the exploration of space. The walls of bias and prejudice that exist today in virtually all aerospace activities—the exclusively male membership in the right stuff club—must systematically be torn down by scientific fact.

### Group Studies Should Be Done in Analog Environments Using the Most Recent Psychological Selection and Evaluation Techniques

Group studies must be done now, on the ground, in environments that simulate salient aspects of the space environment. Mission planners have no real scientific bases on which they can make decisions about crew composition (including personality factors, gender, multicultural makeup, or even the total number of individuals who should comprise crews), leadership styles, or even activity scheduling for long-duration missions. This may be the most important research that NASA can do to plan for long-duration space flights.

### Do Not Rely Entirely on Any Selection Methodology to Prevent Psychological Problems

It is a given that psychological selection—no matter how well it is done—will not entirely eliminate the possibility that problems will develop. This possibility can be further reduced by research on the nature and application of psychological support and training. The programs that exist are laudable, but there is no evidence that these programs or their elements really make a difference in health, performance, and productivity. Some aspects of them are surely beneficial, but it is possible that some aspects are harmful and may actually impair individual and crew performance. Only by requiring scientific study can the humbug and the hype be separated from the essentials.

## AMAZING OURSELVES: THE PSYCHOLOGICAL MEANING OF SPACE EXPLORATION

It is perhaps appropriate at this point in our discussions of the psychological selection of star voyagers that we pause to consider a much more fundamental question. What is it that drives us as a species to expand our horizons, "to go where no one has gone before"? Why would we be impelled to go into this hazardous, inhospitable environment in the first place? What motivates humans to explore the unknown? To constantly expand and redefine the envelope of human experience? If we understand the answers to these questions as they apply to us as a species, might we not unlock the answers as they apply to individuals?

In understanding the psychological evolution of the human species, W. G. Niederland has touched on this fascinating problem in his writing on psychogeography.[14] He views the exploration of space as part of the separation/individuation of our species from Mother Earth. The symbolism and fantasies that motivate people to go beyond their present horizons are rooted in fundamental psychodynamics. Space exploration and the colonization of space by humankind offer a unique opportunity to study the need to explore that drives our species.

During the last great expansion of humankind—first to the New World, then within the New World to the American West—psychiatry and psychology were newly emerging disciplines and not yet ready to tackle such an important task. Space exploration is, after all, the natural extension of an ongoing process that has fascinated humanity ever since evolution began on our planet.

The world has witnessed incredible visual images during thirty years of exploring space—our planet as a blue and living entity, adrift in a sea of darkness; Ed White on the first spacewalk over Mother Earth, his umbilical still firmly attached to the life-giving spacecraft that supplied the mother's gift to her children. These images perhaps reflect not only our past, but also our future. In the movie *2001: A Space Odyssey*, who can forget the child in the womb floating pensively above Earth, the first space child of our species, heralding humanity's destiny.

Lewis Thomas offers us wisdom and comfort in the midst of the world's problems with the thought that we are, in evolutionary terms, still youngsters. He writes in *The Fragile Species*:

> We may all be going through a kind of childhood in the evolution of our kind of animal. Having just arrived, down from the trees and admiring our thumbs, having only begun to master the one gift that distinguishes us from all other creatures it should perhaps not be surprising that we fumble so much. We have not yet begun to grow up. What we call contemporary culture may turn out, years hence, to have been a very early stage of primitive thought on the way to human maturity. What seems to us to be the accident-proneness of statecraft, the lethal folly of nation-states, and the dismaying emptiness of the time ahead may be merely the equivalent of early juvenile delinquency. . . . If we can stay alive, my guess is that we will someday amaze ourselves by what we can become as a species. Looked at as larvae, even as juveniles for all our folly, we are a splendid, promising form of life and I am on our side.[15]

What we may be is unknown. What we are remains under discussion. But there is no doubt that the exploration of space will play a role in the transition of humankind from Now into Tomorrow. In a way, the task of choosing the right stuff is merely the task of identifying the best that is within all of us. And those fortunate individuals who uniquely embody that stuff will carry us as a species to worlds beyond our knowing. Those worlds will challenge us and change us irrevocably from what we were, perhaps to something even better.

## NOTES

1. A. Chaikin, "The Loneliness of the Long-distance Astronaut," *Discover*, February 1985.
2. *NASA Information Summaries: Space Station*, PMS-008A (Hqs) (Houston: Boeing, August 1988). *Space Station Freedom Reference Guide* (Houston: Boeing, 1990).
3. George Bush, Speech delivered at the National Air and Space Museum, Washington, D.C., July 20, 1989.
4. M. B. Duke and P. W. Keaton, eds., *Manned Mars Mission Working Group Summary Report*, NASA M001 (Houston: NASA-JSC, 1986).
5. Ibid. International Space University, *International Mars Mission* (Toulouse, France: ISU, 1991). J. E. Oberg, *Manned Mission to Mars* (Harrisburg, Pa.: Stackpole Books, 1982).
6. Personal communication to author, Houston, Texas, October 1989.
7. R. L. Helmreich, "Applying Psychology in Outer Space: Unfulfilled Promises Revisited," *American Psychologist* 38 (1983): 445–450. N. Kanas, "Psychosocial Factors Affecting Simulated and Actual Space Missions," *Aviation, Space and Environmental Medicine* 56 (1985): 806–811. N. Kanas, "Psychological and Interpersonal Issues in Space," *American Journal of Psychiatry* 144 (1987): 703–709. N. Kanas and A. Fedderson, *Behavioral, Psychiatric and Sociological Problems of Long Duration Space Missions*, NASA Document TMX-58067 (Houston: NASA-JSC, October 1981). R. M. North, "Human Requirements for Long-duration Missions: Antarctic and Arctic Stations, Planetary Surface Operations and Space Transportation Vehicles" (paper presented at the International Biospherics Conference, University of

Alabama, Huntsville, February 20–23, 1992). J. E. Oberg and A. R. Oberg, *Pioneering Space* (New York: McGraw-Hill, 1986).

8. Chaikin, "The Loneliness of the Long-distance Astronaut," 24.

9. Ibid.

10. Moscow World Service, *Psychologists Profile Ideal Mars Crew* (in English), JPRS-USP-89-004 (February 16, 1989).

11. North, "Human Requirements for Long-duration Missions."

12. E. L. Beckman and E. Smith. "Tektite II: Medical Supervision of the Scientist in the Sea," *Texas Rep. Biological Medicine* 30 (1972). E. K. E. Gunderson, "Emotional Symptoms in Extremely Isolated Groups," *Archives of General Psychiatry* (1968): 362–368. A. A. Harrison, Y. A. Clearwater, and C. P. McKay, eds., *From Antarctica to Outer Space: Life in Isolation and Confinement* (New York: Springer-Verlag, 1991). L. A. Palinkas, *Sociocultural Influences on Psychosocial Adjustment in Antarctica,* Report No. 85-49 (San Diego: Naval Health Research Center, 1985). L. A. Palinkas, "Group Adaptation and Individual Adjustment in Antarctica: A Summary of Recent Research," in *From Antarctica to Outer Space: Life in Isolation and Confinement,* edited by A. A. Harrison, Y. A. Clearwater, and C. P. McKay (New York: Springer-Verlag, 1991), 240–251. R. Radloff and R. L. Helmreich, *Groups under Stress—Psychological Research in Sealab II* (New York: Appleton-Century-Crofts, 1968). P. A. Santy, "The Journey Out and In: Psychiatry and Space Exploration," *American Journal of Psychiatry* 140 (1983): 519–527. J. H. Earls, "Human Adjustment to an Exotic Environment," *Archives of General Psychiatry* 20 (1969): 117–122.

13. A. J. W. Taylor, "Antarctica Psychometrika Unspectacular," *New Zealand Antarctic Record* 6 (1978): 36–45.

14. W. G. Niederland, "River Symbolism, Part 1," *Psychoanalytic Quarterly* 25 (1956): 469–504. W. G. Niederland, "River Symbolism, Part II," *Psychoanalytic Quarterly* 26 (1957): 50–75. W. G. Niederland, "The History and Meaning of California: A Psychoanalytic Inquiry," *Psychoanalytic Quarterly* 40 (1971): 485–490. "Origins of Psychogeography and Survivor Syndrome in Experience," in *Frontiers of Psychiatry: Roche Report* (October 1, 1979), 1–2. Santy, "The Journey Out and In."

15. Lewis Thomas, *The Fragile Species* (New York: Charles Scribner's Sons, 1992), 81.

# Appendixes

# APPENDIX 1

## A. DESCRIPTION OF *MERCURY* PSYCHOLOGICAL TESTS

### Projective Personality Tests

***1. Rorschach Ink Blot Test***

The most commonly used projective test at the time was the Rorschach Inkblot Test. Subjects' associations to ten ambiguous inkblots are observed and scored. The Rorschach is a projective psychological test frequently used in psychiatric settings, which yields information about emotional conflicts and defense mechanisms. Ten ambiguous inkblots of various shades and colors are shown to the subject, who is then asked to respond in an unstructured manner. It is the oldest and perhaps most stable of all the projective multidimensional tests. Though research findings about this measure are equivocal, its multifaceted contribution to the assessment profile, plus the considerable experience of the evaluation team in using this instrument with comparable populations, resulted in its inclusion.

***2. Thematic Apperception Test (TAT)***

The TAT, like the Rorschach, is also a projective test consisting of a series of pictures depicting ambiguous situations about which the subject is instructed to make up a story. The Aeromedical Research Laboratory's battery of TAT pictures used eleven of the thirty possible ones (1, 3BM, 4, 6BM, 7BM, 8BM, 12M, 13MF, 16, 18GH, 18BM). Multidimensional analysis of responses is possible. This is the second most widely used projective test and one for which a great deal of comparable data is available.

***3. Draw-A-Person***

The Draw-A-Person Test is a brief projective test. The subject is asked to draw a figure of a person and then one of the opposite sex. From these drawings, inferences about self-concept, ego boundaries, and possible conflict areas can be made. While the data from this test are not contributory to an assessment in every case, the drawings frequently enable one to make significant personality differentiations when other evidence is equivocal.

***4. Sentence Completion Test***

The Sentence Completion Test is also a projective test. Subjects are asked to complete sentences that start with only a few words (e.g., "My mother ____"). Areas of potential conflict can thus be identified from responses.

***5. Who Am I?***

The subject is asked to write twenty answers to the question "Who am I?" The test is considered projective and is an assessment of identity and perception of social roles.

## Objective Personality Tests

**1. *Minnesota Multiphasic Personality Inventory (MMPI)***

One of the most frequently used personality tests, it consists of 566 items for which a subject is asked to respond true or false. This test has probably the most data of any of the personality tests and has been proven reliable and useful in a multitude of clinical and personnel situations. It continues to be used in the 1990s, although it has recently been updated for a more normal population (MMPI-2).

**2. *Gordon Personal Profile***

This test is a self-administered personality inventory consisting of eighteen tetrads of descriptive phrases, from which the subject chooses those "best" and "least" describing himself or herself. It provides a quick assessment of five traits: ascendance, responsibility, emotional stability, sociability, and overall self-evaluation. The main virtues of this test were the small amount of time required for its administration and the availability of comparable Air Force norms.

**3. *Edwards Personal Preference Schedule***

The Edwards Personal Preference Schedule is a 247-item personality inventory in which the candidate must choose from two descriptive phrases the one being more like himself or herself. The test is then scored for sixteen personality domains: achievement, deference, order, exhibition, autonomy, affiliation, introspection, succorance, dominance, abasement, nurturance, change, endurance, heterosexuality, aggression, and consistence. The score on consistency is a measure of profile stability. This test has the virtue of focusing on the relative strengths of *normal* personality variables, rather than concentrating on pathology (i.e., like the MMPI). Comparable Air Force norms are available.

**4. *Shipley Personal Inventory***

This inventory, developed by Shipley, Gray, and Newbert (1946) is a questionnaire prepared originally for the U.S. Navy, but used by other branches of the military establishment for the purpose of screening large numbers of recruits or selectees. A long form contains 145 items and a short form 20 items, which were selected from the long form because of their capacity to distinguish between normal Navy personnel and psychiatric discharges. The items are cast into a forced choice form, and one is required to check the alternative that better describes himself or herself. One alternative is always more characteristic of the normal as revealed by a large-scale case history and analysis. The other characterizes the psychiatrically undesirable. An attempt was made to pair choices that are apparently equal in social desirability.

The subject is asked to choose from twenty pairs of self-descriptive statements, which measure psychosomatic problems. The odd-even reliability of the test was .66 for Navy recruits and .91 among psychiatric discharges.

**5. *Outer-inner Preferences***

This test offers the choice of one alternative from each of fifty-two pairs statements on feelings about activities, things, and other people. It is a measure of interest in and dependency on social groups.

**6. *Pensacola Z***

The subject is asked to choose from sixty-six pairs of statements. The test is a measure of authoritarian attitudes.

*7. Officer Effectiveness Inventory*

This test is a multiple-choice, self-descriptive test of characteristics related to successful Air Force officer performance.

## Performance Tests

*1. Peer Ratings*

Each candidate was asked to indicate which of the other members of the group who accompanied him through the selection program he liked best, which one he would like to accompany him on a two-man mission, and which one he would assign to the mission if he could not go himself.

*2. The Complex Behavior Simulator*

A complex task is used to simulate the job characteristics of a systems operator's tasks. It utilizes a Complex Behavior Simulator (CBS) developed at Brooks Air Force Base in combination with an information-processing task (AUDIT). The information-processing task requires a continuous auditory monitoring and processing of signals by presenting single-letter Morse code signals in random order at a rate of one letter every five seconds. The subject's task is to monitor the different code letters being presented and to signal, by means of push-button switches corresponding to each code letter, whenever he has heard a particular letter a specified number of times. In this particular application, the subject monitored three code letters (a, N, M) and reported whenever he had received three of any one of them. Due to the small amount of practice time available, the subjects were given a mnemonic aid for monitoring the signals.

This stress testing was allocated one hour per subject. The subject received standardized instructions and practice on the task. Practice periods included an opportunity to perform on each task separately and then both together. Practice sessions were carefully monitored to ensure adequate performance by the subject, with coaching where indicated. The criterion for satisfactory base rate performance on the CBS was a subjective evaluation by the examiner based on signal handling rate and observed control movements. The criteria for adequate AUDIT performance were 100 percent signal recognition and five successively correct identifications of randomly sequenced three-signal series.

## Tests of Ability and Intelligence

*1. Wechsler Adult Intelligence Scale (WAIS)*

This test is a well-known measure of verbal and performance functions. It is individually administered and measures intelligence with eleven separate verbal and performance subtests. It is a well-standardized instrument commonly used in the clinical cognitive evaluation of flying personnel. The WAIS measures a broad spectrum of behavior and gives an adequate, though not outstanding, discrimination at the upper ranges of intelligence.

*2. Miller Analogies*

This is a test of intelligence based on the ability to comprehend analogies. It is a timed group test, correlating highly with general intelligence and verbal achievement measures. The test consists of 100 multiple-choice paired analogies. It is well standardized with comparable norms available, permitting differentiation for verbal abilities at a very high level.

3. *Raven Progressive Matrices*
This is a test of non-verbal concept formation.

4. *Doppelt Mathematical Reasoning Test*
The Doppelt Mathematical Reasoning Test is a timed group test consisting of fifty multiple-choice problems requiring the identification of complex mathematical principles. It is another well-standardized instrument whose published norms enable high-level differentiation.

5. *Minnesota Engineering Analogies*
The Engineering Analogies Test is a fifty-item, high-level objective measure of specific engineering knowledge, combining features of an abstract reasoning test with those of engineering achievement. Excellent standardization allows for good separation among candidates at high levels.

6. *Mechanical Comprehension*
This test measures basic comprehension of mechanical principles and the ability to apply them.

7. *Air Force Officer Qualification Test*
This test is used extensively in the U.S. Air Force. The portions of the test used measured verbal ability and quantitative aptitudes.

8. *Aviation Qualification Test (USN)*
This test is used in the U.S. Navy as a measure of academic achievement.

9. *Space Memory*
This test measures the ability to perceive spatial arrangements by assessing the ability to memorize the location of objects in space.

10. *Spatial Orientation/Spatial Visualization*
This test measures the ability to match photographs with maps by measuring the time needed to locate details in aerial photographs.

11. *Gottschaldt Hidden Figures*
The Hidden Figures Test assesses visual perception and analytical ability by asking the subject to locate a certain figure within a mass of irrelevant details. The applicant is asked to find given embedded figures in more complex diagrams.

12. *Guilford-Zimmerman Spatial Visualization*
This test measures the ability to manipulate ideas visually. It is usually used to screen engineers, architects, and draftsmen (or any persons whose work involves mechanical devices).

### Tests of Physical and Psychological Endurance

1. *Isolation*
The isolation chamber was designed to measure ability to adapt to unusual circumstances and to tolerate absence of external stimuli and enforced inactivity as

well as enclosure. Each candidate was confined to a dark, soundproof room for three hours. While this brief period of isolation was not considered particularly stressful for most people, data are obtained on the style of adaptation to isolation. This procedure aided in identifying subjects who could not tolerate enforced inactivity, enclosure in small spaces, or absence of external stimuli. The three-hour period was chosen because a longer time could not be accommodated.

**2. *Pressure Suit Test***

The candidate was dressed in a tightly fitting garment designed to apply pressure to the body to simulate high-altitude flight. Each candidate entered a chamber from which air was evacuated to simulate an altitude of 65,000 feet. This produces severe physical discomfort and a feeling of confinement.

**3. *Acceleration***

The candidate was placed on the human centrifuge in various positions and subjected to different "g" loads. This procedure leads to anxiety, disorientation, and blackouts in susceptible subjects.

**4. *Noise and Vibration***

Each candidate was vibrated at varying frequencies and amplitudes and subjected to high-energy sound while performing different tasks. Efficiency is often impaired under these conditions.

**5. *Heat***

Each candidate spent two hours in a chamber maintained at 130 degrees. This is obviously an uncomfortable experience during which efficiency may be impaired.

## B. CODING OF PSYCHOLOGICAL VARIABLES USED IN *MERCURY* SELECTION

1. Overall Rating of Candidate
2. Drive
3. Freedom from Conflict Anxiety
4. Effective Defense
5. Free Energy
6. Identity
7. Object Relationships
8. Reality Testing
9. Motivation Quantity
10. Motivation Quality
11. Adaptability
12. Social Relationships
13. Freedom from Dependency
14. Freedom from Need for Activity
15. Freedom from Impulsivity

*(From the Wechsler Adult Intelligence Scale)*
16. IQ Total
17. IQ Verbal
18. IQ Performance
19. Information
20. Comprehension
21. Arithmetic
22. Similarities
23. Digit Span
24. Vocabulary
25. Digit Symbol
26. Picture Completion
27. Block Design
28. Picture Arrangement
29. Object Assembly

*(From the MMPI)*
30. Hysteria
31. Depression
32. Hypochondria
33. Psychopathic Deviation

*(From the Rorschach Test)*
34. Total Responses (R)
35. Percent Good Form (F + %)
36. Percent Pure Form (F%)
37. Percent Animal Responses (A%)
38. Number of Popular Responses (P)
39. Number of Human Movement Responses (M)
40. Number of Whole Responses (W)
41. Number of Blank Space Responses (S)
42. Weighted Scorc of Color Responses
    $\frac{(FC + 2CF + 3C)}{2}$
43. Inanimate Movement (Fm + m)
44. Animal Movement (FM)
45. Shade Responses (Fc + c)
46. Percent Responses on Color Cards
    $\frac{(VIII - X (\%)}{R}$
47. Sexual Responses (Sex)
48. Anatomy (At.)

## C. CORRELATION COEFFICIENTS OF FORTY-EIGHT PSYCHOLOGICAL VARIABLES

| | 19 | 20 | 21 | 22 | 23 | 24 | 25 | 26 | 27 | 28 | 29 | 30 | 31 | 32 | 33 | 34 | 35 | 36 | 37 | 38 |
|---|---|---|---|---|---|---|---|---|---|---|---|---|---|---|---|---|---|---|---|---|
| **1** | −860 | 4553 | 1705 | 3030 | 3014 | 3186 | 647 | 1269 | 510 | −62 | 1469 | −3307 | −2567 | −1853 | −833 | 1042 | 3195 | 894 | 107 | −586 |
| **2** | 125 | 3919 | 814 | 3333 | 3620 | 1834 | 2823 | 737 | −343 | 865 | 657 | −2303 | 1700 | −3157 | 885 | 1547 | 116 | −446 | −1051 | −2236 |
| **3** | −3498 | 1921 | −576 | 1026 | 3391 | 3160 | −340 | 1274 | 1126 | 376 | 346 | −3029 | −3970 | −342 | −318 | −798 | 4954 | 954 | 342 | −1335 |
| **4** | −63 | 3935 | 2328 | 1265 | 3519 | 2829 | 1072 | −31 | 683 | −905 | 1064 | −1509 | −2912 | −841 | −1461 | 2150 | 3298 | 1273 | −839 | 997 |
| **5** | 429 | 4457 | 674 | 4098 | 3324 | 3420 | 2280 | 377 | 1598 | 2195 | 550 | −2570 | −735 | −2579 | −189 | 2012 | 1019 | −171 | 846 | −1813 |
| **6** | −468 | 1767 | 1757 | 410 | 1724 | 4432 | −531 | −304 | −150 | −1147 | −661 | −2420 | −5330 | −883 | −1037 | 348 | 3010 | 823 | −513 | 670 |
| **7** | 381 | 3041 | 3835 | 2368 | 1692 | 3510 | −1681 | −197 | −41 | 105 | 2203 | −1335 | −3553 | 594 | −516 | 1232 | 1772 | 948 | −850 | 1753 |
| **8** | 1586 | 4422 | 2085 | 835 | 3256 | 3444 | 1953 | 1291 | −337 | 1144 | 1727 | −3383 | −2514 | −2067 | −2291 | 945 | 1086 | −349 | −1789 | −454 |
| **9** | 1910 | 5043 | 1416 | 4091 | 3788 | 3045 | 1285 | −478 | −424 | 4155 | 2364 | −3957 | −960 | −3380 | −1150 | 1394 | 2392 | −191 | −146 | −641 |
| **10** | 733 | 3596 | 1306 | 4163 | 2107 | −86 | −797 | −146 | −2392 | −108 | −1420 | −3105 | 285 | −1727 | −1327 | 543 | 1842 | 2552 | −2455 | −11 |
| **11** | 241 | 1913 | 2236 | 62 | 3188 | 1923 | 639 | 212 | −1343 | −1460 | 2055 | −1445 | −2335 | −1106 | −1708 | 3621 | 1391 | 2498 | 663 | 2348 |
| **12** | 835 | 2789 | 2469 | 547 | 1846 | 2561 | 659 | −1948 | −1369 | −691 | 3917 | −1846 | −2373 | 146 | 1277 | 325 | 783 | 489 | −3844 | −2657 |
| **13** | −1214 | 2489 | 398 | −1967 | 2277 | 1534 | −1276 | 278 | 635 | −2052 | −1421 | 1505 | −1620 | 2125 | 1774 | 1966 | 1549 | 2004 | −568 | 2987 |
| **14** | −885 | 950 | 1228 | −2117 | 1132 | 1423 | −3224 | −2311 | −2997 | −84 | −1866 | −679 | −3123 | 1276 | −189 | −659 | 1513 | −715 | −897 | 1267 |
| **15** | −2855 | 980 | 992 | 1733 | 456 | 2814 | 483 | 824 | −738 | 1444 | 1032 | −2048 | −1438 | −833 | −1657 | 688 | 929 | 945 | −1083 | 467 |
| **16** | 4399 | 7885 | 6302 | 4520 | 3187 | 5807 | 4713 | 3308 | 3314 | 4096 | 3102 | 361 | −308 | −415 | −1958 | 4407 | −83 | 770 | 1035 | −126 |
| **17** | 5659 | 8097 | 6438 | 5410 | 3827 | 6080 | 1365 | 2755 | 1350 | 2723 | 561 | −144 | −28 | −423 | −1094 | 3872 | 670 | 216 | 1129 | 33 |
| **18** | 1045 | 4776 | 3971 | 2289 | 989 | 3062 | 7454 | 3236 | 5017 | 4860 | 5238 | 602 | −684 | −500 | −2525 | 3853 | −934 | 1760 | 683 | 51 |

| | 19 | 20 | 21 | 22 | 23 | 24 | 25 | 26 | 27 | 28 | 29 | 30 | 31 | 32 | 33 | 34 | 35 | 36 | 37 | 38 |
|---|---|---|---|---|---|---|---|---|---|---|---|---|---|---|---|---|---|---|---|---|
| **19** | 10000 | 3561 | 3590 | 2351 | –315 | 3356 | –11 | –1111 | –635 | 1891 | 417 | 1141 | 1718 | –3286 | –1144 | 300 | –302 | –1771 | 193 | 417 |
| **20** | | 10000 | 4848 | 4823 | 2882 | 3910 | 3282 | 1915 | 2366 | 2530 | 222 | 348 | 6 | 966 | 833 | 2694 | 561 | –352 | –1330 | –1171 |
| **21** | | | 10000 | 2389 | –1169 | 3600 | 1712 | 1157 | 1089 | 786 | 2028 | 685 | 692 | 1077 | –2898 | 2159 | –3832 | 121 | 96 | 487 |
| **22** | | | | 10000 | –399 | 2472 | 8 | 3556 | –544 | 2605 | 1263 | –1031 | 346 | –1765 | –150 | 3875 | 1436 | 1812 | 1141 | 1239 |
| **23** | | | | | 10000 | 1358 | 1871 | 1429 | 1411 | 905 | –1881 | 226 | –875 | 273 | 56 | 2897 | 3043 | 1551 | 289 | –312 |
| **24** | | | | | | 10000 | –193 | 2338 | 1633 | 1279 | 111 | –1010 | –2164 | 161 | –631 | 963 | 745 | –1275 | 2304 | 269 |
| **25** | | | | | | | 10000 | 101 | 4420 | 2075 | 2998 | 2005 | 966 | –640 | –1265 | 1951 | –958 | 140 | –1628 | –2190 |
| **26** | | | | | | | | 10000 | 625 | –381 | 430 | –1257 | –2056 | 83 | –1838 | 4464 | –594 | 3787 | 2809 | 3764 |
| **27** | | | | | | | | | 10000 | 1371 | 959 | 2359 | –1847 | 1283 | 975 | 436 | 955 | –165 | 2319 | 308 |
| **28** | | | | | | | | | | 10000 | 351 | –2761 | –515 | –3492 | –1577 | 286 | –89 | 893 | 1313 | –318 |
| **29** | | | | | | | | | | | 10000 | 750 | 263 | –497 | –2554 | 3175 | –287 | 1528 | 1421 | 319 |
| **30** | | | | | | | | | | | | 10000 | 4029 | 7470 | 5310 | 3400 | –720 | 560 | –772 | –482 |
| **31** | | | | | | | | | | | | | 10000 | 1764 | 1500 | 1237 | –3657 | –1044 | –1451 | –2071 |
| **32** | | | | | | | | | | | | | | 10000 | 4942 | 1746 | –1101 | 1409 | –3249 | –443 |
| **33** | | | | | | | | | | | | | | | 10000 | –534 | 2989 | –553 | –877 | –2183 |
| **34** | | | | | | | | | | | | | | | | 10000 | –1462 | 5165 | 2406 | 4167 |
| **35** | | | | | | | | | | | | | | | | | 10000 | 1558 | 1314 | 659 |
| **36** | | | | | | | | | | | | | | | | | | 10000 | 280 | 4142 |

| | 1 | 2 | 3 | 4 | 5 | 6 | 7 | 8 | 9 | 10 | 11 | 12 | 13 | 14 | 15 | 16 | 17 | 18 |
|---|---|---|---|---|---|---|---|---|---|---|---|---|---|---|---|---|---|---|
| **1** | 10000 | 6328 | 7796 | 7999 | 7636 | 6353 | 6436 | 7554 | 6670 | 6251 | 6962 | 6590 | 4197 | 4174 | 6311 | 3590 | 3815 | 1953 |
| **2** | | 10000 | 2642 | 5150 | 7840 | 1136 | 2265 | 4650 | 6244 | 6430 | 4881 | 4851 | 532 | –595 | 3529 | 3456 | 3508 | 2077 |
| **3** | | | 10000 | 6445 | 5994 | 6813 | 5334 | 4792 | 5159 | 4491 | 5072 | 5238 | 2884 | 4597 | 5421 | 1126 | 1382 | 362 |
| **4** | | | | 10000 | 5125 | 7247 | 6606 | 6561 | 5059 | 6243 | 8297 | 6520 | 5224 | 4761 | 6372 | 3480 | 3688 | 1878 |
| **5** | | | | | 10000 | 4470 | 4326 | 4903 | 6374 | 6199 | 5622 | 4159 | 1434 | 1067 | 4832 | 3725 | 3697 | 2530 |
| **6** | | | | | | 10000 | 6807 | 5643 | 2743 | 3112 | 6576 | 4421 | 5031 | 6045 | 6031 | 1880 | 2487 | 283 |
| **7** | | | | | | | 10000 | 5452 | 4970 | 4166 | 5744 | 5662 | 4518 | 4864 | 5731 | 2902 | 3614 | 878 |
| **8** | | | | | | | | 10000 | 5328 | 3054 | 6019 | 7251 | 3551 | 4518 | 5637 | 4871 | 4260 | 3560 |
| **9** | | | | | | | | | 10000 | 4734 | 4198 | 5773 | 1185 | 2020 | 3428 | 4618 | 4547 | 3092 |
| **10** | | | | | | | | | | 10000 | 5060 | 4309 | 2061 | 1821 | 3503 | 809 | 2557 | –1492 |
| **11** | | | | | | | | | | | 10000 | 6236 | 4596 | 5453 | 5549 | 2312 | 2554 | 996 |
| **12** | | | | | | | | | | | | 10000 | 2592 | 3458 | 4236 | 2830 | 2708 | 1592 |
| **13** | | | | | | | | | | | | | 10000 | 5054 | 3132 | 644 | 1282 | –753 |
| **14** | | | | | | | | | | | | | | 10000 | 4213 | –201 | 1058 | –2222 |
| **15** | | | | | | | | | | | | | | | 10000 | 1808 | 1001 | 2301 |
| **16** | | | | | | | | | | | | | | | | 10000 | 8909 | 7856 |
| **17** | | | | | | | | | | | | | | | | | 10000 | 4254 |
| **18** | | | | | | | | | | | | | | | | | | 10000 |

|  | 39 | 40 | 41 | 42 | 43 | 44 | 45 | 46 | 47 | 48 |
|---|---|---|---|---|---|---|---|---|---|---|
| **1** | 1622 | 3729 | –1081 | 1054 | 385 | 2167 | 26 | 874 | 741 | 2399 |
| **2** | 1123 | 2586 | –212 | 2433 | –43 | 1451 | 2456 | 1028 | –738 | 1218 |
| **3** | 1149 | 744 | –1528 | –2990 | 696 | 1766 | –1245 | 1059 | 1807 | 1496 |
| **4** | 1915 | 3286 | 721 | 1820 | –193 | 3835 | –360 | 12 | 1787 | 3730 |
| **5** | 1825 | 3695 | 218 | 1659 | 1105 | 2386 | 2752 | 2190 | 253 | 1486 |
| **6** | 279 | 1952 | 431 | –99 | 91 | 2119 | –1323 | 155 | 2134 | 2205 |
| **7** | 2764 | 3353 | 60 | –433 | 182 | 2230 | –1091 | –729 | 4253 | 2665 |
| **8** | –160 | 4876 | –22 | 2852 | 1900 | 1477 | 392 | –643 | 866 | 1438 |
| **9** | 1782 | 2182 | –143 | –596 | –561 | 2975 | 1270 | 1306 | 1965 | 1516 |
| **10** | 1524 | –57 | –2482 | 162 | –1639 | –196 | –554 | 1075 | 2952 | 754 |
| **11** | 2767 | 3989 | 1940 | 2685 | 1497 | 4691 | 377 | 230 | 2350 | 4905 |
| **12** | 0 | 2060 | 565 | –249 | 2246 | 2110 | 181 | 978 | 1887 | 3296 |
| **13** | –53 | –310 | 1444 | 1848 | 422 | 2311 | –1158 | –264 | –236 | 4664 |
| **14** | 227 | 2949 | –889 | 660 | 2080 | 1705 | –2983 | –1090 | 2969 | 993 |
| **15** | –1033 | 4194 | –165 | 3394 | 32 | 1651 | 804 | 118 | 553 | 3079 |
| **16** | –408 | 3795 | 3100 | 2308 | 2086 | 3967 | 4244 | 1900 | 1455 | 2264 |
| **17** | 409 | 2545 | 2325 | 1127 | 2192 | 3860 | 3100 | 3446 | 2968 | 1688 |
| **18** | –1213 | 3911 | 2822 | 2883 | 637 | 2551 | 4397 | –297 | –878 | 2014 |

| | 39 | 40 | 41 | 42 | 43 | 44 | 45 | 46 | 47 | 48 |
|---|---|---|---|---|---|---|---|---|---|---|
| **19** | −707 | 296 | −121 | 1622 | 55 | −383 | 242 | 1680 | 2773 | −671 |
| **20** | −793 | 3196 | 1269 | 2117 | 2012 | 1973 | 3678 | 1089 | 787 | 1487 |
| **21** | 1386 | 3234 | −347 | 1782 | −1080 | 2156 | 2593 | 293 | 2159 | 1576 |
| **22** | 19 | 963 | 1715 | 1420 | 382 | 2002 | 3839 | 4437 | 2026 | 946 |
| **23** | 2279 | −503 | 3666 | −2824 | 2676 | 3501 | 617 | 319 | 405 | 781 |
| **24** | 1771 | 2594 | 942 | 238 | 2813 | 3362 | 554 | 3573 | 2293 | 1989 |
| **25** | −1258 | 2227 | 2232 | 2151 | −159 | 736 | 4225 | −3498 | 3094 | 631 |
| **26** | −162 | 1315 | 2602 | 1863 | 1349 | 2059 | 1646 | 3100 | 197 | 863 |
| **27** | −893 | 1197 | 1303 | −804 | −843 | 227 | 1468 | 235 | −3577 | −495 |
| **28** | −1787 | −122 | −945 | −300 | −976 | 727 | 1198 | 2555 | 1852 | −2168 |
| **29** | 1923 | 3201 | 2821 | 1851 | 1524 | 2532 | 2440 | −1282 | 1141 | 4421 |
| **30** | 693 | 1771 | 5145 | 1654 | 4617 | 1500 | 1796 | −2615 | −1496 | 3547 |
| **31** | 1732 | 2024 | 97 | 3668 | 837 | −1578 | 3062 | −1632 | −1618 | 1916 |
| **32** | 412 | 1063 | 2810 | −565 | 5098 | −279 | 433 | −2083 | −40 | 2525 |
| **33** | 582 | −1362 | 773 | −1094 | 3178 | −825 | −896 | −468 | −1550 | 1304 |
| **34** | 3668 | 4578 | 7583 | 4194 | 3799 | 7120 | 5759 | 2652 | 2043 | 6441 |
| **35** | −143 | −4117 | −633 | −4457 | −105 | 267 | −4463 | 791 | 2137 | −49 |
| **36** | 817 | −1849 | 2843 | −812 | 266 | 1637 | −303 | 1721 | 2806 | 3644 |

| | 39 | 40 | 41 | 42 | 43 | 44 | 45 | 46 | 47 | 48 |
|---|---|---|---|---|---|---|---|---|---|---|
| **37** | 1670 | 790 | 1368 | 371 | 76 | 4840 | 571 | 3838 | 29 | 1699 |
| **38** | 829 | 1818 | 1902 | 1959 | –1472 | 2893 | –818 | 1494 | 1800 | 2202 |
| **39** | 10000 | 2153 | 226 | –937 | –80 | 3728 | –237 | –700 | 3456 | 1424 |
| **40** | | 10000 | 2409 | 7456 | 3313 | 3712 | 4344 | –971 | –2 | 4139 |
| **41** | | | 10000 | 2623 | 5345 | 5715 | 5174 | 1060 | –593 | 5560 |
| **42** | | | | 10000 | 1751 | 1995 | 4932 | –695 | –2395 | 4395 |
| **43** | | | | | 10000 | 2881 | 2037 | 1032 | 2291 | 3318 |
| **44** | | | | | | 10000 | 3426 | 3562 | 2464 | 5750 |
| **45** | | | | | | | 10000 | 2061 | –2342 | 4414 |
| **46** | | | | | | | | 10000 | 1999 | 901 |
| **47** | | | | | | | | | 10000 | 1137 |
| **48** | | | | | | | | | | 10000 |

## D. PROJECT *MERCURY* STRESS STUDY: PROGRESS REPORT, 1962*

### Introduction

In July 1960, at the request of the NASA Space Task Group, a study of stress in Mercury flights was undertaken. The goals were:

1. To determine whether significant psychophysiological changes are produced by suborbital or orbital flight;
2. To assess the degree of stress imposed upon the Astronauts;
3. To investigate mechanisms employed for maintaining adaptive behavior under stress;
4. To provide data for application to future pilot selection and training programs.

The stress study arose from the NASA Project Mercury objective of investigating the capabilities of man in a space environment. In addition to determining whether man could contribute to the operation of his spacecraft, and could survive orbital flight, it was felt that an effort should be made to learn the human costs of the mission. Although there was little reason to believe that such costs might be excessive, conclusions regarding them could not be made after flight unless appropriate indices of stress were part of the data to be collected.

The concept of stress used in this study is that of the biological sciences. It involves the idea that when a load is placed upon an organism, a compensatory response occurs. This allows the organism to maintain essential functions and to continue operating in spite of the load. Using such an orienting concept, the present study could be said to ask the following questions: What is the size of the load? What is its nature? How is it handled? What is its effect? What are the individual characteristics which increase stress resistance? And, finally, how do competent individuals differ in their mechanism for coping with stress? To answer these questions, a psychophysiological study was undertaken. This has involved the simultaneous measurement of psychological, physiological, and biochemical variables in response to training and flight activities.

The measures chosen for the study had to meet a series of basic criteria. First, it had to be demonstrated that each measure had been shown by earlier research to be a valid and sensitive index of stress. Second, the tests should be simple, brief, and possible to administer in the field without complex equipment or highly skilled staff. Third, they must be compatible with the operational requirements of Project Mercury. This involved compromises to insure that the necessary data could be obtained with minimum inconvenience and loss of time to the Astronaut, and without disturbing his effectiveness for the mission.

To maintain the confidence of the men, and to assure the reliability of data, it was agreed that individual data would be presented only after the Mercury flights

* This progress report was written by George Ruff, M.D. Every reasonable effort has been made to trace the sources for all references cited in this appendix, but in some instances this has proven impossible. While a selected reference list follows this appendix, the author and publisher will be glad to receive information leading to more complete acknowledgments in subsequent printings of the book and in the meantime extend their apologies for any omissions.

were completed. Hence no material has been disclosed which might have influenced selection for individual flights. This policy also reflected the belief that since the men had been carefully assessed initially, further psychological evaluation would be of limited value in deciding assignments for specific missions. However, if information were obtained which might affect the success of the mission, it was agreed that this should be made available to management without delay. As expected, this was never necessary.

In order to indicate the nature of the work which has been done, this progress report presents separate summaries of the psychological and biochemical findings. For the final report, physiological and performance data recorded during the flights will also be included, and cross-correlational analyses of material from all areas will be carried out to show the interrelationships of changes in psychological, physiological and biochemical functioning. Because of the statistical treatment required, such analyses cannot be undertaken until the data collection phase is completed.

### Study Design and Procedure

The study consists of three general phases: (1) Personality assessment; (2) Repeated measurements during training; and (3) Evaluation of response to flight.

**A. *Orientation***

After initial discussions with Project Mercury staff, the purposes and procedures of the study were discussed in a group session with the seven Astronauts and representatives of management. Both the potential contributions to the space program and the general scientific interest of the work were considered. The number and types of procedures to be used, and the time and other requirements of testing were described. The Astronauts had the opportunity to raise questions, and to protest—if they wished—the personal interviews or the demands of the stress project. The spirit of the first session was one of general agreement on the value of the research and willingness to invest the necessary time and energy. Indeed, in the individual interviews immediately following the orientation session, and generally true also throughout the study, there was evidence of cooperativeness and candor. There have been few signs of conscious resistance. The men have spoken frankly about personal as well as project problems. Where objections subsequently arose, they were directed against the repetitive, dull, or annoying aspects of procedures used in repeated testing and their imposition on the harried pre-flight work of the Astronaut. But at no point, either early or late in the study, does the interview data suggest efforts at intentional deception. These comments are made to underscore the investigators' belief in the reliability of these data.

***B. Phase 1: Personality Assessment***

This phase, undertaken in the summer of 1960, involved studies of the seven Astronauts, in order to investigate the personality mechanisms relevant to later stress behavior. The data of this phase consisted principally of interviews with the Astronauts individually. To maximize reliability and the coverage of different areas, the two psychological investigators each interviewed all men in two one-hour sessions. In addition, the extensive data from the initial assessment program conducted by the Wright Air Development Center were included in the analyses.

In the initial interviews, and amplified throughout the subsequent course of the study, the following general areas have been investigated:

1. *Motivation for project*—The man's original motivation for applying to Project Mercury, and the needs presently served by participation, were explored. His goals and aspirations for the future were similarly examined.
2. *Self concept and concept of the astronaut role*—Self-esteem and self-attitudes, in general and in connection with the astronaut role, were explored. The men's concepts of the "ideal astronaut", and the competence, values, and other aspects of behavior believed relevant were discussed in detail.
3. *Emotional activation and control*—The variety, intensity, conditions for activation and control of various emotional states were studied by exploring behavior in situations of psychological threat and objective danger. The somatic representation and behavioral consequences of anxiety, anger and depression were described. On the more positive side, the sources of pleasure, feelings of competence, mastery, achievement, and other conditions of well-being were also determined.
4. *Type and strength of defenses*—Efforts were made to discover the typical modes of dealing with threats to psychological equilibrium. The efficiency of these mechanisms in coping with stress was studied.
5. *Social behavior*—Relations to family, to the other Astronauts, to management and to other persons were examined. The social organization of the group, as a group, was described.
6. *Other aspects of personality*—Impulsivity and lability, energy level and fatigue, reactions to potentially disruptive agents such as alcohol, and other qualities were reviewed. Various other aspects of the motivation which seemed relevant to pilot behavior were investigated. These included such dimensions as activity-passivity, dependence-independence, trust-distrust, and needs for affiliation, achievement and autonomy.

Overall, the purpose of this phase was to find out what kinds of men these were, what kinds of self and affect organization they had, and how these might have changed in fitting the evolving Astronaut role. Emphasis was placed on the sources of their competence in dealing with stress. Our concern was less with the historical origin of these processes than with the assessment of their present functioning and changes that might have occurred during the project. All interviews were tape-recorded and transcribed for later study. The analysis and findings based on this material derive from independent study of the tapes and transcripts by the two psychological investigators and the convergence of their clinical judgments.

### *C. Phase II: Repeated Measurements during Training*

During training, a small battery of procedures was applied on repeated occasions during the succeeding two years. Where possible, these were paired measurements made before and after the Redstone and Atlas centrifuge simulations, the environmental control systems runs, and selected pre-launch activities. In addition, the battery was used during the more relaxed control occasions. The primary purpose of this phase was to develop a baseline against which the same measures made at the time of flight could be evaluated. Moreover, this provided an opportunity to gather psychophysiological data from a number of systems of stress-relevance, which might indicate more about the relation of stability-variability to adequacy of stress adaptation.

A series of psychological and physiological measures designed to indicate different types or levels of stress response was employed. The area of functioning assessed and the measures used included:

1. *Measures of personality and emotion*—The emotional state on each occasion was described by self-administered adjective checklist (Clyde, 1959), self-rating scales, and a brief questionnaire, which inquired into the Astronaut's perception of the adequacy of his performance and asked him to describe any special circumstances which might have influenced his behavior on that occasion. Where scheduling permitted, interviews exploring these matters were also included. The adjective checklist is one which has been widely used in psychopharmacological research. It seeks to evaluate various aspects of mood. On the basis of factor analytic studies, clusters of items defining six such moods have been distinguished.
2. *Performance measures*—In order to assess possible changes in psychomotor and intellectual functioning caused by training or flight stress, three tests developed for repetitive testing were used (Moran and Mefferd, 1959). These procedures had been administered to the Astronauts at the time of the original selection program, and have otherwise been applied in studies requiring repeated testing of the same individuals. There are twenty alternate forms of matched difficulty. The three tests used were: a) Aiming, which requires that the subject dot the center of connected circles as rapidly as possible; b) Number Facility, a mental arithmetic procedure; and c) Perceptual Speed, a number-cancellation procedure. All of these are timed tests of familiar over learned functions, which should be stable in repeat testing, except for such changes resulting from stress or states of disorganization.
3. *Biochemical measures*—The level of hydrocortisone was measured in plasma, whenever possible. Corticoids and 17-ketosteroids, adrenaline and noradrenaline, and a series of their metabolic products, were measured in urine. Diurnal variations in both corticoids and the catecholamines were investigated in special control experiments. More detailed descriptions of the factors evaluated and details of measurement procedures will be discussed in later sections of this report, in conjunction with the findings in these areas.
4. *Physiological measures*—In addition to the variables studied as part of this study, available data on cardiac and respiratory changes in training and flight situations will be incorporated in subsequent analyses.

These procedures constitute a battery of measures describing selected aspects of the psychophysiological state of each man under different conditions. One may examine the changes in response patterns, as well as in each variable separately. Obviously, there are important questions contained within each realm of data; for example, how adrenal activation during flight might differ from response in control and training conditions. But simultaneous considerations of all these variables should yield a fuller picture of subject-specific or response-specific response profiles.

The design of this portion of the study is of a type described as a P-technique (Cattell, 1943). If the structure of a study such as this is viewed as a tridimensional rectangle, the defining dimensions are subjects, measures and events (or time). This permits more than one kind of analysis. Subjects can be compared, at any point in time, on one or more measures. Measures can be compared across subjects, as they co-vary over time. The time dimension also permits testing the hypothesis that the stability or variability of a measure over time, within each subject, may be an index of general stress resistance. Moreover, it allows the study of correlation between measures, within subjects, which may describe idiosyncratic modes of response. It is clear that persons, tests, and time can be combined in other ways—for example, to study changes in individual response profiles, as we have already suggested. In this presentation, we will present only some of the possible analyses, focusing mainly on the relationship of response changes to the particular stress events, and on comparing men in these stress response findings. More subtle intermeasure analyses and studies of response profiles will be presented in a later report.

### D. *Phase III: Study of Flight Behavior*

The third phase of the study has consisted of more intensive evaluation of the men who have made suborbital or orbital flights. The battery of measures described above was administered immediately before and after each flight. At the same time, there were brief interviews evaluating the Astronaut's condition just prior to and following the flight. There was a longer interview two days earlier to review each man's experiences since the last contact, to explore in greater depth his feelings and anticipations, and to determine fears or doubts he may have had. Similarly, two days after flight, there was another long interview to review the flight experience itself. Observations were also made during debriefing sessions and other less formal contacts. In addition to the primary and back-up pilots, interviews were held with all other Astronauts who were available during the post-flight period. Each of these interviews was at least an hour in length. Where possible, both investigators interviewed the man together, in order to increase the reliability of judgments, and to cover the relevant areas more efficiently. Where this was not possible, one investigator interviewed alone, and the recorded and transcribed interviews were studied by both at a later point.

The present report will be concerned largely with the findings on five Astronauts who have completed flights. Since the major focus of this research is on the evaluation of stress response to flight conditions, separating the data for this group is necessary. However, in certain analyses, particularly those concerned with the personality characteristics of the astronauts in general, all seven are considered.

## Results of the Psychological Studies

### A. *Personality Qualities of the Mercury Astronauts*

Although each of the Astronauts has distinct and separate personality traits, certain features are common to all. In this section we will discuss some of the more salient personality characteristics which are relevant to the understanding of stress behavior.

*General characteristics*—The Astronauts combine high levels of practical intelligence with professional experience. They are concerned with problem solving and have the ability to focus on the essentials of issues which confront them. Their thoughts are quickly organized and effectively communicated. They tend not to be abstract or speculative, but think concretely. Facts are emphasized, rather than theories. What is unknown and uncertain is avoided or handled by efforts to make it known. Details irrelevant to their areas of primary concern are avoided.

These men are not introspective, and tend to have little fantasy life. However, while they seldom dwell on inner processes, they can describe them fully when asked to direct their attention inward. They are oriented toward action rather than thought. They prefer action to inaction, and dislike assuming a passive role. At the same time, they are not overly impulsive, and can refrain from action when it is not appropriate. In most of their behavior, evidence of emotional stability is evident.

Although few of the Astronauts tend to be highly involved in relationships to other people, their techniques for dealing with people are effective. They are quick to see the needs of others, and can usually avoid interpersonal difficulties in getting a job done. While they remain independent when possible, all are comfortable when dependence on others is required.

In most cases, attachments within the Astronaut group itself are not intense. They share a common purpose, and realize that whatever one does affects the others. Yet, their independence of each other is more striking than their existence as a team. They show, for example, few signs of the close inter-relatedness of an infantry squad in combat. On the other hand, they respect one another and have usually been able to keep their competitive feelings from interfering with their joint efforts. They are aware of the strengths and weaknesses of each other. In fact, when divorced from the question of who is better qualified for a flight, their evaluations of each member of the group are almost identical. Although they do not share the same standards of personal behavior, they have common professional values, centering about respect for technical competence.

*Motivation*—All the men are professionally motivated to contribute to the space program. This is part of their strong drive toward mastery and achievement. In each case, this became apparent early in life and was eventually expressed in a desire to fly. Not infrequently, the heightened drive toward mastery once served to reduce self-doubts. However, in most cases, it now functions independently of the need for such reassurance.

At present, the Astronauts feel the challenge of their work and enjoy the opportunity to use all their capacities. They derive a sense of satisfaction from participating in something they consider important—something on the frontier of their field. Along with this, all have the conviction that they are making a social contribution. Since all have a genuine feeling of patriotism, their contribution is more significant by being in the national interest. These feelings are reflected in their answers to hypothetical questions about their future plans. Even if unable to fly and financially independent, they would prefer to continue in the space program.

The men have no special wish to face danger, although they are willing to accept the risks demanded by their work. Where it is present, the desire for danger has not been a motivation for their current work. Although a few men enjoy certain types of risks, this had little to do with volunteering for Project Mercury. None could be expected to take unnecessary chances in the course of his work which might jeopardize its success.

Although a feeling of group identity can be noted, group and personal loyalties have not provided major incentives for work or the Project. There is a genuine belief that this is a "team effort", along with awareness of the contributions of many individuals to its success. But it is as if everyone is working for similar objectives, rather than working for the "team" as such.

The desire for public acclaim and recognition has not been a primary motive, although none of the Astronauts has found it unwelcome. The need for personal sense of status and social approval is found in varying degrees. The extra financial rewards, while pleasant, were not foreseen before volunteering and have not been an important part of the Astronauts' motivation for the Project.

*Affect management*—The question of how the Astronauts handle their emotional reactions is of importance—partly because of uninformed outside opinion that anxiety would necessarily rise to high levels. In all instances affects has been efficiently controlled and expressed through a combination of stable personality organization and prior experience in coping with stress. Each man has faced situations in which fear was appropriate and found that he continued to function in spite of its effects. All have the confidence that if fear begins to build up, they can "get hold of" themselves. This is occasionally done through denial, but more often by conscious effort to divert their attention elsewhere and by focusing on task demands.

*Frustration tolerance*—Since the Astronauts' drive for Achievement is strong, their potential for disappointment is great. Nevertheless, they display striking resilience in the face of frustration. The most important source of disappointment has been delays in the schedule and failure to be selected for a particular flight. In general, both have been dealt with by looking beyond the immediate obstacle and deciding that the problem will be resolved in the future. Whenever a course of action has suggested itself, they have embarked upon it as quickly as possible.

***B. Psychological Response to Training and Flight***

1. Measures of psychological functioning.

a. Comparison of Mercury Astronauts and Non-Selected Candidates During Original Selection Program—The battery of repetitive psychological tests used in the present study was first administered to the Astronauts when they were candidates for Project Mercury. The WADC laboratory responsible for the original selection program has made available the test results for the Astronauts and for the 24 other candidates not selected. All six tests of the Moran battery were administered on seven occasions: 1) a pretest; 2) before, and 3) after performance on a complex and demanding psychological task which presented subjects with "information overloads"; 4) before, and 5) after a period in an altitude chamber, simulating 65,000 feet, during which the subject was in a poorly functioning suit; 6) before, and 7) after a two-hour period in a heat chamber at 140 degrees F. These three conditions are identified as CBS, Altitude, and Heat in the accompanying tables. Since three of the Moran tests have been used in evaluating psychological functioning in response to stress during the present study, it is of interest to look back to the original testing and consider how, if at all, the present Astronauts differed from non-selected candidates, how they compared with each other at the earlier time, and how their stress response then compared with later measures.

In Table 1 are presented the mean scores for all seven measures for the selected and non-selected group on each of the six Moran tests. On inspection, it is clear

Table 1
Mean Scores of Candidates Selected (Seven Mercury Astronauts) and Not Selected on a Battery of Repetitive Psychological Measurements, at Time of Initial Selection Program

| | | | CBS | | Altitude | | Heat | | | |
|---|---|---|---|---|---|---|---|---|---|---|
| **Test** | **Group** | **Pre 1** | **Pre** | **Post** | **Pre** | **Post** | **Pre** | **Post** | **Mean** | **SD** |
| Aiming | Selected | 178 | 187 | 191 | 195 | 203 | 192 | 202 | 193 | 13.80 |
| | Not Selected | 165 | 180 | 185 | 178 | 183 | 185 | 187 | 180 | 23.50 |
| Number Facility | Selected | 38 | 40 | 44 | 37 | 42 | 40 | 43 | 40 | 5.07 |
| | Not Selected | 39 | 44 | 45 | 39 | 43 | 41 | 44 | 42 | 6.54 |
| Perceptual Speed | Selected | 100 | 115 | 107 | 97 | 106 | 106 | 106 | 105 | 11.40 |
| | Not Selected | 95 | 111 | 106 | 102 | 109 | 104 | 107 | 104 | 11.20 |
| Visualization | Selected | 39 | 50 | 50 | 42 | 42 | 46 | 46 | 44 | 2.53 |
| | Not Selected | 39 | 45 | 48 | 44 | 42 | 46 | 45 | 45 | 4.64 |
| Flexibility of Closure | Selected | 11 | 17 | 19 | 13 | 18 | 17 | 16 | 16 | 2.08 |
| | Not Selected | 12 | 16 | 18 | 16 | 19 | 17 | 18 | 16 | 1.30 |
| Speed of Closure | Selected | 28 | 39 | 46 | 39 | 45 | 41 | 40 | 40 | 4.46 |
| | Not Selected | 26 | 36 | 43 | 39 | 42 | 36 | 37 | 37 | 7.53 |

that differences between those who were selected and those who were not are small and insignificant, both in levels of performances and in stress response patterns. Only on the Aiming test is there any noticeable difference in level, with the Astronauts producing somewhat more correct responses to this perceptual-motor task over all occasions. In response to the three experimental stresses, no subject showed a significant change in performance from pre- to post-measures. Since these tests require concentration and focusing of attention, it is expected that potentially disruptive stress might lead to performance deficit. It is thus of interest that the reverse is true. Post-stress measures are generally a bit higher, although the differences are small. Thus, in the 36 possible before-after stress comparisons (6 tests × 2 groups × 3 stresses), in only 6 instances is there a fall. In the remaining 30 cases performance was the same or improved after stress. Though, as noted, these are often slight differences, it is noteworthy that all candidates responded to stress with renewed effort and improved performance.

Tables 2, 3, and 4 present the values for the seven Astronauts for the three tests used in the present study (Aiming, Number Facility, and Perceptual Speed) in response to the three stress situations used in the original selection. These three tests are singled out for more detailed analyses in order to allow comparisons with later functioning. It should be cautioned, however, that later comparison is not possible. This is particularly true for Aiming, where a 2.5 minute time allowance was used in the original study, and 2.0 minutes in the present study. In the cases of the Number Facility and Perceptual Speed Tests, the conditions of administration and method of scoring were similar.

There is little relationship between individual differences in performance on the three procedures. S2 is the best performer on the Aiming test; but the worst on Number Facility. S6 performs best on the Perceptual Speed and worst on Aiming. Each of the three tests, therefore, seems to tap different abilities, which are unrelated to each other, in this sample.

Similarly, there is little concordance among the seven in their patterns of stress response. As already noted, post-stress rise in scores is more likely than stress-induced deficit for all subjects. This generalization is no more or less true of any individual when the men are compared to each other.

Therefore, it appears that at the time of initial selection, the seven Astronauts were not discernibly different in performance level or stress response from non-selected candidates—at least on these measures and in response to these situations. Also, they shared with the others the tendency to improve following stress. The seven did not differ in any consistent way among themselves, either in level or response.

b. Effects of Training and Flight Stress on Measures of Psychological Functioning—Tables 5, 6, and 7 summarize the test findings for the three performance tests administered at various points before and after training and flight activities. The tables are so constructed that the first set of columns, labeled "All Measures," gives the number of occasions, the mean, and the standard deviation for each man. The next block of columns describes the values obtained from non-flight measurements. Since these were usually taken before and after training activities, the averages for all "before" situations and all "after" situations are given in the columns labeled "Mean Pre" and "Mean Post". The final column gives the difference between Pre and Post—with plus indicating an increment of functioning, and minus,

Table 2
Aiming Test: Scores of Seven Mercury Astronauts in Response to Three Stress Situations at Time of Initial Selection Program

| | | Heat | | CBS | | Altitude | | |
|---|---|---|---|---|---|---|---|---|
| Subject | Pre 1 | Pre | Post | Pre | Post | Pre | Post | Mean |
| 1 | 164 | 182 | 190 | 200 | 213 | 180 | 192 | 188.7 |
| 2 | 186 | 222 | 245 | 175 | —* | 233 | 247 | 218.0 |
| 3 | 163 | 184 | 190 | 204 | 203 | 193 | 193 | 190.0 |
| 4 | —* | 189 | 190 | 191 | 176 | 183 | 178 | 184.5 |
| 5 | 169 | 190 | 207 | 166 | 215 | 229 | 230 | 200.9 |
| 6 | 195 | 184 | 199 | 156 | 146 | 158 | 191 | 175.6 |
| 7 | 191 | 191 | 188 | 204 | 195 | 190 | 192 | 193.0 |
| Mean | 178.0 | 191.7 | 201.3 | 185.1 | 191.3 | 195.1 | 203.3 | |

* Data Missing

Table 3
Number Facility Test: Scores of Seven Mercury Astronauts in Response to Three Stress Situations at Time of Initial Selection Program

| | | Heat | | CBS | | Altitude | | |
|---|---|---|---|---|---|---|---|---|
| Subject | Pre 1 | Pre | Post | Pre | Post | Pre | Post | Mean |
| 1 | 29 | 34 | 41 | 39 | 40 | 38 | 36 | 36.7 |
| 2 | 33 | 31 | 37 | 33 | —* | 35 | 41 | 35.0 |
| 3 | 34 | 39 | 37 | 34 | 40 | 30 | 35 | 35.6 |
| 4 | —* | 48 | 50 | 42 | 52 | 36 | 47 | 45.8 |
| 5 | 45 | 44 | 47 | 51 | 46 | 42 | 48 | 46.1 |
| 6 | 39 | 37 | 37 | 37 | 42 | 36 | 37 | 37.9 |
| 7 | 46 | 47 | 49 | 36 | 46 | 41 | 49 | 44.9 |
| Mean | 37.7 | 40.0 | 42.6 | 38.9 | 44.3 | 36.9 | 41.9 | |

* Data Missing

Table 4
Perceptual Speed: Scores of Seven Mercury Astronauts in Response to Three Stress Situations at Time of Initial Selection Program

| | | Heat | | CBS | | Altitude | | |
|---|---|---|---|---|---|---|---|---|
| **Subject** | **Pre 1** | **Pre** | **Post** | **Pre** | **Post** | **Pre** | **Post** | **Mean** |
| 1 | 88 | 88 | 92 | 102 | 92 | 99 | 99 | 94.3 |
| 2 | 97 | 97 | 105 | 111 | —* | 102 | 108 | 103.3 |
| 3 | 99 | 98 | 98 | 111 | 106 | 81 | 99 | 98.9 |
| 4 | —* | 116 | 119 | 120 | 115 | 87 | 108 | 110.8 |
| 5 | 126 | 126 | 119 | 143 | 125 | 125 | 124 | 126.9 |
| 6 | 101 | 117 | 112 | 118 | 105 | 98 | 108 | 108.4 |
| 7 | 91 | 97 | 96 | 98 | 97 | 87 | 96 | 94.6 |
| Mean | 100.3 | 105.6 | 105.9 | 114.7 | 106.7 | 97.0 | 106.0 | |

* Data Missing

a decrement. The last block of columns presents the values obtained during the flight period. The measurements made at the time of the medical examination, preceding the orbital flights, were done two days prior to flight. These are in the column "T-2". Where measurements were made at the time of scrubbed flights, these are indicated in the next two columns. Finally, those measurements made at the time of flight itself are given, along with the differences between them. Decrement is indicated by negative values; improvement by positive. Each of the tests yields two scores: 1) the number of items correctly completed within the time limit; and 2) the number of errors made. These two scores are essentially independent (Moran & Mefferd, 1959). In these analyses, therefore, both are considered as separate measures of behavioral efficiency.

On the Aiming Test, it is clear that there are no great differences in ability when all measures of the number of items correctly completed are considered (Table 5). There are greater differences in the error scores. Since this test requires the subject to place a dot within each of the circles as quickly as possible, an error is counted if the dot is on or beyond the boundary of the printed circle. Three of the subjects made fewer than two errors in approximately 200 attempts (per trial), while two of the subjects made 30 or more errors on the average. One of these, S5, was also the most variable on the Number Facility Test.

The Aiming Test shows the clearest effects of flight (Table 5). For all subjects there was some improvement in functioning after the training situations, such as centrifuge runs and simulator trials. By contrast, from before to after flight there was, in every case, a distinct drop in performance and a corresponding rise in errors. This measure thus seems to be the most sensitive measure of the impact of the flight experience on the ability to maintain efficient psychomotor performance.

The findings on the other two tests are less clear. The mean scores for all five Astronauts on the Number Facility Test are generally parallel to those of the Aiming procedure—improvement following training events and deficit following flight—although to a smaller degree (Table 6). The differences between men in the Number Facility Test were small. There was a tendency for faster performance to be accompanied by a greater number of errors. Thus, S3 had both the smallest number of correct items, but also the smallest number of errors.

The Perceptual Speed Test (Table 7) showed the greatest range of individual differences in overall performance, and the least systematic response to flight. Subjects tended to make the same type of response to the flight as they did to training. For example, S1 improved after training events, and improved after flight; while S6 showed a decrement after both training and flight. But in general, pre to post differences were small. One finding is worthy of special comment. The pre-flight measure for all tests was above the general level of performance. The reverse might have been expected, since this measure was made at a time when pre-flight anxiety might be highest. Furthermore, it measures performance in the early hours of the morning, when most subjects are less efficient. That performance was so good at this point suggests a state of activation. The anticipatory anxiety such as might exist not only did not lessen, but seems to have facilitated psychological functioning in the immediate pre-flight hours.

In summary, these measures suggest fairly small effects of flight on performance. The men tended to improve following training stress (as already noted in the stress

Table 5
Aiming Test: Pre and Post Measures of Psychological Functioning at the Times of Training and Flight Events (Five Astronauts Who Have Completed Flights by October 1962)

| | | All Measures | | | Non–Flight Events | | | | Flight Events | | | | | |
|---|---|---|---|---|---|---|---|---|---|---|---|---|---|---|
| | | | | | | Mean | Mean | Diff. | | Scrub | | Flight | | Diff. |
| Ss | Score | N | Mean | SD | Mean | Pre | Post | Post-Pre | T-2 | Pre | Post | Pre | Post | Post-Pre |
| 1 | Correct | 23 | 208.7 | 12.5 | 210.8 | 208.7 | 211.9 | + 3.2 | 201 | — | — | 202 | 184 | –18 |
| | Errors | 23 | 1.7 | 2.0 | 1.6 | 1.8 | 1.4 | – .4 | 0 | — | — | 0 | 7 | + 7 |
| 3 | Correct | 22 | 198.8 | 28.1 | 190.1 | 190.6 | 193.7 | + 3.1 | 227 | 240 | — | 226 | 214 | –12 |
| | Errors | 22 | 1.3 | 2.6 | 2.1 | 3.0 | 1.6 | – 1.4 | 3 | 1 | — | 1 | 3 | + 2 |
| 4 | Correct | 10 | 201.2 | 12.5 | 203.2 | 200.7 | 205.7 | + 5.0 | — | 195 | 218 | 204 | 176 | –28 |
| | Errors | 10 | 28.6 | 13.4 | 38.2 | 39.0 | 37.3 | – 1.7 | — | 12 | 7 | 16 | 22 | + 6 |
| 5 | Correct | 13 | 207.6 | 34.2 | 202.4 | 188.6 | 216.2 | +27.6 | 236 | — | — | 229 | 210 | –19 |
| | Errors | 13 | 36.6 | 30.7 | 45.0 | 36.6 | 53.4 | +16.8 | 17 | — | — | 1 | 8 | + 7 |
| 6 | Correct | 13 | 196.5 | 11.9 | 199.0 | 196.3 | 205.0 | + 8.7 | — | 188 | — | 193 | 183 | –10 |
| | Errors | 13 | .8 | .8 | .9 | .3 | 2.0 | + 1.7 | — | 0 | — | 0 | 1 | + 1 |
| Mean Correct | | | 202.6 | 19.8 | 201.1 | 197.0 | 206.5 | + 9.5 | 221 | 208 | 218 | 211 | 193 | –17 |
| Mean Errors | | | 13.8 | 9.9 | 17.6 | 16.1 | 19.1 | + 3.0 | 7 | 4 | 7 | 4 | 8 | + 5 |

Table 6
Number Facility: Pre and Post Measures of Psychological Functioning at the Times of Training and Flight Events (Five Astronauts Who Have Completed Flights by October 1962)

| | | All Measures | | | Non-Flight Events | | | | Flight Events | | | | | |
|---|---|---|---|---|---|---|---|---|---|---|---|---|---|---|
| | | | | | | | | | | Scrub | | Flight | | |
| | | | | | | Mean | Mean | Diff. | | | | | | Diff. |
| Ss | Score | N | Mean | SD | Mean | Pre | Post | Post-Pre | T-2 | Pre | Post | Pre | Post | Post-Pre |
| 1 | Correct | 23 | 30.1 | 2.35 | 30.2 | 30.2 | 30.0 | − .2 | 29 | — | — | 30 | 30 | 0 |
| | Errors | 23 | .8 | .85 | .7 | 1.0 | .4 | − .6 | 2 | — | — | 0 | 2 | + 2 |
| 3 | Correct | 22 | 28.4 | 2.13 | 28.1 | 28.4 | 28.1 | − .3 | 28 | 29 | — | 32 | 30 | − 2 |
| | Errors | 22 | .4 | .58 | .5 | .1 | .4 | + .3 | 0 | 0 | — | 0 | 0 | 0 |
| 4 | Correct | 10 | 36.3 | 2.50 | 36.0 | 35.3 | 36.7 | +1.4 | — | 37 | 33 | 39 | 38 | − 1 |
| | Errors | 10 | 1.8 | .90 | 2.0 | 2.3 | 1.7 | − .6 | — | 2 | 1 | 1 | 2 | + 1 |
| 5 | Correct | 13 | 30.3 | 3.68 | 30.1 | 28.2 | 32.0 | +3.8 | 31 | — | — | 32 | 30 | − 2 |
| | Errors | 13 | 2.1 | 1.80 | 2.2 | 2.6 | 1.8 | − .8 | 0 | — | — | 4 | 1 | − 3 |
| 6 | Correct | 13 | 36.9 | 2.63 | 37.2 | 37.3 | 36.3 | −1.0 | — | 40 | — | 36 | 32 | − 4 |
| | Errors | 13 | .7 | .85 | .8 | .7 | .3 | − .4 | — | 0 | — | 0 | 1 | + 1 |
| Mean Correct | | | 32.4 | 2.66 | 32.3 | 31.9 | 32.6 | + .7 | 39.3 | 35.3 | 33.0 | 33.8 | 32.0 | − 1.8 |
| Mean Errors | | | 1.2 | 1.00 | 1.2 | 1.3 | .9 | − .4 | .7 | .7 | 1.0 | 1.0 | 1.2 | + .2 |

Table 7
Perceptual Speed: Pre and Post Measures of Psychological Functioning at the Times of Training and Flight Events (Five Astronauts Who Have Completed Flights by October 1962)

| | | | | | Non-Flight Events | | | | Flight Events | | | | | |
|---|---|---|---|---|---|---|---|---|---|---|---|---|---|---|
| | | All Measures | | | | Mean | Mean | Diff. | | Scrub | | Flight | | Diff. |
| Ss | Score | N | Mean | SD | Mean | Pre | Post | Post-Pre | T-2 | Pre | Post | Pre | Post | Post-Pre |
| 1 | Correct | 23 | 75.0 | 6.02 | 75.3 | 74.0 | 76.8 | +2.8 | 71 | — | — | 71 | 77 | + 6 |
| | Errors | 23 | 2.6 | 1.60 | 2.8 | 2.9 | 3.0 | + .1 | 3 | — | — | 0 | 1 | + 1 |
| 3 | Correct | 22 | 82.2 | 5.70 | 81.8 | 83.1 | 81.3 | −1.8 | 79 | 75 | — | 93 | 94 | + 1 |
| | Errors | 22 | .7 | .90 | .8 | 1.1 | .7 | − .4 | 0 | 0 | — | 2 | 0 | − 2 |
| 4 | Correct | 10 | 89.9 | 8.05 | 88.5 | 89.0 | 88.0 | −1.0 | — | 83 | 91 | 99 | 95 | − 4 |
| | Errors | 10 | 1.3 | .95 | 1.0 | 1.0 | 1.0 | 0 | — | 3 | 2 | 0 | 2 | + 2 |
| 5 | Correct | 13 | 102.3 | 5.81 | 102.8 | 102.2 | 103.4 | +1.2 | 90 | — | — | 101 | 111 | +10 |
| | Errors | 13 | 3.1 | 1.65 | 2.9 | 3.4 | 2.4 | −1.0 | 1 | — | — | 6 | 4 | − 2 |
| 6 | Correct | 13 | 92.6 | 5.59 | 92.0 | 92.3 | 90.7 | −1.6 | — | 94 | — | 98 | 92 | − 6 |
| | Errors | 13 | 2.3 | 2.06 | 2.5 | 4.3 | 1.3 | −3.0 | — | 0 | — | 1 | 4 | + 3 |
| Mean Correct | | | 88.0 | 6.20 | 88.1 | 88.1 | 88.0 | − .1 | 79.7 | 84.0 | 91.0 | 92.4 | 93.8 | + 1.4 |
| Mean Errors | | | 2.0 | 1.40 | 1.9 | 2.5 | 1.7 | − .8 | 1.3 | 1.0 | 2.0 | 1.8 | 2.2 | + .4 |

tests of the selection program). It is noteworthy that the pre-flight levels were above the subjects' norms. Following flight, there were mild deficits, particularly in Aiming, and somewhat less in Number Facility. The psychomotor integration and effort of the pre-flight period were not maintained. Differences between subjects in level of performance and stress response were small.

c. Emotional Reactions—As part of the test battery, along with the psychometric procedures, a mood scale consisting of 53 adjectives was administered. The Astronaut was asked to what degree (on a four-point scale) each adjective described his mood at the moment. On factor analysis, Clyde (1959) had extracted six factor variables from this list: 1) "Friendly"; 2) "Energetic"; 3) "Clear Thinking"; 4) "Aggressive"; 5) "Jittery"; and 6) "Depressed". The score for each of these is the sum of the ratings for the adjectives found loaded on the factor. Since each factor score is based on the sum of a different number of adjective ratings in this original scoring system, they have been recalculated on a mean-per-adjective basis to facilitate comparison. In this section, we will discuss the changes in these scores derived from the adjective check list. Other evidences of the subjects' emotional state—from interviews, questionnaires, and other ratings—will be summarized in the next section.

In Table 8 are the mean scores for each of the six factor variables for each subject, organized in the same fashion as scores of the performance measures. It can be seen that there are mean-Pre and mean-Post nonflight events, and differences between them, as well as Pre- and Post-flight events, and differences between them. As noted, in all cases entries are based on mean rating per adjective to allow more direct comparison between variables.

To facilitate consideration of these data, two combinations of variables are used. On logical analysis, the scores for Energetic and Clear Thinking are combined and contrasted with those for Jittery and Depressed. The first of these pairs is named "Effectance"; the second "Disturbance". It is expected that the well-functioning person would normally be higher on Effectance and lower on Disturbance, and that emotionally stressful states should elevate the Disturbance rating and lower the Effectance. The residual two variables—"Friendly" and "Aggressive"—again suggest a polarity. Both are concerned with relations to other persons—Friendly of a warm, cooperative type, and Aggressive of an active, conflictual type. Whereas Effectance vs. Disturbance has its referents largely within the person, Friendly vs. Aggressive is more concerned with social behavior. Table 9 presents the Effectance vs. Disturbance analysis.

On all occasions, the men described themselves more in terms of the positive than the negative emotional states. Thus, the highest ratings were obtained for Friendly, Energetic, and Clear-Thinking; the lowest ratings were for Aggressive, Jittery, and Depressed (Table 8). As with the performance variables, Pre-to-Post changes were rather small, although flight-induced changes were greater than those produced by training events. Following flight, there was a general tendency for the Effectance variables to drop and for greater anxiety to be admitted. There was a parallel increase in Friendly. Thus, compared to their state prior to suiting, after flight they tended to be somewhat less energetic and clear thinking, somewhat more anxious, and to feel more warmly related to people. However, it should be noted that most of these change scores are small, and that both before and after flight the men describe themselves as alert and attentive and generally without fear or other disturbing affect.

Table 8
Clyde Mood Scale: Factor Variables: Pre and Post Measures of Emotional State at the Times of Training and Flight Events (Five Astronauts Who Have Completed Flights by October 1962)

| | | All Measures | | | Non-Flight Events | | | | Flight Events | | | | | |
|---|---|---|---|---|---|---|---|---|---|---|---|---|---|---|
| | | | | | | | | | Scrub | | | Flight | | Diff. |
| Ss | Variable | N | Mean | SD | Mean | Mean Pre | Mean Post | Diff. Post-Pre | T-2 | Pre | Post | Pre | Post | Post-Pre |
| | Friendly | 22 | 3.00 | .13 | 2.98 | 2.96 | 3.00 | + .04 | 3.00 | — | — | 3.00 | 3.50 | + .50 |
| | Energetic | 22 | 2.85 | .28 | 2.84 | 2.98 | 2.78 | – .20 | 2.80 | — | — | 3.00 | 3.00 | 0 |
| 1 | Clear Thinking | 22 | 3.00 | 0 | 3.00 | 3.00 | 3.00 | 0 | 3.00 | — | — | 3.00 | 3.00 | 0 |
| | Aggressive | 22 | 1.23 | .21 | 1.19 | 1.21 | 1.16 | – .05 | 1.64 | — | — | 1.28 | 1.36 | + .08 |
| | Jittery | 22 | – .40 | .16 | – .07 | – .01 | – .11 | – .10 | .29 | — | — | 0 | – .14 | – .14 |
| | Depressed | 22 | 1.15 | .19 | 1.13 | 1.12 | 1.11 | – .01 | 1.50 | — | — | 1.30 | 1.20 | – .10 |
| | Friendly | 22 | 3.28 | .50 | 3.28 | 3.40 | 3.24 | – .16 | 3.30 | 3.40 | — | 3.0 | 3.30 | + .30 |
| | Energetic | 22 | 3.34 | .39 | 3.34 | 3.49 | 3.04 | – .45 | 3.56 | 3.75 | — | 3.63 | 2.75 | – .88 |
| 3 | Clear Thinking | 22 | 3.70 | .25 | 3.68 | 3.67 | 3.67 | 0 | 3.83 | 3.83 | — | 3.83 | 3.50 | – .33 |
| | Aggressive | 22 | 1.86 | .09 | 1.89 | 1.87 | 1.88 | + .01 | 1.77 | 1.82 | — | 1.73 | 1.64 | – .09 |
| | Jittery | 22 | – .07 | .26 | – .86 | – .24 | – .06 | + .18 | 0 | .29 | — | 0 | .57 | + .57 |
| | Depressed | 22 | 1.24 | .22 | 1.25 | 1.20 | 1.27 | + .07 | 1.16 | 1.30 | — | 1.30 | 1.10 | – .20 |
| | Friendly | 10 | 3.15 | .17 | 3.13 | 3.23 | 3.03 | – .20 | — | 3.50 | 3.00 | 3.10 | 3.10 | 0 |
| | Energetic | 10 | 3.10 | .23 | 3.09 | 3.04 | 3.13 | + .09 | — | 3.63 | 2.75 | 3.13 | 3.00 | – .13 |
| 4 | Clear Thinking | 10 | 3.30 | .11 | 3.25 | 3.28 | 3.22 | – .07 | — | 3.50 | 3.30 | 3.30 | 3.30 | 0 |
| | Aggressive | 10 | 1.91 | .15 | 2.02 | 1.97 | 2.06 | + .09 | — | 1.73 | 1.73 | 1.73 | 1.82 | + .09 |
| | Jittery | 10 | – .11 | .27 | – .26 | – .24 | – .29 | – .05 | — | .43 | – .29 | .14 | .14 | 0 |
| | Depressed | 10 | 1.08 | .15 | 1.07 | 1.00 | 1.13 | + .13 | — | 1.10 | 1.30 | 1.00 | 1.00 | 0 |

| | | All Measures | | | Non-Flight Events | | | | Flight Events | | | | | |
|---|---|---|---|---|---|---|---|---|---|---|---|---|---|---|
| | | | | | | | | | Scrub | | | Flight | | |
| Ss | Variable | N | Mean | SD | Mean | Mean Pre | Mean Post | Diff. Post-Pre | T-2 | Pre | Post | Pre | Post | Diff. Post-Pre |
| | Friendly | 13 | 3.76 | .32 | 3.72 | 3.70 | 3.74 | + .04 | 3.90 | — | — | 3.90 | 3.90 | 0 |
| | Energetic | 13 | 3.89 | .15 | 3.86 | 3.88 | 3.85 | – .03 | 4.00 | — | — | 4.00 | 3.90 | – .10 |
| 5 | Clear Thinking | 13 | 3.87 | .16 | 3.83 | 3.83 | 3.83 | 0 | 3.00 | — | — | 3.00 | 3.00 | 0 |
| | Aggressive | 13 | 2.44 | .28 | 2.34 | 2.35 | 2.33 | – .02 | 2.73 | — | — | 2.64 | 3.00 | + .36 |
| | Jittery | 13 | – .31 | .18 | – .30 | – .31 | – .29 | + .02 | – .43 | — | — | – .43 | – .29 | + .14 |
| | Depressed | 13 | 1.00 | 0 | 1.00 | 1.00 | 1.00 | 0 | 1.00 | — | — | 1.00 | 1.00 | 0 |
| | Friendly | 12 | 2.28 | .25 | 3.10 | 3.10 | 3.30 | + .20 | — | 3.10 | — | 2.90 | 3.40 | + .50 |
| | Energetic | 12 | 3.48 | .45 | 3.35 | 3.29 | 3.09 | – .20 | — | 4.00 | — | 3.88 | 3.75 | – .13 |
| 6 | Clear Thinking | 12 | 3.62 | .39 | 3.57 | 3.50 | 3.50 | 0 | — | 4.00 | — | 3.70 | 3.50 | – .20 |
| | Aggressive | 12 | 1.86 | .21 | 1.78 | 1.57 | 1.70 | + .13 | — | 1.64 | — | 1.82 | 1.73 | – .09 |
| | Jittery | 12 | .20 | .26 | .16 | .33 | – .04 | – .37 | — | .43 | — | .29 | .29 | 0 |
| | Depressed | 12 | 1.09 | .10 | 1.09 | 1.10 | 1.03 | – .07 | — | 1.30 | — | 1.00 | 1.00 | 0 |
| M | Friendly | | 3.09 | .27 | 3.24 | 3.28 | 3.26 | – .02 | 3.40 | 3.33 | 3.00 | 3.18 | 3.44 | + .26 |
| e | Energetic | | 3.33 | .30 | 3.30 | 3.34 | 3.18 | – .16 | 3.45 | 3.79 | 2.75 | 3.53 | 3.28 | – .25 |
| a | Clear Thinking | | 3.50 | 3.50 | 3.47 | 3.46 | 3.44 | – .02 | 3.28 | 3.78 | 3.30 | 3.37 | 3.26 | – .11 |
| n | Aggressive | | 1.86 | 1.86 | 1.84 | 1.79 | 1.83 | + .04 | 2.05 | 1.73 | 1.73 | 1.84 | 1.91 | + .07 |
| s | Jittery | | – .14 | .14 | – .27 | – .09 | – .16 | – .07 | – .05 | .38 | – .29 | 0 | .11 | + .11 |
| | Depressed | | 1.11 | 1.11 | 1.11 | 1.08 | 1.11 | + .03 | 1.22 | 1.23 | 1.30 | 1.12 | 1.06 | – .06 |

Table 9

Clyde Mood Scale: Effectance vs. Disturbance: Pre and Post Measures of Psychological Functioning at the Times of Training and Flight Events (Five Astronauts Who Have Completed Flights by October 1962)

| | | | | | Non-Flight Events | | | | Flight Events | | | | | |
|---|---|---|---|---|---|---|---|---|---|---|---|---|---|---|
| | | All Measures | | | | | | | Scrub | | | Flight | | |
| Ss | Variable | N | Mean | SD | Mean | Mean Pre | Mean Post | Diff. Post-Pre | T-2 | Pre | Post | Pre | Post | Diff. Post-Pre |
| 1 | Effectance | 22 | 2.9 | .12 | 2.8 | 2.9 | 2.8 | − .1 | 2.8 | — | — | 2.9 | 2.9 | 0 |
| | Disturbance | 22 | .7 | .09 | .6 | .7 | .6 | − .1 | 1.0 | — | — | .8 | .7 | − .1 |
| 3 | Effectance | 22 | 3.4 | .27 | 3.4 | 3.5 | 3.3 | − .2 | 3.6 | 3.7 | — | 3.6 | 3.0 | − .6 |
| | Disturbance | 22 | .7 | .18 | .7 | .6 | .7 | + .1 | .7 | .9 | — | .8 | .9 | + .1 |
| 4 | Effectance | 10 | 3.1 | .16 | 3.1 | 3.1 | 3.1 | 0 | — | 3.5 | 2.9 | 3.2 | 3.1 | − .1 |
| | Disturbance | 10 | .6 | .14 | .5 | .5 | .6 | + .1 | — | .8 | .7 | .7 | .7 | 0 |
| 5 | Effectance | 13 | 3.8 | .13 | 3.8 | 3.8 | 3.8 | 0 | 3.9 | — | — | 3.9 | 3.9 | 0 |
| | Disturbance | 13 | .5 | .03 | .5 | .5 | .5 | 0 | .4 | — | — | .4 | .5 | + .1 |
| 6 | Effectance | 12 | 3.5 | .39 | 3.4 | 3.3 | 3.2 | − .1 | — | 3.9 | — | 3.7 | 3.6 | − .1 |
| | Disturbance | 12 | .7 | .15 | .7 | .8 | .6 | − .2 | — | 1.0 | — | .7 | .7 | 0 |
| Mean | Effectance | | 3.3 | .21 | 3.3 | 3.3 | 3.2 | − .1 | 3.4 | 3.7 | 2.9 | 3.5 | 3.3 | − .2 |
| Mean | Disturbance | | .6 | .12 | .6 | .6 | .6 | 0 | .7 | .9.7 | .7 | .7 | 0 | |

The Astronauts did not differ greatly from one another in these patterns of emotional self-ratings. On the whole, S3 tends to show greater changes from before to after both flight and non-flight situations. S5 seems most deviant from the general pattern—showing a more distinct rise in the Aggressive category after flight than the other men. Moreover, except for S5, all of the men showed some tendency to be more Jittery immediately before flight and to rise somewhat higher after flight, compared to their general levels.

Overall, it is quite clear that the men generally feel vigorous, alert, and positively oriented toward others, and that neither flight or non-flight training stress importantly changed these mood states. Careful study of individual curves reveals particular occasions when there was more disturbance than at the time of flights. On closer review of such occasions, there were usually understandable and appropriate frustrations for explaining the mood changes. On other ratings, made at the same time as the mood ratings, the men most often rate their performance as good, with anticipatory and during-event anxiety described as low.

2. *Clinical summary of stress behavior*—Throughout the study, adaptive behavior has been recorded under various conditions. As in any job situation, there have been day-to-day problems. Because of the particular nature of this project, the number of delays and frustrations has necessarily been large. Furthermore, responsibilities and procedures have often been unlike those in the military services where their professional lives have been spent. Both the novel technology and the unusual social organization were potential sources of difficulty. In spite of this, few adverse reactions were noted during the development and training phase of the program. The Astronauts took a practical, realistic approach to these problems.

None of their training activities, as such, were emotionally stressful, except where performance fell below aspiration. Such "failures" produced disappointments, but usually served to stimulate increased effort. The men differ in the extent to which self-esteem is bound up with achievement. Some of them set their levels of aspiration high and then struggle strenuously to attain them. Others accept a lower level of accomplishment. In such cases, performance at objectively lower levels, which may reflect less effort in training, is not always viewed as personal failure.

For most of the men, not being chosen for the first flight was the hardest problem to master. The immediate disturbing implication was that the men not selected were not doing as well as they had thought. They tried to learn in what ways their performance might be lacking, and eventually each was able to decide that his day would come. Once a man knew that he was chosen for a flight, further delays had little impact. From there on, the issue of "when?" was of secondary importance.

During the period between selection and flight, the major concern has been achieving a sense of readiness. Until the men have felt "on top" of things, a feeling of discomfort has been evident. This has arisen from such matters as changes in the flight plan or insufficient simulator experience with particular emergency procedures. Conscious thought of danger and possible death has been of secondary importance.

The Astronauts have achieved a sense of readiness at different intervals before the flight. Where "everything has fallen into place" well ahead of time, the final phase of the training period has provided an opportunity to relax and polish details. But where readiness has been achieved only during the final days before the flight, more strain has been evident.

During the immediate pre-flight period, all the men have felt ready to go. They have been preoccupied with operational details, and show little anticipatory anxiety. When this has appeared, it is described as a mild feeling of tension or of being on edge. It is similar to that felt in combat or other stressful situations and—being familiar—is not disturbing. In most cases, anxiety seems more related to intense concern that the flight will be successful than to the fear of injury or death.

All of the Astronauts have considered the risks. They are convinced that as a result of their past experience and intensive training, they are prepared to handle any emergency. Much of the ability to control anticipatory anxiety comes from their confidence in this preparation. Having considered each eventuality and having done all they can, they feel there is little point in worrying any further. When thoughts of danger arise, they are put aside in favor of a review of the flight plan or other technical aspects of the flight. As one man said, "Whenever I think of something that may go wrong, I think of a plan to take care of it."

During the period from insertion to launch, the men have reported being on edge, which they view as a positive sign. If tension begins to build up, their response is to stop, take stock of themselves, and decide what to do to bring matters under control. No man has been unable to do this. Although there have been evidences of excitement, no instance of severe anxiety has been observed.

The same pattern has emerged during flight. Thoughts have centered on procedures to be carried out and on the experiences of being in orbit. The successful launch and subsequent course of the flight induce a feeling of exhilaration, which may be reinforced by the pleasant sensation of weightlessness. Anxiety levels have not been high. Even in the instances where the possibility of death arose, emotional reactions were maintained within normal limits.

Responses after the flight have uniformly involved some degree of elation, coupled with fatigue. Elation has come from both a sense of a difficult job well done and a sense of relief that the long-anticipated flight is over. Concern with public relations functions and other post-flight activities is not great. The Astronauts appropriately feel that they have mastered the major challenge and that nothing else will bother them now.

### C. Summary of the Psychological Findings

1. *Performance*—At the time of selection, both the Astronauts and the non-selected candidates showed a high level of performance, with improvement following experimental stresses. During the stress study, acceleration, ECS tests, simulator runs, and prelaunch activities produced mild changes in performance, often in the direction of improvement. Immediately prior to flight, performance was above baseline values. The flights themselves led to a small but consistent deficit in one test, but had little effect on the others. These measures reveal overall stability of functioning, a state of heightened efficiency prior to flight, and a slight decline of performance following flight. Thus, neither training nor flight activities produced important psychological deficit.

2. *Emotional state*—Varied emotional states were observed during the study; however, no disabling responses were noted in the nonflight and flight situations studied.

3. *Personality factors*—In general, the Astronauts are characterized by task orientation, strong achievement needs, and satisfaction in professional accomplishment. They are practical, intelligent, individualistic and have clear needs for active control of situations and inner feelings. They approach flight with confidence in the technical support and in their own training and ability. By intensified concern with goal-directed activity and by blocking negative feelings, they effectively cope with the potential psychological effects of flight.

### Results of Biochemical Studies

[*Author's Note:* The biochemical findings were collected by S. Korchin and were never published. Unfortunately, the data are lost and cannot be included.]

### Discussion and Theoretical Implications of the Study

Although the study was designed to study the stress of Mercury flights, and the capacities of the Astronauts to cope with it, the findings obtained can extend our knowledge of the psychophysiology of stress behavior, in general. Further data collection and analyses are still to be done, but some of the broader issues to which our data relate can be discussed.

#### A. *Pattern of Stress Responses*

Within the sphere of human behavior, stress may be viewed as a state involving "a sufficiently potent danger to psychological or physical well-being as to require extraordinary measures for the maintenance of organized behavior or—these failing—which may lead to disordered behavior, anxiety, or other emotional disturbance." (Korchin, in press) Conceptually, there are important parallels in the processes of stress, defense and breakdown at various levels of biological, psychological and social functioning. It is clear that there are stresses of different orders, and that each may affect the mechanisms of regulation and/or disorganization in different ways. Thus, a threat to vital functioning (e.g., deprivation of oxygen supply) may occur without conscious representation of danger or anxiety, while neurotic conflicts without life-threatening consequences may profoundly alter physiological functioning.

In advance, it was conceivable that space flight with its psychological and physical demands might have resulted in serious disturbances. Measures were thus made at different levels of biological organization, so that the extent as well as the locus of stress reaction could be assessed. It was possible that flight stress would lead to emotional arousal, deficits in psychological functioning, and activation of both adrenomedullary and adrenocortical systems.

In fact, the extent of arousal and the consequent effects have been small. Rather than indicating stress disorganization, a state of heightened energization is suggested by the emotional, psychological, and catecholamine findings. The capacity for effective performance—at least as indicated by simple psychometric measures—was facilitated rather than disturbed. Astronaut reports describe the flight experiences as exhilarating. The required performance in the capsule was at high levels. The stress effect was seen primarily in adrenocortical output.

Recent theory suggests that stress input and psychological performance or physiological output are related in curvilinear rather than monotonic fashion (Leeper, 1948; Hebb, 1955; Korchin, 1962). In this conceptualization, lower orders of arousal lead to facilitation up to an optimum. Beyond this, increased stress results in psychological deficit and psychophysiological disturbances. The point of optimum functioning, as well as the shape of the curve generally, are functions of individual characteristics as well as of the nature of the stress.

In the present study, we would propose that the flight led to activation or energization in men who were both initially competent and well-prepared for the oncoming event. The emotional responses, test and actual performance, and the output of adrenaline and its metabolic products all support the view of the situation as one producing facilitative activation. The cost to the bodily economy of maintaining this state is seen in the heightened level of adrenocortical activity, of a sort found in response to other conditions. Even in this realm, the extent of response is not comparable to that found in extreme anxiety or high levels of physical stress.

One need not expect that measures of stress behavior at different biological levels should operate in a convergent manner. It is conceivable and supported by experimental evidence (Korchin and Herz, 1960), that psychologically stressful events may produce proportionately more adrenal activation than emotional change. The effectively functioning person has adaptive mechanisms which can block psychological disturbance without at the same time preventing physical response. In this view, it may be more characteristic of the less adequate person to have greater concordance between psychological and physiological measures of stress response.

The findings of this study are based on focal stresses of relatively short duration. If the same performance were required over a longer period, one could not be certain whether the adaptive mechanisms would operate efficiently. Although all of the men performed well and were able to stay within their capacities, more sustained stress of the same intensity might have led to a breakdown of functioning. The evidence from the adrenocortical findings is that the men differed in the degree to which they depleted their reserve for further adaptive stress response, even during the time involved in these suborbital and orbital flights.

In some sense, the flights studied might be likened more to the situation of the sprinter than to that of the long-distance runner. High output is required for relatively short period. Even with depleted reserves, successful performance may be possible. Whether the same capacities, but in greater amount, are required for operation under continued stress; or whether the "long-distance runner" requires other abilities, cannot be determined. But it would seem reasonable to suggest that a stress of greater duration should be less intense, in order to permit adequate functioning without premature depletion of resources. In present terms, this suggests the need for simplified flight plans, fewer in-flight activities, and slower pacing in longer flights.

One additional quality should be noted. The focal stress—the flight itself—is actually the terminal experience of a series of events which may be as or more stressful. The days and weeks preceding flight are harried for the Astronaut, as competing demands on the flight plan are balanced and compromised. All of this occurs in an atmosphere of heightened excitement and public attention. The flight itself may thus come as an emotional relief. Since a social stress is imposed on a background of the going general stress, the response is a function of how it cumulates or contrasts with the background level.

### *B. Sources of Ego Strength*

This study was undertaken on the assumption that the Astronauts were men of more than average competence, and that the study could, therefore, contribute to the understanding of the conditions of effective functioning in psychologically-healthy persons. The results presented bear out the initial assumption. There is little question but that the Astronauts, as a group, are men of particular psychological competence. In some respects they are like the "homoclites" recently described by Grinker (1962). The term describes psychologically healthy men who are realistically oriented to their world, without unattainable ambitions and consequent neurotic concerns, who are adequate to the demands on them and able to function well without emotional distress. In some respects, however, the Astronauts seem different from, and more than, the homoclite. If, as Grinker suggests, the homoclites are the solid core of mentally healthy Americans, the Astronauts are at least "uncommon common men". On the whole, they are more ambitious, more committed to career advancement, more concerned with success and sensitive to failure, as well as having a higher order of ability and intelligence. They share, however, in having little consuming self-doubt. In general, they have strong egos, capable of dealing with psychological events without undue disturbance, with good reality testing, and control of potentially disturbing inner processes. Many of these qualities were described as characteristic of the present Astronauts and other candidates in the original selection program (Ruff & Levy, 1959).

What are the sources of this ego strength? Although fuller answers will depend on further study of the available material and data yet to be collected, some general speculations are possible.

In general, the Astronauts are men with firm identities. They know who they are, and where they are going. This appears related to development in well-organized families, with considerable solidarity. They grew up in stable communities, usually smaller ones, where the family's social position was secure. Within the family, there was unusually strong identification with a competent father. From childhood on, their lives flowed in relatively smooth progressions. There were few crises or turning-points in their lives. Each phase led naturally into the next. Where there were obstacles, and they may have been inadequate to the demands of a realistically difficult situation, compensatory mechanisms were called into play. Although these may once have been of neurotic origin, they became integrated syntonically into the larger organization of a healthy personality. Subsequently, such behaviors became functionally autonomous, and, freed of neurotic functions, remained as sources of personal gratification and social utility.

The high order of innate capacities possessed by these men should not be overlooked. They started with ample ability and have been exposed to situations which could be mastered within the repertory of their capacities. This has led to success, and success to heightened self-esteem. One can picture life histories which start with good abilities and favoring childhood environment which provided basic emotional security and firm identification. Thereafter, there was a recycling progression going through successive phases of appropriate aspiration, success, increased self-esteem, increased aspiration, etc. Examination of their professional careers indicates smooth progression without major set-backs. This type of life history, we believe, is an important source of their present personal competence and of their probable adaptation to future stress.

### Recommendations

**A. *Further Research***

1. *Completion of present study*—It is recommended that stress data be collected for the MA-9 flight. The conclusions drawn so far are based on results of five flights. Their significance will be increased by the addition of a sixth. Because of its longer duration, this flight will be of special importance in indicating the stress effects of the space environment.

In studying individual differences in psychophysiological response characteristics of the Astronauts, it is important to gain flight data on as many of the men as possible. This will be of potential importance in determining future selection techniques. Major Cooper is the only eligible pilot for whom flight data are not yet available. Since his baseline values have already been established, little more of his time will be required to complete this study.

After the final flight data have been examined, it will be desirable to re-interview each Astronaut to bring the material up to date and to answer questions raised by analysis of the results. A short series of interviews similar to those conducted in July 1960 is contemplated. To minimize the time required, as much of the necessary information as possible will be gathered through a questionnaire. Interviews can be scheduled at any time the astronauts' schedules permit.

Certain cardiovascular and respiratory data obtained during training and flight periods will be required for incorporation into the final data analysis. Ratings of performance are also desirable. If possible, it would be helpful to have interviews with Mr. Gilruth and Mr. Williams. This would allow their observations to be included in the analysis and would insure that questions they may have are covered in the final report.

2. *Future Astronaut studies*—Since it is recognized that work involving Astronauts should be held to a minimum, careful consideration must be given to follow-up stress studies for the Gemini and Apollo programs. Insofar as these may be desirable, they should be conducted by an in-house group within the Manned Spacecraft Center.

3. *Related laboratory studies*—Where further data are needed to supplement and extend the psychological findings of the present study, attempts will be made to use non-astronaut subjects. Whether or not such data are required will not be known until final analyses of the results have been completed.

**B. *Future Selection and Training Programs***

As a result of the data gathered during the present study, it is apparent that recommendations for future selection and training programs can be made. However, these conclusions must remain tentative until results for each of the men are available.

## E. CHRONOLOGY OF *MERCURY* ASTRONAUT SELECTION

| Number of Candidates | Event |
|---|---|
| 225 | USAF personnel records screened |
| 225 | Navy personnel records screened |

| | |
|---|---|
| 23 | Marine Corps personnel records screened |
| 35 | Army personnel records screened |
| 508 | TOTAL records screened (Jan. 1959) |
| 110 | Total met minimum standards, in terms of test pilot school, jet hours, age, height, technical education, medical record (end of Jan. 1959) |
| 69 | Reported to briefing in Washington (2 groups, Feb. 2 and 9, 1959) |
| −6 | Too tall |
| −8 | Declined invitation to volunteer |
| 55 | TOTAL given written tests, technical interviews, psychiatric interviews, medical history reviews |
| −8 | Declined invitation to continue |
| −15 | Disqualified (medical, psychological or technical reasons) |
| 32 | TOTAL sent to Lovelace Clinic |
| −1 | Eliminated for medical reasons |
| 31 | TOTAL sent to Wright ADC for psychological & stress testing |
| 0 | Eliminated for psychological reasons |
| 31 | TOTAL left for final selection |
| −24 | Not selected |
| 7 | TOTAL selected as *Mercury* astronauts (April 1, 1959) |

## SELECTED REFERENCES

Cattell, R. B. 1943. "The Description of Personality: Basic Traits Resolved into Clusters." *Journal of Abnormal Psychology,* 475–506.

Grinker, Sr., R. R., and R. R. Grinker, Jr. 1962. "Mentally Healthy Young Males (Homoclites): A Study." *Archives of General Psychiatry* 6: 305–353.

Korchin, S. J., and H. A. Heath. 1961. "Somatic Experience in the Anxiety State: Some Sex and Personality Correlates of 'Autonomic Feedback.'" *Journal of Consultative Psychology* 25: 398–404.

Korchin, S. J., and M. Herz. 1960. "Differential Effects of Shame and Disintegrative Threats on Emotional and Adrenocortical Functioning." *Archives of General Psychiatry* 2: 640–651.

Oken, D., R. R. Grinker, H. A. Heath, M. Herz, S. J. Korchin, M. Sabshin, and N. B. Schwartz. 1962. "Relation of Physiological Response to Affect Expression." *Archives of General Psychiatry* 6: 405–453.

Ruff, G. E., and E. Z. Levy. 1959. "Psychiatric Evaluation of Candidates for Space Flight." *American Journal of Psychiatry* 116: 385–391.

# APPENDIX 2

## A. SUMMARY OF USAF ISOLATION/CONFINEMENT STUDIES (1953–1966)

| Laboratory | Duration | Subjects | Facility | Primary Variables | Additional Psychological Components |
|---|---|---|---|---|---|
| AMRL (1953) | 24 Hours | 16 – MIL & CIV Staff | F-84 Cockpit | Partial Immobilization<br>Drugs (d-Amphetamine)<br>Prolonged Confinement (1 Man) | Psychomotor Measures<br>Psychophysiologic Measures |
| AMRL (1955) | 56 Hours | 4 – 2 Pilots 2 CIV | F-84 Cockpit | Partial Immobilization<br>Prolonged Confinement (1 Man) | Psychomotor Performance<br>Subjective States |
| SAM (1957) | 24 Hours | 1 Airman | Small Altitude Chamber | Same | Biomedical Measures |
| AMRL (1958) | 3 Hours to 7 Days | 100 + – Mixed | Soundproof Room | Prolonged Confinement (1 Man)<br>Prolonged Isolation | Subjective States |
| AMRL (1958) | 5 Days | 5 | "Mission Simulator" | Same | Psychophysiologic Measures<br>Subjective States |
| SAM (1958) | 10 Days | 2 – 1 NCO 1 CIV Staff | Small Altitude Chamber | Same (Except 2 Men) | Psychomotor Measures<br>Physiological Measures |
| SAM (1958–59) | 7 Days | 7 – 1 Airman 6 Pilots | 1-Man Space | Space Environment and Atmosphere ("Pure" $0_2$ At Altitude)<br>Partial Immobilization<br>Prolonged Confinement (1 Man) | Psychomotor Performance<br>Psychophysiologic Measures<br>Physiologic Measures<br>Biochemical Measures<br>Subjective States |

| Laboratory | Duration | Subjects | Facility | Primary Variables | Additional Psychological Components |
|---|---|---|---|---|---|
| SAM (1958–1959) | 30 Hours | 8 – 4 Airman 4 Pilots | 1-Man Space Cabin | Same | Same |
| AMRL (1959) | Intermittent | 15 – College Students | Crew – Stations Simulator | Simulator "Shakedown" | Not Applicable |
| AMRL (1959) | 4 Days | 16 – College Students | Crew – Stations Simulator | Prolonged Confinement (4 Men)<br>Work/Rest Ratios (2/2,4/4,6/6,8/8) | Psychophysiologic Measures<br>Subjective Impressions |
| AMRL (1960) | 48 Hours | 20 – 10 Officers 10 Controls | Escape Capsule | Partial Immobilization<br>Prolonged Confinement (1 Man)<br>Experimental Diet | Psychomotor Measures<br>Intellectual Funct. Measures<br>Subjective States |
| AMRL (1960) | 4 Days | 20 – College Students | Crew Stations Simulator | Prolonged Confinement (5 Men)<br>Work/Rest Ratios (4/2,6/2) | Psychophysiologic Measures<br>Subjective Impressions |
| SAM (1960) | 14 Days | 2 – NCO'S | 2-Man Space Cabin | Simulated Space Environment<br>Simulated Space Atmosphere ("Pure" $O_2$ At Altitude)<br>Prolonged Confinement (2 Men)<br>Manned Space Flight Logistics | Subjective Impressions<br>Limited Psychiatric Monitoring |

| Laboratory | Duration | Subjects | Facility | Primary Variables | Additional Psychological Components |
|---|---|---|---|---|---|
| Sam (1960) | 30 Days | 2 – Pilots | 2-Man Space Cabin | Same | Psychomotor Measures<br>Prolonged Confinement Effects<br>Diurnal Effects<br>Work/Rest Cycle Effects<br>Workload Effects<br>Limited Psychiatric Monitoring<br>Subjective Reports |
| AMRL (1961) | 15 Days | 11 – 2 B-52 Crews | Crew Stations Simulator | Prolonged Confinement (5 Men)<br>Work/Rest Ratios (4/2) | Psychophysiologic Measures<br>Sleep Impressions<br>Subjective Impressions |
| SAM (1961) | 17 Days | 10 – Pilots | 2-Man Space Cabin | Simulated Space Environment<br>Simulated Space Atmosphere ("Pure" $0_2$ At Altitude)<br>Prolonged Confinement (2 Men)<br>Manned Space Flight Logistics | Psychomotor Measures<br>Diurnal Effects<br>Confinement Effects<br>Work/Rest Effects<br>Workload Effects<br>Subjective Reports (Diaries)<br>Psychiatric Monitoring<br>Interpersonal Interaction<br>Biomedical Measures |
| SAM (1961) | 30 Days | 2 – Pilots | 2-Man Space Cabin | Same | Same |

| Laboratory | Duration | Subjects | Facility | Primary Variables | Additional Psychological Components |
|---|---|---|---|---|---|
| AMRL (1962) | 15 Days | 6 – AF Cadets | Crew Stations Simulator | Prolonged Confinement (6 Men)<br>Work/Rest Ratios (4/2) | Psychophysiologic Measures<br>Subjective Impressions<br>Sleep Impressions |
| AMRL (1962) | 30 Days | 10 – Pilots | Crew Stations Simulator | Same (Except 4/4 And 5 Men) | Same |
| SAM (1962) | 12 Days | 2 – Pilots | 2-Man Space Cabin | Simulated Space Environment<br>Simulated Space Atmosphere/ Altitude/$C0_2$<br>Prolonged Confinement (2 Men)<br>Manned Space Flight Logistics | Biomedical Measures<br>Psychomotor Measures |
| SAM (1962) | 14 Days | 4 – Pilots | 2-Man Space Cabin | Same | Same |
| SAM (1962) | 17 Days | 6 – Pilots | 2-Man Space Cabin | Same (Except No $C0_2$ Variations) | Same |
| AMRL (1963) | 12 Days | 32 –<br>6 Cadets<br>9 Officers<br>17 Pilots | Crew Stations Simulators | Prolonged Confinement (5–6 Men)<br>Work/Rest Ratios (4/2,4/4 Plus Sleep Deprivation) | Psychophysiologic Measures<br>Subjective Impressions<br>Sleep Impressions |

| Laboratory | Duration | Subjects | Facility | Primary Variables | Additional Psychological Components |
|---|---|---|---|---|---|
| SAM (1963) | 17 Days | 4 – Pilots | 2-Man Space Cabin | Simulated Space Environment<br>Simulated Space Atmosphere/Altitude/$C0_2$<br>Prolonged Confinement (2 Men)<br>Manned Space Flight Logistics | Biomedical Measures<br>Psychomotor Measures |
| SAM (1963) | 30 Days | 8 – Airmen | 2-Man Space Cabin | Same | Same |
| SAM (1964) | 30 Days | 4 – Airmen | 2-Man Space Cabin | Simulated Space Environment<br>Simulated Space Atmosphere<br>Prolonged Confinement (2 Men)<br>Manned Space Flight Logistics | Biomedical Measures<br>Psychomotor Measures |
| SAM (1964) | 15 Days | 4 – Airmen | 4-Man Space Cabin | Simulated Space Environment<br>Simulated Space Atmosphere ($He$-$0_2$)<br>Prolonged Confinement (4 Men)<br>Manned Space Flight Logistics | Same |
| SAM (1964) | 30 Days | 2 – Engineers | Minuteman Launch Complex | Minuteman Crew Systems Evaluation<br>Prolonged Confinement (2 Men)<br>Missile System Logistics | Psychomotor Measures<br>Sleep Study<br>Subjective Reports (Diaries) |

| Laboratory | Duration | Subjects | Facility | Primary Variables | Additional Psychological Components |
|---|---|---|---|---|---|
| SAM (1964) | 15 Days | 2 – Airmen | Small Altitude Chamber | Prolonged Confinement (2 Men)<br>Nutritional Studies | Psychomotor Measures |
| SAM (1965) | 15 Days | 4 – Airmen | 4-Man Space Cabin | Simulated Space Environment<br>Simulated Space Atmosphere ($He-0_2$)<br>Prolonged Confinement (4 Men)<br>Manned Space Flight Logistics | Biomedical Measures<br>Psychomotor Measures |
| SAM (1965) | 14 Days | 4 – Airmen | 4-Man Space Cabin | Same (Except $N_2-O_2$) (Plus Trace Contaminants Analysis) | Same |
| SAM (1965) | 57 Days | 4 – Rated Officers | 4-Man Space Cabin | Full Scale Simulation ($He-0_2$) With Augmented Biomedical, Nutritional, Logistic, Etc. | Same (Plus Sleep Reports) |
| SAM (1965) | 28 Days | 8 – Airmen | Small Altitude Chamber | Prolonged Confinement (2 Men)<br>Nutritional Studies | Psychomotor Measures |
| SAM (1966) | 12 Days | 14 – Airmen | Double Decker Isolator | Prolonged Confinement (2 Men)<br>Work/Rest Schedules (4/2, 4/4, 16/8 Plus Sleep Deprivation) | Subjective Changes<br>Sleep Reports |

| Laboratory | Duration | Subjects | Facility | Primary Variables | Additional Psychological Components |
|---|---|---|---|---|---|
| SAM (1966) | 5 Days | 8 – Airmen | 4-Man Space Cabin | Prolonged Confinement (4 Men)<br>Simulated Space Environment<br>Variations in $C0_2$<br>Nutritional Studies<br>Manned Space Flight Logistics | Psychomotor Measures<br>Biomedical Measures<br>Sleep Reports |
| SAM (1966) | 21 Days | 4 – Airmen | 4-Man Space Cabin | Same (Plus Trace Contaminants) | Same |
| SAM (1966) | 5 Days | 16 – Airmen | 4-Man Space Chamber | Same (Except 3% Or 4% $C0_2$ In Place Of Trace Contaminants) | Same |
| SAM (1966) | 28 Days | 4 – Airmen | Small Altitude Chamber | Prolonged Confinement (2 Men)<br>Nutritional Studies | Psychomotor Measures<br>Subject Impressions |
| SAM (1966) | 16 Days | 4 – Airmen | Small Altitude Chamber | Same | Subject Impressions |
| SAM (1966) | 32 Days | 6 – Airmen | Small Altitude Chamber | Same | Same |
| SAM (1966) | 32 Days | 4 – Airmen | Small Altitude Chamber | Prolonged Confinement (4 Men)<br>Nutritional Studies | Same |

B. Psychometric Test Variable Means

| | Variable | Mercury | | Gemini/Apollo | | |
|---|---|---|---|---|---|---|
| | | Selected | Not Selected | Selected | Not Selected | Control |
| 1. | Clin. Psych. Rating | 7.6 | 6.5 | 4.3 | 3.8 | – |
| 2. | WAIS FSIQ | 135.1 | 131.8 | 134.9 | 131.0 | 118.6 |
| 3. | WAIS VIQ | 136.0 | 131.2 | 133.4 | 129.4 | 118.9 |
| 4. | WAIS PIO | 129.9 | 128.6 | 132.4 | 129.3 | 115.8 |
| 5. | Ror. # Response | 32.0 | 31.0 | 37.3 | 26.0 | 35.5 |
| 6. | Ror. F + % | 90.0 | 6.5 | 87.7 | 90.1 | 72.8 |
| 7. | Ror. F % | 39.0 | 37.5 | 51.6 | 45.1 | 44.8 |
| 8. | Ror. A % | 31.7 | 36.5 | 42.6 | 42.3 | 39.6 |
| 9. | Ror. # Popular | 4.3 | 5.3 | 7.4 | 7.1 | 6.2 |
| 10. | Ror. # M | 2.9 | 3.2 | 4.6 | 2.7 | 2.5 |
| 11. | Ror. # W | 16.1 | 14.7 | 11.7 | 9.4 | 12.0 |
| 12. | Ror. Σ C | 3.6 | 3.1 | 4.0 | 2.7 | 4.6 |
| 13. | Ror. # m | 1.6 | 1.4 | 2.8 | 1.6 | 1.5 |
| 14. | Ror. # FM | 4.0 | 4.8 | 5.3 | 4.0 | 3.6 |
| 15. | Ror. # Shading Re | 3.9 | 2.5 | 5.4 | 4.0 | 4.3 |
| 16. | Ror. % Resp. to Chrom Cards | 36.1 | 35.0 | 58.6 | 54.0 | – |

*Source:* Bryce O. Hartman and Richard C. McNee, "Psychometric Characteristics of Astronauts." Paper presented at the NATO AGARD Conference on Recent Advances in Space Medicine, Athens, Greece, October 1976.

C. Psychometric Test Variable Standard Deviations

| | Variable | Mercury | | Gemini/Apollo | | |
|---|---|---|---|---|---|---|
| | | Selected | Not Selected | Selected | Not Selected | Control |
| 1. | Clin. Psych. Rating | 0.60 | 0.88 | 1.52 | 0.79 | – |
| 2. | WAIS FSIQ | 3.7 | 6.8 | 6.8 | 6.3 | 7.1 |
| 3. | WAIS VIQ | 4.3 | 7.3 | 6.2 | 5.4 | 8.6 |
| 4. | WAIS PIO | 4.8 | 7.7 | 7. | 10.5 | 8.0 |
| 5. | Ror. # Response | 15.6 | 18.2 | 15.5 | 13.8 | 27.8 |
| 6. | Ror. F + % | 8.8 | 8.3 | 11.5 | 9.4 | 19.8 |
| 7. | Ror. F % | 4.9 | 17.4 | 7.7 | 18.7 | 17.6 |
| 8. | Ror. A % | 12.7 | 9.6 | 11.7 | 10.6 | 12.0 |
| 9. | Ror. # Popular | 1.8 | 2.2 | 3.6 | 1.7 | 2.9 |
| 10. | Ror. # M | 3.2 | 1.9 | 3.2 | 1.6 | 2.7 |
| 11. | Ror. # W | 6.8 | 6.7 | 6.0 | 3.9 | 8.0 |
| 12. | Ror. Σ C | 1.9 | 1.9 | 1.8 | 2.0 | 3.4 |
| 13. | Ror. # m | 1.3 | 1.3 | 1.8 | 1.7 | 2.1 |
| 14. | Ror. # FM | 2.4 | 3.3 | 2.7 | 2.3 | 2.8 |
| 15. | Ror. # Shading Response | 2.9 | 2.5 | 2.8 | 2.8 | 4.5 |
| 16. | Ror. % Resp. to Chrom Cards | 8.0 | 7.8 | 7.0 | 7.9 | – |

*Source:* Bryce O. Hartman and Richard C. McNee, "Psychometric Characteristics of Astronauts." Paper presented at the NATO AGARD Conference on Recent Advances in Space Medicine, Athens, Greece, October 1976.

## D. SAMPLE RORSCHACH EVALUATION SUMMARIES IN NINE "TYPICAL" ASTRONAUT CANDIDATES FROM *MERCURY, GEMINI,* AND *APOLLO*

| Candidate | Responses to Rorschach |
|---|---|
| 1 | Highly intellectualized responses, and in some cases over-controlled. There was a slightly immature quality to many of the responses. Anxiety indices were somewhat high demonstrating some anxiety in affection and fulfillment. Other ratios computed on the total scoring of the protocol suggested a marked tendency to withdraw from the flux of normal interpersonal relationships. |
| 2 | The Rorschach was well-integrated, and well-balanced. It indicates good emotional resources which are well-mobilized. There appeared to be good sensitivity to the needs of self and others. There were some signs of underlying anxiety, affectional in origin, but the tension involved is channeled into productive outlets. There were indications of continuing emotional maturation and possibility of achieving even greater emotional depth. |
| 3 | Rorschach analysis revealed an individual with very strong status needs who attempted to demonstrate his conceptual adequacy somewhat beyond the point of diminishing returns. Though highly intellectual, the productivity of responses was suggestive of intellectual over-compensation, perhaps for unconscious needs in the affectional area. The subject was quite able to adequately utilize control factors and to maintain both intellectual and emotional equilibrium throughout. His concentration on anatomical percepts was suggestive of some concern over his physical condition though this may well have been situationally induced. A rigidity of conceptualization was noted throughout, and little evidence of creativity was obtained. |
| 4 | Rorschach analysis indicated an unusual degree of creative ability coupled with well-integrated and well-mobilized intellectual and emotional resources. The subject appeared quite emotionally reactive to his environment and demonstrated a high level ability to structure ambiguous stimuli into meaningful and consistent patterns. Indices of ego strength and general emotional maturity were high. |
| 5 | The subject gave a superior Rorschach protocol; there were creative percepts well-tempered by periodic return to a more conventional approach. This is a well-integrated individual, though somewhat over-mobilized. There is considerable emphasis upon immediate gratification of needs and impulses; this is balanced by mature modes of gratification. There were signs of some underlying generalized anxiety channeled productively. High energy level and drive. |
| 6 | This was a highly productive protocol revealing good organizational ability with an emphasis on the practical and everyday kinds |

of percepts but yet an ability to perceive the unusual and different. There was a sensitivity to and an interest in his environment. His need for immediate gratification was not out of proportion to his overall structure. He may tend to react somewhat impulsively on occasion but has the capacity to delay and shift his approach. There was evidence of rather intense underlying affectional anxiety with some awareness of this need. This may in some respect act as a drive mechanism, which motivates the subject to a heightened level of activity. These tensions or anxieties were not detrimental to his overall functioning.

7 This subject's approach was rather avoidant and cautious. His goals are relatively well-defined and there was no disproportionate emphasis upon immediate satisfaction at the expense of long-range planning. He has potential for good abstractive and integrative ability but his efforts are somewhat strained and commonplace. His level of aspiration is greater than his ability. The emphasis on animal and shading responses is suggestive of a heightened level of anxiety. The large number of space responses are suggestive of some oppositional tendencies. There were also some indications of generalized feelings of inadequacy. In general, this was a rather stereotyped, unimaginative protocol suggestive of a cautious, inhibited, toned-down hesitant approach to external stimuli.

8 A well-balanced Rorschach, indicated good emotional resources which are well-mobilized. There was a good balance between immediate and more mature gratification of impulses. He is sensitive to and tolerant of the needs of himself and others. The subject can function in a matter-of-fact manner, but prefers a more creative approach.

9 Responses reflected an unusually efficient mobilization of intellectual resources in both a creative and immediately practical manner. Other indices reflected a high degree of emotional responsiveness to his environment. This may, in part, be due to slight but pervasive anxiety stemming from unmet affectional needs. Response analysis further indicates a tendency to withdraw somewhat from associates and maintain a somewhat detached but individually meaningful perception of his environment.

# APPENDIX 3

## A. 1988 ASTRONAUT CANDIDATE PSYCHIATRIC EVALUATION CLINICAL INTERVIEW FORMAT

### Personal Information

Name; Age; Applicant for which position (pilot astronaut or mission specialist astronaut); Sex; Birth date; Address and phone; Marital status; Names and ages of children; Highest academic level and degree; Occupation; Employer; Current living situation.

### Current Primary Problems

Are there any current or ongoing problems of a psychological, medical, or social nature? History and impact, if appropriate.

### Past Psychiatric History

Past psychiatric illness or treatment of any kind; Family history of psychiatric illness or treatment.

### History of Childhood and Adolescence

Parents (including health status, age at death, marital status, etc.); Siblings; Applicant's development; Applicant's assessment of parenting; Applicant's assessment of siblings; School history; Trouble with police or authority.

### Adult Personal and Social History

Educational history (high school, college, graduate school; field of study; participation in school activities); Employment history; Sexual and marital history; Current status of important personal relationships; Current life satisfaction; Why is applicant applying to be an astronaut; Reaction of significant others to his/her application.

### Applicant's Responses to Stress

Applicant is asked to describe or give an anecdote of the most stressful period and the most life-threatening situation in the applicant's life. How did he or she deal with it? What methods were used to cope with the situation? How did the situation end up?

### Formal Mental Status Examination

Appearance; Behavior; Speech; Mood and affect; Sensorium and intellectual functions; Thought process; Thought content; Perception.

### Diagnostic Formulation

Presence or history of Axis I or II psychiatric disorders according to the third edition of the *Diagnostic and Statistical Manual* of the American Psychiatric Association (DSM-III).

## B. MEMBERS OF THE NASA IN-HOUSE WORKING GROUP ON PSYCHIATRIC AND PSYCHOLOGICAL SELECTION OF ASTRONAUTS (1988–1990)

### Chair

Patricia A. Santy, M.D., M.S.
Medical Officer/Psychiatry
NASA Johnson Space Center

### Participants

Ellen Baker, M.D.
Astronaut
NASA Johnson Space Center

Manley "Sonny" Carter, M.D.
Astronaut
NASA Johnson Space Center

George E. Ruff, M.D.
Professor
University of Pennsylvania

Don E. Flinn, M.D.
Professor
Texas Tech University

Carlos Perry, M.D.
Professor
University of Texas, San Antonio

David R. Jones, M.D.
Chief of Neuropsychiatry
Brooks Air Force Base School of Aviation Medicine

John C. Patterson, Ph.D.
Psychologist
Brooks Air Force Base School of Aviation Medicine

Robert M. Rose, M.D.
Chairman, Department of Psychiatry
University of Texas Medical Branch, Galveston

Harry C. Holloway, M.D.
Chairman, Department of Psychiatry
Uniformed Health Services University

Robert L. Helmreich, Ph.D.
Professor
University of Texas, Austin

E. K. Eric Gunderson, Ph.D.
Department of the Navy
San Diego, California

Terrence McGuire, M.D.
Consultant
NASA Johnson Space Center

Clay Foushee, Ph.D.
Psychologist
NASA Ames Research Center

Chiaki Mukai, M.D.
Payload Specialist Astronaut
National Space Development Agency of Japan

Chiharu Sekiguchi, M.D.
Flight Surgeon
National Space Development Agency of Japan

Dr. Minoru Kume
Psychological Consultant
National Space Development Agency of Japan

Shigenobu Kanba, M.D.
Keio University School of Medicine
Psychiatric Consultant
National Space Development Agency of Japan

Klaus-Martin Goeters, Ph.D.
Psychologist
German Space Agency (DLR)

Anke L. Putzka, M.D.
Flight Surgeon
Institute for Aerospace Medicine
German Space Agency (DLR)

Alv Dahl, M.D.
Professor
University of Oslo
Consultant
European Space Agency

Gerald Post, M.D.
Professor
George Washington University

Ed Foulks, M.D.
Professor
Tulane University

Jean Endicott, Ph.D.
Professor
Columbia College of Physicians and Surgeons

Deborah Hasan, Ph.D.
Assistant Professor
Columbia College of Physicians and Surgeons

William Ravelle, Ph.D.
Professor
Northwestern University

David Buss, Ph.D.
Center for Advanced Studies in Behavioral Sciences
Stanford University

Auke Tellegen, Ph.D.
Professor
University of Minnesota

Janet Spence, Ph.D.
Professor
University of Texas, Austin

Lewis Goldberg, Ph.D.
Professor
Oregon Research Institute

## C. RECOMMENDED PSYCHIATRY STANDARDS FOR SELECTION

1. General
   a. Criteria for the diagnosis of current or previous psychiatric disorders will be based on the most recent edition of the *Diagnostic and Statistical Manual* (DSM) of the American Psychiatric Association.
   b. Any psychiatric, psychological, mental, or behavioral disorder that, in the opinion of the examiner, makes the candidate a hazard to flying safety or mission success. These disorders are suggested by (but not limited to) the following, and may require further evaluation if present:
      (1) History of one or more suicide gestures or attempts;
      (2) History of an adjustment disorder, or presence of an adjustment disorder, or symptoms of an adjustment disorder;
      (3) History of any interpersonal problem, or presence of an interpersonal problem which is sufficiently severe to interfere with the performance of duties, or to suggest inability to adapt to stressful situations (e.g., marital problems, parent-child problem or other V-code diagnoses);
      (4) Character and behavior disorders that are evident by history and objective examination, such that the degree of immaturity, instability, personality inadequacy, or dependency will seriously interfere with performance of duties;
      (5) History of repeated behavior characterized by lack of impulse control, or lack of judgment or concern for the consequences of their behavior.

2. Substance Use
   a. Alcohol dependence or abuse, or history of dependence or abuse.
   b. Drug dependence or abuse, or history of dependence or abuse, characterized by:
      (1) Drug use, other than that prescribed by a recognized health care practitioner, of any narcotic drug;
      (2) The repeated use of any drug or chemical substance with such frequency that it appears that the examinee has accepted the use of, or reliance on, the substance as part of his pattern of behavior.
3. Psychiatric Disorders with Psychotic Symptoms
   a. While psychosis or psychotic symptoms are not specific to any particular psychiatric disorder, any current evidence of, or documented history of, a psychosis, whatever its etiology (other than those of brief duration associated with a toxic or infectious process) is disqualifying.
   b. Documented history of psychosis or psychotic disorders which have evidence of genetic loading in two or more members of the family of origin (father, mother, siblings) including:
      (1) Schizophrenia
      (2) Delusional (paranoid) disorder
      (3) Brief reactive psychosis
      (4) Schizophreniform disorder
      (5) Schizoaffective disorder
      (6) Atypical psychotic disorder (not due to exposures to any toxic or infectious process)
      (7) Bipolar disorder (any variant)
      (8) Major depression
4. Personality
   a. Any personality disorder or history of personality disorder as described by the most recent edition of the *Diagnostic and Statistical Manual* of the American Psychiatric Association
5. Other Psychiatric Disorders
   a. Any psychiatric disorder or documented history of a psychiatric disorder which was clinically significant in the opinion of the examiner. These shall include:
      (1) Mood disorders (including major depression, bipolar disorder or any variant, atypical depression, and cyclothymia)
      (2) Anxiety disorders (including panic, phobic, obsessive-compulsive disorders, and post-traumatic stress disorder)
      (3) Somatiform disorders
      (4) Dissociative disorders
      (5) Sexual disorders
      (6) Sleep disorders
      (7) Factitious disorders
      (8) Disorders of impulse control
      (9) Conduct disorders
      (10) Eating disorders (bulimia or anorexia nervosa)
   b. The use, or history of any use, of psychotropic medication will require further work-up and evaluation by the examiner, and may lead to disqualification under section 1.a.

# APPENDIX 4

## RECOMMENDATIONS TO NASA ON TENTATIVE PSYCHOLOGICAL "SELECT-IN" CRITERIA FOR ASTRONAUT SELECTION (WHITE PAPER, JULY 1988)

### Background

A subgroup of the NASA In-House Working Group on Psychiatric and Psychological Selection of Astronauts was formed to develop specific guidelines for NASA Astronaut selection.

The task of the subgroup was to define the "best" psychological make-up for the job of Astronaut; and the "best" crew psychological mix, particularly for extended-duration space flights. These "best" characteristics (referred to as "select-in" criteria) will be used in selecting the most qualified candidates for the job of Astronaut, and possibly for Space Station Crew selection.

The task is perceived by the Group to be a longitudinal one, since there has never been validation of the criteria utilized in previous Astronaut psychiatric and psychological evaluations (from Mercury through Shuttle). The components of the task were defined to be the following:

1. Develop a set of "tentative" select-in criteria, based on the combined experiences of the group, and on data presented in the psychological and psychiatric literature.
2. Develop an operational definition of each specific select-in criterion.
3. Determine a method of assessment for that criterion.
4. Determine the reliability of the assessment method.
5. Validate the criteria using data on behavior, performance and emotions.

A final task will be to determine the transcultural validity of the select-in criteria, since it is anticipated that Space Station crews will be culturally and ethnically diverse.

### Assumptions

The following assumptions were made in determining the tentative select-in psychological criteria:

1. Astronaut selection is considered to be an ongoing process, and not a specific occurrence at one point in time. This assumption is crucial, since it will be nearly impossible to select crews for extended missions without ongoing evaluation of performance and interpersonal interactions.
2. Both individual Astronaut training and Crew training is considered to be part of the selection process, since it is the milieu in which performance and interpersonal dynamics can be observed and evaluated.

3. Extended duration missions necessitate that crews develop a common group experience during training; and that research to define the nature, length and possible content of that experience is essential.
4. Further programmatic research is required to define the entire selection process, from individual Astronaut selection through Crew selection and actual Space Station missions.
5. There is currently no structured method of evaluating individual or crew performance. Finding a measure will be crucial to validating any psychological criteria developed. Any measure of performance will have to be determined with the concurrence of the Astronaut office and Flight Crew Operations Directorate. Decisions about confidentiality and operational use of such information are very important, and must be decided on prior to the initiation of the research.
6. The method by which research on the assessment and validation processes is done must be scientifically sound, as well as acceptable to current astronauts.

As we move into the era of long-duration space missions such as Space Station, the focus of psychiatric and psychological evaluation of candidates for the job of Astronaut shifts from the individual concerns to the question of how people learn to work together effectively and harmoniously as a group in the isolated and hazardous environment of space.

Individual personality traits may be more or less conducive to group functioning, but some traits may be developed or enhanced to some extent in the context of a training process which emphasizes group and mission orientation. It is also important to consider when defining specific select-in criteria, that some characteristics used to select-out candidates may turn out to be a select-in factor at a later phase of the process.

The political and social reality is that there are likely to continue to be payload specialists whose motivations and backgrounds may be significantly different from professional Astronauts. Some of these may be from other cultures and nations. For these individuals in particular, the training paradigm may be absolutely essential—that is, they must be assigned to a crew as early in the training process as possible. We would recommend, in fact, that all Payload Specialists (particularly those from other countries) be required to undergo the majority of the same training experiences as professional Astronauts and spend an extended time at the Johnson Space Center working directly with the Astronaut Office. The recommended length of duty is approximately 1 year, but the exact time is yet to be determined. This approach is suggested to most effectively facilitate integration of these individuals into the Astronaut Corps, as well as into long-duration mission crews. It is possible that a shortened training experience may suffice, but to justify such an approach when the evidence in many other environments is to the contrary, considerable research would have to be done. An extended training with other Astronauts does not take on the same degree of importance for Shuttle crews as it would for Space Station crews, but will become increasingly important for 14–28 day Shuttle flights, if these are initiated by NASA.

**Recommendations**

The following points need to be stressed prior to discussing the specific recommendations on psychological select-in criteria:

First, behavioral scientists may list specific characteristics, but that is not the same thing as the ability to measure those characteristics in individual applicants with any degree of reliability.

Second, even if specific traits can be measured reliably, it is not certain that utilization of the measure, or any group of measures, will ultimately be more effective than the current management Selection Committee is at selecting the "best" applicants.

Third, it is understood that no one individual can possibly have all of these characteristics listed; nor are they likely to have them in the same degree. In other words, each individual will have strengths and weaknesses which will need to be evaluated in the context of the entire Program objectives. This cannot be done just on the basis of psychological testing.

Finally, it is unknown at this point in time if these tentative criteria accurately reflect those characteristics which will predict success as an Astronaut or a crewmember on an extended-duration space mission. Validation of these criteria through ongoing performance and behavioral measures is essential. The criteria put forth in this paper are considered to be a "best guess" and reflect our understanding of the job of Astronaut and the environment of space, as well as our own knowledge and expertise in the area of Psychiatry and Psychology.

Keeping these points in mind, three major areas, or clusters, of attributes were identified. The three areas are considered of equal importance and are: Aptitude for the Job; Motivation; and Sensitivity to Self and Others. These three areas and the specific criteria encompassed within each one are discussed in the following paragraphs. At the present stage of development the criteria are merely listed with a brief description when necessary. These descriptions are not specific operational definitions of the criteria, the writing of which will be a task at subsequent meetings of the Working Group.

### Aptitude for Job

This general cluster of attributes involves both general and specific select-in criteria identified relating to the job description of the Astronaut. It is assumed that Astronauts are being evaluated in the context of their ability to function effectively in an extended-duration mission environment (as we currently understand it).

**General Criteria:**

Intelligence and Technical Aptitude

History of Success Professionally

Adaptability and Flexibility

Team Player

Ability to Represent NASA Effectively

**Specific Criteria:**

Stress or Discomfort Tolerance

Ability to Function Despite Imminent Catastrophe or Personal Danger

Ability to Compartmentalize (i.e. temporarily put aside personal feelings or problems)

Ability to Tolerate Separation from Loved Ones

Ability to Tolerate Isolation

### Motivation

This cluster of attributes describes the category of motivation. Specifically motivation is the degree of interest and enthusiasm for the job. It assumes that one has high energy levels in pursuing one's goals. Generally, there must be a high degree of "mission-oriented" motivation and a lesser degree of "personal" motivational factors. In particular, the latter factors must not be neurotically overdetermined. This means that self-destructive wishes and attempts to compensate for identity problems or feelings of inadequacy are clearly undesirable.

**Specific traits**

Achievement/Goal-Orientation

Hard Working/Self-Starting

Mastery

Persistence

Optimism

No Excessive Extrinsic or Unhealthy Motivation

Healthy Sense of Competition

Capacity to Tolerate Boredom and Low Levels of Stimulation

Mission-Orientation

Healthy Risk-Taking Behavior

### Sensitivity to Self and Others

This cluster of attributes reflects the emotional maturity and sensitivity of the individual—an area of particular importance for extended duration space missions. However, it is also critical for effective functioning and performance of routine duties and responsibilities of the Astronaut job.

**Specific Traits**

Overall Emotional Maturity and Stability

Self-Esteem

Ability to Form Stable and Quality Interpersonal Relationships

Openness and Warmth

Psychological-Mindedness

Sense of Humor

Ability to Relate to Others with Sensitivity, Regard, and Empathy

No Overt Hostility, Irritability, or Irascibility

Fairness and Sense of Proportion

Appropriate Assertiveness

### Conclusion

The above select-in psychological criteria for Astronaut Selection are recommended based on our current knowledge of the expectations of Astronauts for future space missions. Specific operationally oriented definitions of each criterion will be defined and a method to assess the criteria determined over the next several months. It is strongly recommended to NASA management that they consider sponsoring a scientific study which validates the operationally defined criteria against Astronaut Performance.

# APPENDIX 5

## A. 1989 ASTRONAUT CANDIDATE MULTIDIMENSIONAL APTITUDE BATTERY (MAB) SCORES (N = 106)

| Mab Scale | Mean | Standard Deviation |
|---|---|---|
| Information | 69.5 | 5.87 |
| Comprehension | 62.9 | 2.88 |
| Arithmetic | 67.4 | 7.24 |
| Similarities | 66.4 | 5.04 |
| Vocabulary | 66.6 | 5.43 |
| Digit Symbol | 63.6 | 6.10 |
| Picture Completion | 64.3 | 6.82 |
| Spatial | 62.9 | 7.24 |
| Picture Assembly | 67.1 | 7.34 |
| Object Assembly | 63.7 | 5.54 |
| Verbal IQ | 125.9 | 5.93 |
| Performance IQ | 127.3 | 9.10 |
| Full Scale IQ | 127.3 | 6.63 |

## B. 1989 ASTRONAUT CANDIDATE MILLON CLINICAL MULTIAXIAL INVENTORY (MCMI-II) SCORES (N = 101, 5 TESTS INVALID)

| MCMI Scale | Mean | Standard Deviation |
|---|---|---|
| Disclosure (X) | 15.5 | 12.3 |
| Desirability (Y) | 61.1 | 16.0 |
| Debasement (Z) | 12.3 | 6.5 |
| 1- Schizoid | 42.1 | 16.6 |
| 2- Avoidant | 12.7 | 9.85 |
| 3- Dependent | 53.6 | 23.7 |
| 4- Histrionic | 57.0 | 16.1 |
| 5- Narcissistic | 52.4 | 16.9 |
| 6A- Anti-Social | 31.4 | 16.0 |
| 6B- Aggressive/Sadistic | 48.6 | 16.2 |
| 7- Compulsive | 68.4 | 16.6 |
| 8A- Passive-Aggressive | 9.82 | 9.77 |
| 8B- Self-Defeating | 15.2 | 13.7 |
| S- Schizotypal | 24.4 | 13.1 |
| C- Borderline | 13.6 | 8.38 |
| P- Paranoia | 35.2 | 19.1 |
| A- Anxiety | 23.0 | 5.62 |
| H- Somatoform | 37.5 | 16.1 |
| N- Bipolar/Manic | 45.9 | 13.2 |
| D- Dysthymic | 24.3 | 6.69 |
| B- Alcohol Dependence | 8.71 | 9.28 |
| T- Drug Dependence | 23.2 | 13.4 |
| SS- Thought Disorder | 6.75 | 12.9 |
| CC- Major Depression | 8.33 | 13.4 |
| PP- Delusional Disorder | 26.2 | 17.4 |

C. MCMI: GENERAL PROFILE OF 1989 CANDIDATES

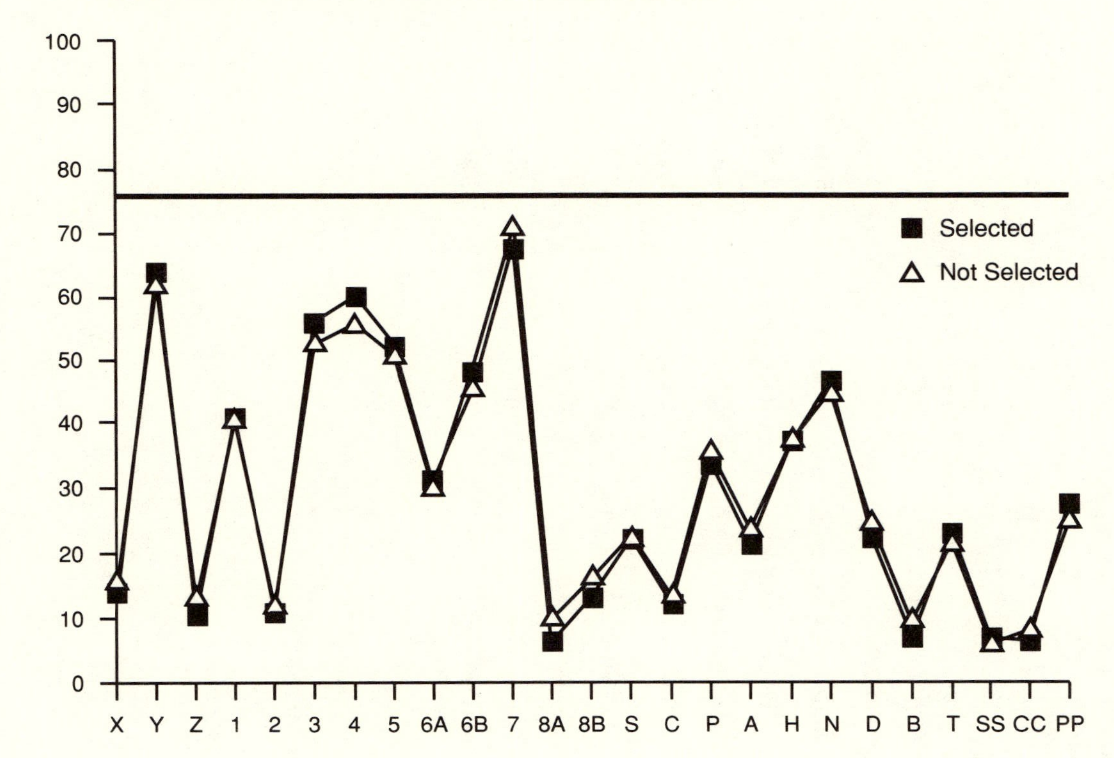

## D. 1989 ASTRONAUT CANDIDATE MMPI T-SCORES (N = 106)

| MMPI Scale | Mean | Standard Deviation |
|---|---|---|
| L | 52.6 | 8.21 |
| F | 46.0 | 3.43 |
| K | 64.0 | 5.84 |
| Hysteria | 49.8 | 4.24 |
| Depression | 51.5 | 6.42 |
| Hypochondriasis | 57.7 | 5.38 |
| Psychopathic Deviation | 55.8 | 6.85 |
| Masculinity/Femininity | 58.1 | 7.99 |
| Paranoia | 55.4 | 5.05 |
| Psychcasthenia | 52.7 | 5.49 |
| Schizophrenia | 53.7 | 5.01 |
| Mania | 54.2 | 8.06 |
| Social Introversion | 43.0 | 5.87 |
| MacAndrew | 19.5 | 3.07 |

# APPENDIX 6

## A. PERSONAL CHARACTERISTICS INVENTORY (PCI) DESCRIPTION OF SCALES

| Low Score Meaning | Trait | High Score Meaning |
|---|---|---|
| Not self assertive<br>"Feminine"<br>Socially not desirable | Instrumentality<br>(I +) | Self assertive<br>"Masculine"<br>Socially desirable, both sexes |
| Does not express interpersonal relations<br>"Masculine"<br>Socially not desirable | Expressivity<br>(E+) | Expresses interpersonal relations<br>"Feminine"<br>Socially desirable, both sexes |
| Socially desirable, both sexes | Negative Instrumentality<br>(I-) | Affects environment but uses socially undesirable methods/strategies<br>"Masculine" |

| Low Score Meaning | Trait | High Score Meaning |
| --- | --- | --- |
| Low verbal aggressivity | Verbal Aggressivity (Eva) | Neurotic, passive-aggressive “Feminine” Socially undesirable, both sexes |
| Socially desirable | Negative Communion (Ec-) | Feminine counterpart to I- strategies |
| Low achievement motivation | Mastery | High motivation for undertaking new and challenging tasks |
| Low needs or satisfaction for working well | Work Orientation | High satisfaction, need for and pride in working well |
| Low competitiveness | Competition Orientation | Desire to surpass others in all areas |
| Low motivation for achievement | Achievement Motivation (A/S) | High motivation for achievement |
| Patient, does not get irritated | Impatience Irritation (I/I) | Impatient, Irritable |

## B. PERSONALITY CLUSTERS DERIVED FROM HELMREICH'S PERSONAL CHARACTERISTICS INVENTORY (PCI)

| Cluster | | Component Traits | |
|---|---|---|---|
| IE + | ("right stuff") | High | Instrumentality |
| | | High | Expressivity |
| | | High | Mastery |
| | | High | Work Orientation |
| Ec - | ("no stuff") | High | Negative Communion |
| | | Low | Instrumentality |
| | | Low | Achievement Motivation |
| I - | ("wrong stuff") | High | Verbal Aggressivity |
| | | High | Negative Instrumentality |
| | | High | Competition Orientation |

## C. NEO-PERSONALITY INVENTORY DESCRIPTION OF SCALES

| SCALE/SUBSCALE | GENERAL DESCRIPTION |
|---|---|
| Neuroticism | Adjustment versus instability and psychological distress |
| Anxiety | |
| Hostility | |
| Depression | |
| Self-Consciousness | |
| Impulsiveness | |
| Vulnerability | |
| Extroversion | Quantity and intensity of interpersonal interaction; activity level; and capacity for joy |
| Warmth | |
| Gregariousness | |
| Assertiveness | |
| Activity | |
| Excitement-Seeking | |
| Positive Emotions | |

| | |
|---|---|
| Openness | Proactive seeking and appreciation of experience for its own sake; toleration for and exploration of the unfamiliar |
| Fantasy | |
| Aesthetics | |
| Feelings | |
| Actions | |
| Ideas | |
| Values | |
| Agreeableness | Quality of interpersonal orientation |
| Conscientiousness | Degree of organization, persistence, motivation, and goal-directed behavior |

# Bibliography


Abe, M., K. Sumita, and M. Kuroda. A *Manual of the MMPI* (in Japanese). Japanese standard ed. Kyoto: Sankyobo, 1963.

Aldrich, C. K. "The Clouded Crystal Ball: A 35-year Follow-up of Psychiatrists' Predictions." *American Journal of Psychiatry* 143 (1986): 45–49.

Allport, G. W. *The Use of Personal Documents in Psychological Science.* S.S.R.C. Bulletin No. 49. 1942.

American Psychiatric Association. *Diagnostic and Statistical Manual of Mental Disorders.* 3rd ed. Washington, D.C.: American Psychiatric Association Press, 1980.

American Psychiatric Association. *Diagnostic and Statistical Manual of Mental Disorders.* 3rd ed. revised. Washington, D.C.: American Psychiatric Association Press, 1987.

Amir, Y., Y. Kovarsky, and S. Sharron. "Peer Nominations as Predictors of Multistage Promotions in a Ramified Organization." *Journal of Applied Psychology* 54 (1970): 462–469.

Amthauer, R. *Intelligenz-Struktur-Test.* Gottingen, Germany: Hogrefe, 1953.

Astrachan, B. M., D. J. Levinson, and D. A. Adler. "The Impact of National Health Insurance on the Tasks and Practice of Psychiatry." *Archives of General Psychiatry* 33 (1976): 785–794.

Atkinson, Joseph D., and Jay M. Shafritz. *The Real Stuff: A History of NASA's Astronaut Recruitment Program.* New York: Praeger, 1985.

Barling, J., S. D. Bluen, and V. Moss. "Type A Behavior and Marital Dissatisfaction: Dissecting the Effects of Achievement Striving and Impatience/Irritability." *Journal of Psychology* 124 (1990): 311–319.

Beckman, E. L., and E. Smith. "Tektite II: Medical Supervision of the Scientist in the Sea." *Texas Reports on Biology and Medicine* 30 (1972).

Beregovoy, G. T., I. V. Davydov, N. V. Krylova, and I. B. Solovyeva. "Psychological Training—One of the Most Important Factors of Enhancing the Safety of

Space Flights." Paper presented at the Thirtieth Congress of the International Astronautical Federation, IAF79, Munich, September 17–22, 1979.

Beregovoy, G. T., V. N. Grigorenko, R. B. Bogdasherskiy, and I. N. Pochkayev. "Characteristics of Psychological Selection." In *Kosmicheskaya Akademiya (Space Academy)*, 16–25. Translated by Lydia Stone. Moscow: Mashinostroyeniye, 1987.

Berry, Charles A. "View of Human Problems to Be Addressed for Long-duration Space Flights." *Aerospace Medicine* 44 (1973): 1136–1146.

Bishop, S., D. Faulk, J. Patterson, and P. Santy. "The Use of IQ Assessment in Astronaut Screening and Evaluation." Paper presented at the Aerospace Medical Association Meeting, Toronto, May 23–26, 1993.

Bloch, S., and P. Reddaway. *Psychiatric Terror: How Soviet Psychiatry Is Used to Suppress Dissent.* New York: Basic Books, 1977.

Bluen, S. D., J. Barling, and W. Burns. "Predicting Sales Performance, Job Satisfaction and Depression by the Achievement Striving and Impatience/irritability Dimensions of Type A Behavior." *Journal of Applied Psychology* 75 (1990): 212–216.

Bluth, B. J. "Soviet Space Stress." *Science 81* 2 (1981): 30–35.

Bluth, B. J., and M. Helppie. *Soviet Space Station Analogs.* 2d ed. NASA Grant NAGW-659. Washington, D.C.: NASA Headquarters, 1987.

Bodrov, V. A., V. B. Malkin, B. L. Pokrovskiy, and D. I. Shpachenko. *Psychological Selection of Pilots and Cosmonauts* (in Russian). Vol. 48 of *Problems in Space Biology.* Edited by B. F. Lomov. Moscow: Nauka Press, 1984.

Bodrov, V. A., V. B. Malkin, B. L. Podrovskiy, and D. I. Shpachenko. "Development and Status of Psychological Selection of Pilots and Cosmonauts: A Short Historical Review." In *Psychological Selection of Pilots and Cosmonauts*, 9–37. Vol. 48 of *Problems in Space Biology.* Translated by Lydia Stone. Edited by B. F. Lomov. Moscow: Nauka Press, 1984.

Bodrov, V. A., V. B. Malkin, B. L. Podrovskiy, and D. I. Shpachenko. "Psychological Selection and Professional Performance of Pilots and Cosmonauts." In *Psychological Selection of Pilots and Cosmonauts*, 192–217. Vol. 48 of *Problems in Space Biology.* Translated by Lydia Stone. Edited by B. F. Lomov. Moscow: Nauka Press, 1984.

Bodrov, V. A., V. B. Malkin, B. L. Podrovskiy, and D. I. Shpachenko. "Research Methods of Experimental Psychology." In *Psychological Selection of Pilots and Cosmonauts*, 90–108. Vol. 48 of *Problems in Space Biology.* Translated by Lydia Stone. Edited by B. F. Lomov. Moscow: Nauka Press, 1984.

Bond, D. D. *The Love and Fear of Flying.* New York: International University Press, 1952.

Brosin, H. W. "Psychiatry Experiments with Selection." *Social Service Review* 22 (1948): 461–468.

Cattell, R. B. *The Description and Measurement of Personality.* Yonkers-on-Hudson, N.Y.: World Book Co., 1946.

Cattell, R. B. *Personality and Motivation Structure and Measurement.* Yonkers-on-Hudson, N.Y.: World Book Co., 1957.

Chaikin, A. "The Loneliness of the Long-distance Astronaut." *Discover,* February 1985, pp. 20–31.

Chidester, T. R. "Trends and Individual Differences in Response to Short Haul Flight Operations." Ph.D. diss., University of Texas, Austin, 1988.

Chidester, T. R., and H. C. Foushee. "Leader Personality and Crew Effectiveness: Factors Influencing Performance in Full-mission Air Transport Simulation." In *Proceedings of the 66th Meeting of the Aerospace Medical Panel on Human Stress: Situations in Aerospace Operations.* The Hague: Advisory Group for Aerospace Research and Development, 1988.

Chriss, N. C. "Psychological Training for the Cosmonauts." *Houston Chronicle,* January 16, 1988, 3.

Comet, B. "Medical Standards for Selection and Annual Certification of French Astronauts." Unpublished manuscript, 1986.

Cooper, C. L. *The Stress Check.* Englewood Cliffs, N.J.: Prentice-Hall, 1981.

Cordes, C. "Mullane: Tests Are Grounded." *APA Monitor* (October 1983): 24.

Costa, P. T., and R. R. McCrae. "Objective Personality Assessment." In *The Clinical Psychology of Aging,* 119–143. Edited by M. Storandt, I. C. Siegler, and M. F. Elias. New York: Plenum Press, 1978.

Costa, P. T., and R. R. McCrae. *The NEO Personality Inventory Manual.* Odessa, Fla.: Psychological Assessment Resources, 1985.

Costa, P. T. and R. R. McCrae. "Concurrent Validation after 20 Years: Implications of Personality Stability for Its Assessment." In *Advances in Personality and Assessment,* 31–54. Vol. 4. Edited by J. N. Butcher and C. D. Spielberger. Hillsdale, N.J.: Erlbaum, 1985.

Costa, P. T., and R. R. McCrae. "Major Contributions to Personality Psychology." In *Hans Eysenck: Consensus and Controversy.* Edited by S. Modgil and C. Modgil. Philadelphia: Falmer Press, 1986, 63–72.

Crowne, D. P., and D. Marlow. "A New Scale of Social Desirability Independent of Psychopathology." *Journal of Consulting Psychology* 24 (1960): 349–354.

Curtis, A. R. *Space Almanac.* Woodsboro, Md.: Arcsoft Publishers, 1990.

Diaz-Loving, R., R. Diaz-Guerrero, R. L. Helmreich, and J. T. Spence. "Comparacion transcultural y analysis psicometrico de una medida de rasgos masculinos (instrumentales) y femeninos (expresivos)." *Revista de la Asociacion Latinoamericana de Psicologia Social* 1 (1981): 3–37.

Digman, J. M. "Personality Structure: Emergence of the Five-factor Model." *Journal of Personality and Social Psychology* 55 (1988): 258–265.

Downey, R. G., and P. J. Duffy. *Review of Peer Evaluation Research.* Technical Paper No. 342. U.S. Army Research Institute for the Behavioral and Social Sciences. Fort Knox, Ky., 1987.

Duke, M. B., and P. W. Keaton, eds. *Manned Mars Mission Working Group Summary Report.* NASA M001. Houston: NASA Johnson Space Center, May 1986.

EAPCWG. "Proposed Psychological Criteria for the Selection of European Astronaut Candidates." Final draft, 1990.

Earls, J. H. "Human Adjustment to an Exotic Environment." *Archives of General Psychiatry* 20 (1969): 117–122.

Endicott, J., N. C. Andreasen, and R. L. Spitzer. *Family History—Research Diagnostic Criteria (FH-RDC).* 3d ed. New York: New York State Psychiatric Institute, 1978.

Endicott, J., and R. L. Spitzer. "A Diagnostic Interview: The Schedule for Affective Disorders and Schizophrenia." *Archives of General Psychiatry* 35 (1978): 837–844.

Eysenck, S., and H. J. Eysenck. "An Experimental Investigation of 'Desirability' Response Set in a Personality Questionnaire." *Life Sciences* 5 (1953): 343–345.

Fassbender, C., and K. M. Goeters. "Psychological Evaluation of European Astronaut Applicants: Results of the 1991 Selection Campaign." *Aviation, Space and Environmental Medicine*. In press, 1993.

Fausto-Sterling, A. *Myths of Gender: Biological Theories about Men and Women*. New York: Basic Books, 1985.

Fichtbauer, S. *Ein mobiler Kleinsimulator fur Tracking und Mehrfacharbeitsversuche*. DFVLR Internal Report IB-355-75-06. Cologne: DFVLR, 1975.

Flaherty, B. F., D. E. Flinn, G. T. Hauty, and G. R. Steincamp. *Psychiatry and Space Flight*. Report 60-80. Brooks Air Force Base, Texas: U.S. Air Force School of Aviation Medicine, 1960.

Flinn, Don E. "Psychiatric Factors in Astronaut Selection." In *Psychophysiological Aspects of Space Flight*. Edited by B. E. Flaherty. New York: Columbia University Press, 1961.

Flinn, Don E. "Behavior and Communication during Space Flights." In *Communication and Social Interaction*, 247–261. Edited by P. F. Ostwald. New York: Grune & Stratton, 1977.

Flinn, Don E., E. S. Flyer, and F. E. Holdredge. "Behavioral and Psychological Studies in Aerospace Medicine." *Annals of the New York Academy of Science* 107 (1963): 613–634.

Flinn, Don E., Bryce O. Hartman, D. H. Powell, and R. E. McKenzie. "Psychiatric and Psychological Evaluation." In *Aeromedical Evaluations for Space Pilots*, 199–230. Edited by L. E. Lamb. Brooks Air Force Base, Texas: U.S. Air Force School of Aerospace Medicine, Aerospace Medical Division (AFSC), 1963.

Flinn, Don E., J. T. Monroe, E. H. Cramer, and D. H. Hagen. "Observations in the SAM Two-man Space Cabin Simulator. Behavioral Factors in Selection and Performance." *Aerospace Medicine* 36 (1961): 610–615.

Forer, B. R. *The Forer Structured Sentence Completion Test: Manual*. Los Angeles: Western Psychological Services, 1988.

Foushee, H. C., and R. L. Helmreich. "Group Interactions and Flightcrew Performance." In *Human Factors in Modern Aviation*, 189–277. Edited by E. Weiner and J. Nagel. New York: Academic Press, 1988.

Gagarin, Y. A. *The Road to Space*. Moscow: Nauka Press, 1978.

Goeters, Klaus-M. "The Recruitment and Organizational Integration of Space Personnel." *Acta Astronautica* 17 (1988): 227–229.

Goeters, Klaus-M., and C. Fassbender. *Definition of Psychological Testing of Astronaut Candidates for Columbus Missions*. ESA 8730/90/NL/IW. 1991.

Goeters, Klaus-M., E. Schwartz, C. Budczinski, M. Nordhausen, and B. Repp. *Psychological Selection of Spacelab Payload Specialists: The Evaluation of German Applicants*. ESA Technical Translation ESA-TT-586. Hamburg: ESA, 1979.

Goldberg, D. P., and V. F. Hiller. "A Scaled Version of the General Health Questionnaire." *Psychological Medicine* 9 (1979): 139–145.

Goldberg, L. R. "Language and Individual Differences: The Search for Universals in Personality Lexicons." In *Review of Personality and Social Psychology,* 141–165. Vol. 2. Edited by L. Wheeler. Beverly Hills, Calif.: Sage, 1981.

Goldberg, L. R. "An Alternative Description of Personality: The Big-five Factor Structure." *Journal of Personality and Social Psychology* 59 (1990): 1216–1229.

Grechko, G. Comments made during a visit to the NASA Johnson Space Center, May 1989.

Gregorich, S., R. L. Helmreich, and J. A. Wilhelm. *The Structure of Cockpit Management Attitudes.* NASA/University of Texas Technical Memo 89-1. Austin, Texas: NASA/UT Project, 1989.

Gregorich, S., R L. Helmreich, J. A. Wilhelm, and T. R. Chidester. "Personality Based Clusters as Predictors of Aviator Attitudes and Performance." In *Proceedings of the Fifth International Symposium on Aviation Psychology,* edited by R. S. Jensen. Columbus: Ohio State University, 1989, 686–691.

Gunderson, E. K. E. "Emotional Symptoms in Extremely Isolated Groups." *Archives of General Psychiatry* 9 (1968): 362–368.

Hackman, J. R., and R. L. Helmreich. "Assessing the Behavior and Performance of Teams in Organizations: The Case of Air Transport Crews." In *Assessment for Decision.* Edited by D. Peterson and D. B. Fishman. New Brunswick, N.J.: Rutgers University Press, 1987, 283–313.

Harrison, A. A., Y. A. Clearwater, and C. P. McKay, eds. *From Antarctica to Outer Space: Life in Isolation and Confinement.* New York: Springer-Verlag, 1991.

Hartman, Bryce O. Presentation to the National Academy of Sciences Committee on Manned Space Flight, April 1967 (unpublished manuscript).

Hartman, Bryce O., and Don E. Flinn. "Crew Structures in Future Space Missions." In *Lectures in Aerospace Medicine.* Brooks Air Force Base, Texas: U.S. Air Force School of Aerospace Medicine, 1963.

Hartman, Bryce O., and Richard C. McNee. "Psychometric Characteristics of Astronauts." Paper presented at the NATO AGARD Conference on Recent Advances in Space Medicine, Athens, Greece, October 1976.

Hathaway, S. R., and J. C. McKinley. *MMPI-2: Manual for Administration and Scoring.* Minneapolis: National Computer Systems, 1989.

Helmreich, R. L. "Social Psychology on the Flight Deck." In *Proceedings of the NASA Workshop on Resource Management Training for Airline Flight Crews.* Edited by G. Cooper. NASA CP-2120. NASA Ames Research Center, 1979, 1–18.

Helmreich, R. L. *Pilot Selection and Training.* Washington, D.C.: American Psychological Association, 1982.

Helmreich, R. L. "What Changes and What Endures: The Capabilities and Limitations of Training and Selection." Paper presented at the Aer Lingus/Irish Airline Pilots Association Flight Symposium. Dublin, 1983.

Helmreich, R. L. "Applying Psychology in Outer Space: Unfulfilled Promises Revisited." *American Psychologist* 38 (1983): 445–450.

Helmreich, R. L. "Pilot Selection and Performance Evaluation: A New Look at an Old Problem." In *Proceedings of the Tenth Symposium: Psychology in the Department of Defense.* Edited by G. E. Lee. USAFA-TR-1. Colorado Springs: U.S. Air Force Academy, 1986.

Helmreich, R. L. "Exploring Flight Crew Behavior." *Social Behavior* 21 (1987): 583–589.

Helmreich, R. L. "Cockpit Management Attitudes." *Human Factors* 26 (1987): 583–589.

Helmreich, R. L. "Theory Underlying CRM Training: Psychological Issues in Flightcrew Performance and Crew Coordination. In *Cockpit Resource Management Training: Proceedings of the NASA/MAC Workshop*. Edited by H. W. Orlady and H. C. Fouschee. CP2455. San Francisco: NASA Ames Research Center, 1987, 14–21.

Helmreich, R. L., W. E. Beane, G. W. Lucker, and J. T. Spence. "Achievement Motivation and Scientific Attainment." *Personality and Social Psychology Bulletin* 4 (1978): 222–226.

Helmreich, R. L., T. R. Chidester, H. C. Foushee, S. E. Gregorich, and J. A. Wilhelm. *How Effective Is Cockpit Resource Management Training? Issues in Evaluating the Impact of Programs to Enhance Crew Coordination*. NASA/University of Texas Technical Report 89-2. Austin, Texas: NASA/UT Project, 1989.

Helmreich, R. L., H. C. Foushee, R. Benson, and R. Russini. "Cockpit Management Attitudes: Exploring the Attitude Performance Linkage." *Aviation, Space and Environmental Medicine* 57 (1986): 1198–1200.

Helmreich, R. L., J. R. Hackman, and H. C. Foushee. *Evaluating Flight Crew Performance: Policy, Pressures, Pitfalls and Promise*. NASA/University of Texas Technical Report 86-1. Austin, Texas: NASA/UT Project, 1986.

Helmreich, R. L., and P. A. Santy. "Preliminary Psychological Data Analysis of 1989 Astronaut Selection." Presentation to NASA Life Sciences Management, 1989.

Helmreich, R. L., L. L. Sawin, and A. L. Carsrud. "The Honeymoon Effect in Job Performance: Delayed Predictive Power of Achievement Motivation." *Journal of Applied Psychology* 71 (1986): 1085–1088.

Helmreich, R. L., and J. T. Spence. "The Work and Family Orientation Questionnaire: An Objective Instrument to Assess Components of Achievement Motivation and Attitudes toward Family and Career." *JSAS Catalog of Selected Documents in Psychology* 8 (1978): 35, ms. 1677.

Helmreich, R. L., J. T. Spence, W. E. Beane, G. W. Lucker, and K. A. Matthews. "Making It in Academic Psychology: Demographic and Personality Correlates of Attainment." *Journal of Personality and Social Psychology* 39 (1980): 896–908.

Helmreich, R. L., and J. A. Wilhelm. "When Training Boomerangs: Negative Outcomes Associated with Cockpit Resource Management Training." In *Proceedings of the Fifth International Symposium on Aviation Psychology*. Edited by R. S. Jensen. Columbus: Ohio State University, 1989, 692–697.

Helmreich, R. L., J. A. Wilhelm, and S. E. Gregorich. *Notes on the Concept of LOFT: An Agenda for Research*. NASA/University of Texas Technical Manual 88-1. Austin, Texas: NASA/UT Project, 1988.

Helmreich, R. L., J. A. Wilhelm, and T. R. Runge. "Study of achievement motivation and job satisfaction in supertanker crews." Paper presented at the California Maritime Academy, Vallejo, California Seventh Annual Marine Industry Symposium: Motivation, Organization and Satisfaction Aboard Ship, 1981.

Horn, W. *Das Leistungsprufsystem.* Gottingen, Germany: Hogrefe, 1969.

Hyler, S. E., R. O. Rieder, J. B. W. Williams, R. L. Spitzer, M. Lyons, and J. Hendler. "A Comparison of Clinical and Self-report Diagnoses of DSM-III Personality Disorders in 552 Patients." *Comprehensive Psychiatry* 30 (1989): 170–178.

International Space University. *International Mars Mission.* Toulouse, France: ISU, 1991.

Jackson, D. N. *Multidimensional Aptitude Battery (MAB).* Menlo Park, Calif.: Sigma Assessment Systems, 1984.

Jennings, Richard T., and Patricia A. Santy. "Reproduction in the Space Environment: Part II—Concerns for Human Reproduction." *Obstetrical and Gynecological Survey* 45 (1990): 7–17.

John, O. P. "The 'Big Five' Factor Taxonomy: Dimensions of Personality in Natural Language and in Questionnaires." In *Handbook of Personality Theory and Research,* 66–100. Edited by L. A. Perrin. New York: Guilford, 1990.

Jones, D. R., and C. A. Annes. "The Evolution and Present Status of Mental Health Standards for Selection of USAF Candidates for Space Missions." *Aviation, Space and Environmental Medicine* 54 (1983): 730–734.

Kanas, N. "Psychosocial Factors Affecting Simulated and Actual Space Missions." *Aviation, Space and Environmental Medicine* 56 (1985): 806–811.

Kanas, N. "Psychological and Interpersonal Issues in Space." *American Journal of Psychiatry* 144 (1987): 703–709.

Kanas, N. "Psychosocial Support for Cosmonauts." *Aviation, Space and Environmental Medicine* 62 (1991): 353–355.

Kanas, N., and A. Fedderson. *Behavioral, Psychiatric and Sociological Problems of Long Duration Space Missions.* NASA Document TMX-58067. Houston: Johnson Space Center, October 1981.

Kelly, A. D., and N. Kanas. "Crew Member Communication in Space: A Survey of Astronauts and Cosmonauts." *Aviation, Space and Environmental Medicine* 63 (1992): 721–726.

Kirsch, H. *Selektions-strategie und psychologische tests bei der eignungsuntersuchung von Bewerbern fur die fliegerische Ausbildung bei der DLH.* DFVLR Internal Report IB-355-76-04. Hamburg: DLR, 1976.

Kirsch, H., K. M. Goeters, and R. Ewe. *Faktorenanalyse eines neuen mehrdimensionalen Personlichkeits-Fragebogens.* DLR FB 75-20. Hamburg: DLR, 1975.

Kitamura, T. *Psychiatric Selection Procedure of Japanese Astronauts for the NASA Space Station: I. The Selection Framework.* Japan National Institute of Mental Health, 1989.

Kitamura, T., and S. Kanba. *Psychiatric Selection Procedure of Japanese Astronauts for the NASA Space Station: II. Suggested Select-out Procedures.* Tokyo: National Institute of Mental Health, 1989.

Kragh, U. "The Defense Mechanism Test: A New Method for Diagnosis and Personnel Selection." *Journal of Applied Psychology* 44 (1960): 303–309.

Kragh, U. "Predictions of Success of Danish Attack Divers by the Defense Mechanism Test (DMT)." *Perceptual and Motor Skills* 15 (1962): 103–106.

Kruger, C., K. M. Goeters, and H. Eissfeldt. *Geschlechtsspezifische Unterschiede im Leistungs—und Personlichkeitsbereich bei Bewerbern fur den Gehobenen Flug—Verketirskontrolldienst.* DFVLR FB-316-88-01. Cologne: DFVLR, 1988.

Kume, M. "Present Status of Japanese Psychological Astronaut Selection." Report presented at the joint NASA/NASDA Working Group Meetings, Tokyo, February 9–10, 1989.

Kuznetsov, O. N., and V. I. Lebedev. *Psychology and Psychopathology of Solitude.* Moscow, 1972.

Logsdon, John M. "U.S.-European Cooperation in Space: A 25-year Perspective." *Science* 223 (January 6, 1984): 11–16.

Logsdon, John M. *Together in Orbit: The Origins of International Participation in Space Station Freedom.* Washington, D.C.: Space Policy Institute, George Washington University, December 1991.

Lukyanova, N. F. "Personality Tests." In V. A. Bodrov, V. B. Malkin, B. L. Pokrovskiy, eds., *Psychological Selection of Pilots and Cosmonauts,* 108–125. Vol. 48 of *Problems in Space Biology.* Translated by Lydia Stone. Edited by B. F. Lomov. Moscow: Nauka Press, 1984.

Manzey, D., and A. Schiewe. "Psychological Training of German Science Astronauts." *Acta Astronautica* 27 (1992): 131–138.

Matthews, K. A., R. L. Helmreich, W. E. Beane, and G. W. Lucker. "Pattern A, Achievement-striving, and Scientific Merit: Does Pattern A Help or Hinder?" *Journal of Personality and Social Psychology* 39 (1980): 962–967.

McCrae, R. R., and P. T. Costa. "Clinical Assessment Can Benefit from Recent Advances in Personality Psychology." *American Psychologist* 41 (1986): 1001–1003.

McCrae, R. R., and P. T. Costa. "More Reasons to Adopt the Five-factor Model." *American Psychologist* 44 (1989): 451–452.

McCrae, R. R., and P. T. Costa. "Validation of the Five Factor Model of Personality across Instruments and Observers." *Journal of Personality and Social Research.* In press, 1992.

McFadden, T. J., and R. L. Helmreich. "Personality Predictors of Performance in Astronauts." Presentation to the Aerospace Medical Association, Miami, 1992.

McFadden, T. J., R. L. Helmreich, R. M. Rose, and L. Fogg. "Predicting Astronaut Effectiveness: A Multivariate Approach." Submitted for publication to *Aviation, Space and Environmental Medicine,* 1993.

Meehl, Paul E. *Clinical vs. Statistical Prediction: A Theoretical Analysis and a Review of the Evidence.* Minneapolis: University of Minnesota Press, 1954.

Mezzich, J. E., H. Fabrega, G. A. Coffman, and R. Haley. "DSM-III Disorders in a Large Sample of Psychiatric Patients: Frequency and Specificity of Diagnoses." *American Journal of Psychiatry* 146 (1989): 212–219.

Millon, T. *Millon Clinical Multiaxial Inventory–II.* Minneapolis: National Computer Systems, 1987.

Moscow World Service. *Psychologists Profile Ideal Mars Crew* (in English). JPRS-USP-89-004. February 16, 1989.

Mukai, C., and C. Sekiguchi. "Japanese Medical Selection." Presented to the NASA In-House Working Group on Psychiatric and Psychological Selection of Astronauts, 1988.

Muller, A. *Das Aufmerksamkeitsprufgerat.* Hamburg, 1956.

Mumford, M. D. "Social Comparison Theory and the Evaluation of Peer Evaluations: A Review and Some Applied Implications." *Personnel Psychology* 36 (1983): 867–881.

Myasnikov, V. I., and O. P. Kozerenko. "Prevention of Psychoemotional Disorders in Prolonged Space Flight by Method of Psychological Support." *Space Biology and Aerospace Medicine* 2 (1981): 25–29.

Myasnikov, V. I., O. P. Kozerenko, A. A. Gerasimovich, and E. V. Ryabov. *Psychological Support and Psychological Adaptation of Crewmembers in Flights of Long Duration.* Moscow: Kaluga, 1979.

Myasnikov, V. I., O. P. Kozerenko, and F. N. Ukov. "Actual Problems in Providing Medico-Psychological Support in Flights of Long Duration." In *Aerospace Medicine.* Dresden, 1980, 41–42.

Myasnikov, V. I., E. F. Panchenkova, and F. N. Uskov. "Prospects of Using Radio and TV Communication Data in the Medical Supervision of Cosmonauts In-flight." Paper presented at the Twenty-fifth International Congress of Aviation and Space Medicine, Helsinki, September 4–9, 1977.

Nardini, J. E., R. Herrmann, and J. E. Rasmussen. "Navy Psychiatric Assessment Program in the Antarctic." *American Journal of Psychiatry* 119 (1962): 97–105.

NASA *Information Summaries: Space Station.* PMS-008A (Hqs). Washington, D.C.: NASA Headquarters, August 1988.

NASA *Medical Standards: NASA Class I Pilot Astronaut Selection and Annual Medical Certification.* JSC-11569. Houston: Johnson Space Center, 1984.

Niederland, W. G. "River Symbolism, Part I." *Psychoanalytic Quarterly* 25 (1956): 469–504.

Niederland, W. G. "River Symbolism, Part II." *Psychoanalytic Quarterly* 26 (1957): 50–75.

Niederland, W. G. "The History and Meaning of California: A Psychoanalytic Inquiry." *Psychoanalytic Quarterly* 40 (1971): 485–490.

Nikonov, A. V. "Reflections of the Dynamics of the Human Operators' Psychophysiological State in Radio Exchange." In *Psychological Problems of Space Flight,* 335–343. Moscow: Nauka Press, 1979.

North, R. M. "Human Requirements for Long-duration Missions: Antarctic and Arctic Stations, Planetary Surface Operations and Space Transportation Vehicles." Paper presented at the International Biospherics Conference, University of Alabama, Huntsville, February 20–23, 1992.

Novikov, M. A. "Psychophysiological Selection, Crew Manning, and Training for Space Flight." In *Psychological Problems of Space Flight,* 196–204. Moscow: Nauka Press, 1979.

Novikov, M. A. "Principles and Methods of Studying Psychophysiological Compatibility." Paper presented at the Twelfth U.S./U.S.S.R. Joint Working Group Meeting on Space Biology and Medicine, Washington, D.C., November 9–22, 1981.

Oberg, J. E. *Manned Mission to Mars.* Harrisburg, Pa.: Stackpole Books, 1982.

Oberg, J. E. *Uncovering Soviet Disasters.* New York: Random House, 1988.

Oberg, J. E., and A. R. Oberg. *Pioneering Space.* New York: McGraw-Hill, 1986.

"Origins of Psychogeography and Survivor Syndrome in Experience." In *Frontiers of Psychiatry: Roche Report,* 1–2. October 1, 1979.

Olff, M., G. Godaert, and H. Ursin, eds. *Quantification of Human Psychological Defence.* Berlin: Springer-Verlag. In press, 1993.

Palinkas, L. A. *Sociocultural Influences on Psychosocial Adjustment in Antarctica.* Report No. 85-49. San Diego: Naval Health Research Center, 1985.

Palkinas, L. A. "Group Adaptation and Individual Adjustment in Antarctica: A Summary of Recent Research." In *From Antarctica to Outer Space: Life in Isolation and Confinement,* 240–251. Edited by A. A. Harrison, Y. A. Clearwater, and C. P. McKay. New York: Springer-Verlag, 1991.

Perry, Carlos J. G. "Psychiatric Selection of Candidates for Space Missions." *Journal of the American Medical Association* 194 (1965): 841–844.

Perry, Carlos J. G. "Psychiatric Support for Man in Space." *International Psychiatry Clinics* 4 (1967): 197–221.

Pfohl, B. "Structured Interview for DSM-III-R Personality Disorders (SIDP-R)." Draft ed. Iowa City: Department of Psychiatry, University of Iowa, 1989.

Pred, R. S., J. T. Spence, and R. L. Helmreich. "The Development of New Scales for the Jenkins Activity Survey Measure of the Type A Construct. *Social and Behavioral Science Documents* 16 (1986): 51–52.

Radloff, R., and R. L. Helmreich. *Groups under Stress—Psychological Research in Sealab II.* New York: Appleton-Century-Crofts, 1968.

Reich, J., R. Crowe, R. Noyes, and E. Troughton. *Family History of DSM-III Personality and Anxiety Disorders (FHPD).* Iowa City: Department of Psychiatry, University of Iowa, 1982.

Remek, V. "Communication Problems of International Crews." Paper presented at the Thirtieth Congress of the International Astronautical Federation, Munich, September 17–22, 1979.

Rivolier, J., C. Bachelard, G. Cazes, H. Mathian, C. Shaw, and G. Veron. "WP630 and WP730: University of Reims." In *Definition of Psychological Testing of Astronaut Candidates for Columbus Missions,* 117–150. Hamburg: German Aerospace Research Establishment, Department of Aviation and Space Psychology, 1991.

Robins, L. N., J. E. Helzer, M. M. Weissman, H. Orvaschel, E. Gruenberg, J. D. Burke, and D. A. Regier. "Lifetime Prevalence of Specific Psychiatric Disorders in Three Sites." *Archives of General Psychiatry* 41 (1984): 949–958.

Rose, R. M. "Predictors of Psychopathology in Air Traffic Controllers." *Psychiatric Annals* 12 (1982): 925–933.

Rose, R. M. "Memo to DLR: Preliminary Data Analysis on TOM Variables." October 26, 1990.

Rose, R. M., L. Fogg, R. L. Helmreich, and T. McFadden. "Psychological Predictors of Astronaut Effectiveness." *Aviation, Space and Environmental Medicine.* In press, 1993.

Rose, R. M., R. L. Helmreich, L. Fogg, and T. J. McFadden. "Assessments of Astronaut Effectiveness." *Aviation, Space and Environmental Medicine.* In press, 1993.

Rose, R. M., C. D. Jenkins, and M. W. Hurst. *Air Traffic Controller Health Change Study.* FAA Contract FA73WA-3211. 1978.

Ruff, George E. "Psychological Tests." In *Project Mercury Candidate Evaluation Program,* 81–86. Edited by Charles L. Wilson. WADC Technical Report 59-505. Wright-Patterson Air Force Base, Ohio: Wright Air Defense Center, 1959, 81–86.

Ruff, George E. "Selection of Crews for Space Flight." Unpublished paper, 1962.

Ruff, George E. "Report on the NASA In-House Working Group on Psychiatric and Psychological Selection of Astronauts Meeting with the NASDA Psychological Standards Working Group, Tokyo, February 9–10, 1989." Unpublished, 1989.

Ruff, George E., and Edwin Z. Levy. "Psychiatric Evaluation of Candidates for Space Flight." *American Journal of Psychiatry* 116 (1959): 385.

Santy, Patricia A. "The Journey Out and In: Psychiatry and Space Exploration." *American Journal of Psychiatry* 140 (1983): 519–527.

Santy, Patricia A. "Women in Space: A Medical Perspective." *Journal of the American Medical Women's Association* 39 (1984): 13–16.

Santy, P. A. "Psychiatric Support for a Health Maintenance Facility (HMF) on Space Station." *Aviation, Space and Environmental Medicine* 58 (1987): 1219–1224.

Santy, Patricia A., ed. "The NASA In-House Working Group on Psychiatric and Psychological Selection of Astronauts: Summary and Transcripts." Unpublished paper, 1988.

Santy, P. A., J. Endicott, D. R. Jones, R. M. Rose, J. Patterson, A. W. Holland, D. M. Faulk, and R. Marsh. "Results of a Structured Psychiatric Interview to Evaluate NASA Astronaut Candidates." *Military Medicine* 158 (1993): 5–9.

Santy, Patricia A., Al W. Holland, and Dean M. Faulk. "Psychiatric Diagnoses in a Group of Astronaut Applicants." *Aviation, Space and Environmental Medicine* 62 (1991): 969–973.

Santy, P. A., A. W. Holland, L. Looper, and R. Marcondes-North. "Multicultural Factors in the Space Environment: Results of an International Shuttle Crew Debrief." *Aviation, Space and Environmental Medicine* 64 (1993): 196–200.

Sarbin, T. R. "A Contribution to the Study of Actuarial and Individual Methods of Predictions." *American Journal of Sociology* 48 (1942): 593–602.

Schmale, H., and H. Schmidtke. *Der Berufseignungstest.* Stuttgart: Huber, 1966.

Seifert, R. "Neue gerate zur untersuchung der psychomotorik." *Diagnostica* 12 (1966): 6–16.

Sekiguchi, C., and M. Kume. "Cultural Characteristics of Japanese People." Presentation to the NASA In-House Working Group on Psychiatric and Psychological Selection of Astronauts. Houston, 1988.

Sekiguchi, C., S. Yumikura, M. Kume, and N. Okada. "Japanese Astronaut Selection Psychological Results." Paper presented at the Aerospace Medical Assocation Meeting. Miami, May 10–14, 1992.

Senkevich, Y. A., and M. M. Korotayev. "Medical Selection Methods and Criteria for Initial Qualification of Cosmonauts and Placement in Group Training for the Purpose of Standardizing Psychological and Physiological Examinations." Translated by N. Timacheff. Unpublished paper, 1991.

Sledge, W. H., and J. A. Boydstun. "The Psychiatrist's Role in Aerospace Operations." *American Journal of Psychiatry* 137 (1980): 956–959.

Sobchik, L. N. *Manual for Use of the MMPI Psychological System.* Moscow: RSFSR, 1971.

"Soviets Gain WPA Readmission with Conditions." *Psychiatric News,* November 3, 1989, p. 1.

"Soviets Reenter World Psychiatric Society." *Science News* 136 (1989): 278.

"Space Psychological Support: An Interview with Victor Blagov." *Soviet Life* 3 (March 1989).

*Space Station Freedom Reference Guide*. Houston: Boeing, 1990.

Spence, J. T., and R. L. Helmreich. *Masculinity and Femininity: Their Psychological Dimensions, Correlates and Antecedents*. Austin: University of Texas Press, 1978.

Spence, J. T., and R. L. Helmreich. "Achievement-related Motives and Behavior." In *Achievement and Achievement Motives: Psychological and Sociological Approaches*. Edited by J. T. Spence. San Francisco: W. H. Freeman, 1983, 10–74.

Spence, J. T., R. L. Helmreich, and C. K. Holahan. "Negative and Positive Components of Psychological Masculinity and Femininity and Their Relationships to Self Reports of Neurotic and Acting-out Behaviors." *Journal of Personality and Social Psychology* 37 (1979): 1673–1644.

Spence, J. T., R. L. Helmreich, and R. S. Pred. "Impatience versus Achievement Strivings in the Type A Pattern: Differential Effects on Students' Health and Academic Achievement." *Journal of Applied Psychology* 72 (1987): 522–528.

Spitzer, R. L., and J. Endicott. *Schedule for Affective Disorders and Schizophrenia*. Biometrics Research, New York: New York State Department of Mental Hygiene, 1975.

Spitzer, R. L., J. B. W. Williams, M. Gibbon, and M. B. First. *Structured Clinical Interview for DSM-III-R—Patient Version* (SCID-P, 6/1/88). Biometrics Research Department, New York State Psychiatric Institute, 1988.

Spitzer, R. L., J. B. W. Willliams, M. Gibbon, and M. B. First. *Instruction Manual for the Structured Clinical Interview for DSM-III-R* (SCID, 5/1/89 Revison). Biometrics Research Department, New York State Psychiatric Institute, 1989.

Steinkamp, G. R. *Human Experimentation in the Space Cabin Simulators*. Report 59-101. Brooks Air Force Base, Texas: U.S. Air Force School of Aviation Medicine, 1959.

Taylor, A. J. W. "Antarctica Psychometrika Unspectacular." *New Zealand Antarctic Record* 6 (1978): 36–45.

Taylor, J. A. "A Personality Scale of Manifest Anxiety." *Journal of Abnormal and Social Psychology* 48 (1953): 285–290.

Thomas, Lewis. *The Fragile Species*. New York: Charles Scribner's Sons, 1992.

Thurstone, L. L., and T. M. Thurstone. *SRA Primary Mental Abilities*. Chicago: Science Research Associates, 1949.

Tyrer, P., M. S. Alexander, D. Cicchettti, M. S. Cohen, and M. Remington. "Reliability of a Schedule for Rating Personality Disorders." *British Journal of Psychiatry* 135 (1979): 168–174.

Tyrer, P., J. Alexander, and B. Ferguson. *Personality Assessment Schedule (PAS)*. 5th revision. London, 1987.

Ursano, R. J., and D. R. Jones. "The Individual's vs. the Organization's Doctor: Value Conflict in Psychiatric Aeromedical Evaluation." *Aviation, Space and Environmental Medicine* 52 (1981): 704–706.

Vaernes, R. "The Defense Mechanism Test Predicts Inadequate Performance under Stress." *Scandinavian Journal of Psychology* 23 (1982): 37–43.

Veron, G. "French Selection Criteria for Astronaut Candidates." Paper presented to the European Astronaut Psychological Criteria Working Group, Paris, 1989.

Voas, Robert B. "Operations Part of the Mercury Technical History." Preliminary draft, NASA-JSC Archives, 1963.

Wechsler, D. *Die Messung der Intelligenz Erwachsener.* Stuttgart: Huber, 1956.

Wiggins, J. S., and P. D. Trapnell. "Personality Structure: The Return of the Big Five." In *Handbook of Personality Psychology.* Edited by S. R. Briggs, R. Hogan, and W. H. Jones. In Press, 1993.

Wilson, Charles L., ed. *Project Mercury Candidate Evaluation Program.* WADC Technical Report 59-505. Wright-Patterson Air Force Base, Ohio: Wright Air Development Center, 1959, 3.

Wolfe, Tom. *The Right Stuff.* New York: Farrar, Straus & Giroux, 1979.

Zamaledtinov, I. S., N. V. Drylova, S. A. Kiselev, and Y. V. Trufanova. "The Possibility of Identifying Levels of Emotional Stress According to Speech Indices in Space Flight Activity." In *Psychological Problems of Space Flight,* 131–148. Moscow: Nauka Press, 1979.

# Index

## About the Author

PATRICIA A. SANTY is Associate Professor in the Department of Psychiatry and Behavioral Science, University of Texas Medical Branch, Galveston. She was formerly a Medical Officer at NASA Johnson Space Center and was the crew surgeon for a number of shuttle missions, including *Challenger*.